Structures

Structures
SI UNITS

The late W. T. Marshall
B.SC., PH.D., F.I.C.E., F.I.STRUCT.E.
Regius Professor of Civil Engineering
University of Glasgow

H. M. Nelson
B.SC., A.R.T.C., M.I.C.E.
Senior Lecturer in Civil Engineering
University of Glasgow

Second Edition

Pitman

PITMAN BOOKS LIMITED
128 Long Acre, London WC2E 9AN

PITMAN PUBLISHING INC
1020 Plain Street, Marshfield, Massachusetts 02050

Associated Companies
Pitman Publishing Pty Ltd, Melbourne
Pitman Publishing New Zealand Ltd, Wellington
Copp Clark Pitman, Toronto

© The late W. T. Marshall and H. M. Nelson 1969, 1977

First published in Great Britain 1969
Reprinted (with corrections) 1972
Second Edition 1977
Reprinted 1978 (twice), 1982

Text set in 10/11 pt Monotype Times New Roman, printed and bound
in Great Britain at The Pitman Press, Bath

ISBN 0 273 00827 7

Preface to Second Edition

When this book was first published the use of the digital computer had altered the presentation of structural theory and while students had to be familiar with that presentation it remained more important that the fundamental principles of structural analysis were understood. Books had been published dealing with the application of the computer and the use of matrices in structural analysis, but the Authors endeavoured to write a book for undergraduate students of engineering which would be useful to them at all stages in their course. The understanding of basic principles requires practice by hand calculation of simple problems and the calculation techniques are those used in practice. At the same time, the more general approaches best suited to computer calculations are demonstrated.

The Authors also hoped to present a concise introduction to matrix methods of value to the older engineer making the transition from the 'least work' of his student days to the modern language of the Theory of Structures.

Books on this subject all have a problem with nomenclature and there is still no generally accepted system. The Authors tried to avoid inventing new symbols, but this leads to some minor inconsistencies. The simple title Structures was chosen for two main reasons: one that all the available permutations of analysis, fundamental structures and theory seemed to have been used up, and that this expression is generally used by students themselves. Since the book was written principally with the undergraduate in view, it appeared to be consistent to think mainly of him in the choice of title. A book for undergraduates cannot by its very nature contain original work and is a matter of selection and presentation. The Authors were very conscious of this and of their indebtedness to the many sources used.

The second edition retains nearly all the original but some introductory material, now judged to be unnecessary, has been deleted. The chapter dealing with the flexibility approach has been largely re-written and brief sections on stiffened plates and stability functions added to the chapter on compression members.

Professor Marshall died suddenly just after the revisions were drafted but not fully checked so that remaining errors are entirely the responsibility of the second author.

Glasgow 1977   H.M.N.

Contents

Contents

Chapter 8
Stiffness Method

Chapter 9
Successive Approximations

Chapter 10
The Energy Theorems

1

Introduction

1.1 Types of Structures

A structure can be defined as a body capable of resisting applied loads without any appreciable deformation of one part relative to another. The last part of this sentence is necessary in order to eliminate from the field of structures those mechanisms in which kinematic and dynamic effects are important.

The function of a structure, therefore, is to transmit forces from one point in space to another.

There are many bodies which do satisfy the definition given. A common example is the framed building, but the aeroplane and the ship are also structures. These three structures all have a common make-up. In each case the load, be it office loading on the floor, uplift pressure on the wings, or hydrostatic force on the hull, is carried by a slab or plate and transmitted to beams. In the framed building the load from the beams is transferred to the columns and then to the foundations; in the aeroplane and ship the beams transfer the load to bulkheads.

These framed structures may be differentiated from structures such as a dam, which relies on its weight to resist the forces applied to it by the water which it is holding back.

There are two broad sub-divisions of structures, therefore: those which resist applied loads by virtue of their geometry—these are known as *framed structures*; and those which resist applied loads by virtue of their weight—these are known as *mass structures*. This book will concentrate on the former.

1.2 Forces in the Members of a Framed Structure

The three basic elements of a framed structure are the *rod*, the *beam* (or *shaft*), and the *slab* (or *plate*). Each is characterized by a dominant form of load-carrying capacity and deformation.

A rod is essentially subject to load along the axis (tensile for a *tie*, compressive for a *column* or *strut*) and the deformation under load is a simple change in length (*see* Fig. 1.1).

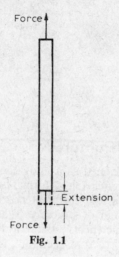

Fig. 1.1

Beams are supported at one or more points in their length and carry loads normal to the member axis (Fig. 1.2); their characteristic deformation by bending action is a plane curvature. A shaft carries torsional moments giving rise to a twist without deformation of the axis of the member.

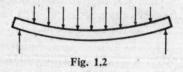

Fig. 1.2

In a plate it is necessary to consider two primary patterns: one is produced by a load normal to the plate (then usually called a *slab*) which deforms it like a beam but into a dished shape having two-way curvature; the other pattern is made by edge loads in the plane of the plate (*see* Figs. 1.3(*a*) and (*b*)).

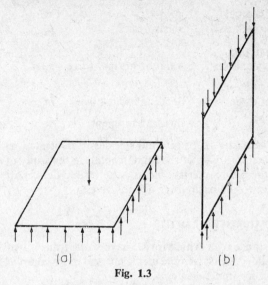

Fig. 1.3

In practice, a member is seldom subject to simple loadings. Often bending actions and axial forces are mixed. With plates the problem is further altered fundamentally if the plate is initially curved to form a shell structure (Fig. 1.4).

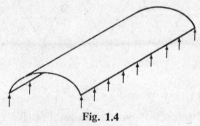

Fig. 1.4

In general, the force actions on any structural element may be six in number (as shown in Fig. 1.5)—

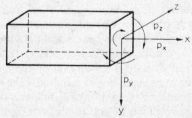

Fig. 1.5

Forces normal
   to element axis: Shear forces p_y, p_z

Force parallel
   to element axis: Axial or normal force p_x

Bending actions
   about three axes: Bending moments m_y, m_z

                           Twisting moment m_x

A characteristic set of internal stresses is associated with each of the basic actions. One of the fundamentals in the study of structures is to find these characteristic stress patterns and deformations and to relate them to the basic force actions given.

1.3   Structural Forms

As the distance over which forces have to be transmitted increases, the elementary forms become inefficient and more complex forms are built up from simple elements.

As the span of a beam increases it becomes uneconomical to use a solid beam and then an open beam or *truss* similar to that shown in Fig. 1.6 is used. Just as for a simple beam under vertical loading, the

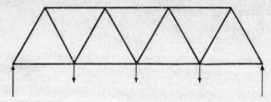

Fig. 1.6

forces in the upper chord members are compressive and those in the lower chord tensile. Shear forces are resisted by the inclined (*web*) members and the forces in these may be either tensile or compressive.

Figure 1.7 shows a tall building where the loads from the floor slabs are transmitted to beams and then transferred to columns. The wall loads are also carried by beams. Figure 1.8 shows a similar building in which the beams and columns have been omitted and the structure built of horizontal and vertical slabs.

The variety of structural form is very large and, although the two examples given are assemblies of essentially the same basic elements, other forms combine the actions of different basic elements. For example, the suspension bridge combines a flexible cable and a stiff beam.

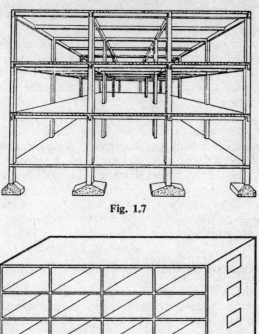

Fig. 1.7

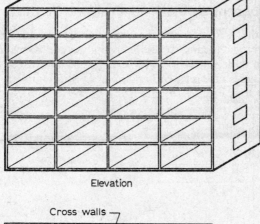

Elevation

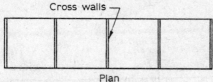

Plan

Fig. 1.8

A structure having all its members and loads in one plane is termed a *plane structure* and plane frames will be used in most of the examples here in order to avoid complicated arithmetic.

The more general case is the three-dimensional or *space structure*. The analysis of such structures follows the same fundamental principles as that for plane frames but the arithmetic is usually more troublesome.

The final object of structural analysis is to enable the engineer to design and construct a structure which is satisfactory in service. This means that it must have (1) strength, i.e. it must not collapse when loads are applied, and (2) stiffness, i.e. the deformations must not be excessive.

The major part of this book deals with the analysis of structures in order to determine the force actions produced by a given loading system. Chapter 3 deals with the stresses produced in members by different force actions. A knowledge of such stresses is necessary as a first step in ensuring that a structure has adequate strength. The type and magnitude of deformations resulting from given force actions are dealt with in Chapters 4 and 5. Applications of the principles given therein is required when determining whether the structure possesses adequate stiffness.

1.4 Types of Joint in a Framed Structure

It is clear that a framed structure must consist of a number of members joined together and so before an analysis can be made of such a structure it is necessary to know the type of joint used. There are two basic types of joint: *stiff* and *pinned*. Stiff joints are similar to those shown in Fig. 1.9 (*a*) and (*b*). Fig. 1.9 (*a*) shows a joint in a

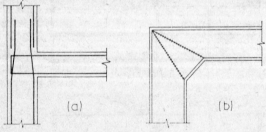

(a)   (b)

Fig. 1.9

reinforced concrete frame and Fig. 1.9 (*b*) one in a welded steel frame. The feature of the stiff joint is that a flexure of one member meeting at the joint has an effect on the other members as shown in Fig. 1.10; if a joint is perfectly stiff then the angle between the members is unaltered while the joint rotates.

A true pinned joint (Fig. 1.11) is now virtually an obsolete form of construction, but there are some structures where the effects of stiff connexions between the members do not appreciably affect the behaviour of the structure and the joint can be assumed to behave

Fig. 1.10   Fig. 1.11   Fig. 1.12

as a pinned joint. The feature of the pinned joint is shown in Fig. 1.12, where flexure of a member on one side of the joint does not affect other members meeting at that joint.

Hence a stiff joint can transmit bending from one side to another whilst a pinned joint cannot do so.

1.5 Types of Support in a Framed Structure

The load applied to a framed structure is transmitted to supports which will supply the necessary reactive forces to maintain equilibrium. In the plane frame the different types of support are shown in Fig. 1.13 (*a*), (*b*), (*c*) and (*d*). The encastré, fixed-end or built-in

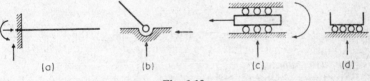

(a)   (b)   (c)   (d)

Fig. 1.13

support shown in Fig. 1.13 (*a*) is capable of supplying three reactive forces: (1) horizontal, (2) vertical, and (3) a fixing moment. The supports shown in Fig. 1.13 (*b*) and (*c*) can supply two reactive forces, whilst the roller bearing shown in Fig. 1.13 (*d*) can only supply a vertical reaction.

In the space frame there are additional types of support. If the horizontal axes are referred to as x, and y, and the vertical axis as z, then for a space frame a pin provides x, y and z reactive forces, a groove (as shown in Fig. 1.14) y and z, or x and z, and a roller support

Fig. 1.14

a z reaction only. Built-in supports in a space frame can be designed to supply moments about both the x- and y-axes or about one only of these axes.

1.6 Equations of Equilibrium

A structure is a body in static equilibrium without any dynamic
effects. Any force action in space can be replaced by a force at the
origin of the three reference axes shown in Fig. 1.5, together with a
moment about one of these axes.

For example, the force P acting at $(x, y, 0)$ can be replaced by P_x
at $(0, 0, 0)$ and a moment about the z-axis $M_z = Py$.

Then for equilibrium the resultant action in any direction must be
zero giving the 6 equations of equilibrium—

$$\sum P_x = 0 \qquad \sum M_x = 0$$
$$\sum P_y = 0 \qquad \sum M_y = 0 \qquad (1.1)$$
$$\sum P_z = 0 \qquad \sum M_z = 0$$

In the case of a plane frame lying in the xy-plane these equations
reduce to

$$\sum P_x = 0$$
$$\sum P_y = 0 \qquad (1.2)$$
$$\sum M_z = 0$$

Some structures can be completely analysed by the use of these
equations. Such structures are known as *statically determinate* or
simply *determinate*.

Structures which cannot be analysed solely by the application of
these equations are known as *statically indeterminate, hyperstatic* or
redundant.

1.7 Conditions for Determinancy

In the simple beam the condition for determinancy is that the supports must be such that there are not more than three reactive forces
In other words, they must be either built-in at one end with no

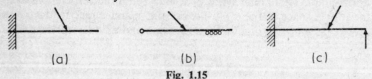

(a)   (b)   (c)

Fig. 1.15

support whatsoever at the other (i.e. a *cantilever*), or the supports
must consist of a pin at one end and a roller bearing at the other.
These two cases are shown in Fig. 1.15 (*a*) and (*b*). On the other
hand, the beam shown in Fig. 1.15 (*c*), which is built-in at one end
and simply supported at the other, is a statically-indeterminate structure since these are four reactive forces and these cannot be obtained

from the three equations of static equilibrium for a plane frame.

Trusses similar to that shown in Fig. 1.6 present a two-fold problem. First, as already mentioned, no more than three reactive forces must be given; and secondly there must be sufficient bars present to ensure that under any applied load system the frame is stable and not a mechanism. The first assumption with such trusses is that the joints are pinned; consequently, if the loads are applied at the joints, the forces in the members are direct tensions or compressions.

The simplest plane frame is the triangle shown in full lines in Fig. 1.16. This satisfies both the previously-mentioned conditions in that: firstly it will not change its shape when load is applied to it; and secondly, since not more than two members meet at any joint,

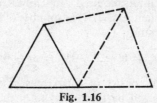

Fig. 1.16

the forces in all the members can be obtained by the application of the equations of static equilibrium.

The frame can be extended beyond this triangle by joining the first point outside the triangle to it by two bars, shown dotted in Fig. 1.16, thus giving once again a statically-determinate frame; a further extension to this frame can be made by a further two bars, shown chain-dotted in Fig. 1.16.

Thus if there are j joints in the frame the first 3 joints will require 3 bars, and the remaining $(j - 3)$ joints will require 2 bars each, hence the number of bars, n, is given by

$$n = 2(j - 3) + 3 = 2j - 3 \tag{1.3}$$

This gives the relationship between the number of bars and joints for a statically-determinate plane frame. This relationship will of course only hold good if the correct number of reactive forces has been provided, i.e. if the frame is supported on a pin at one end and a roller bearing at the other.

If the supports provided for the frame consist of two pins, then this foundation itself replaces the original triangle of Fig. 1.16, and in order to give a statically-determinate frame only 2 bars must be used for each joint outwith the original pinned supports. Thus if j_f denote the number of joints outwith the original pinned supports, the number of bars n_p for a statically-determinate frame is given by the equation

$$n_p = 2j_f \tag{1.4}$$

In a space frame the reactions must normally provide six forces, a minimum of one being along each axis. One system which supplies this is a pin (three forces), a groove (two forces) and a roller bearing (one force).

Provided that there are only six reactive forces, then the simplest space frame is a tetrahedron, which has three members meeting at a joint. The forces in this can be obtained by considering the equilibrium in the x-, y- and z-directions. Any additional joints will require to be connected to the original tetrahedron by three bars only in order to give a statically-determinate frame.

Thus the first four joints required six members, and each additional joint requires three members. If n_s = number of bars for a statically-determinate space frame

$$n_s = 6 + 3(j - 4) = 3j - 6 \tag{1.5}$$

Where the supports consist of three pins, then each additional joint must be connected by three bars to enable the forces in the members to be obtained from the equations of static equilibrium. If there are j_f such joints then for a space frame

$$n_{sp} = 3j_f \tag{1.6}$$

Example 1.1
Examine the sufficiency of bracing for the plane frames shown in Fig. 1.17 (a), (b), (c) and (d). Where the frame is deficient, sketch a suitable arrangement of bars to make good the deficiency, and if

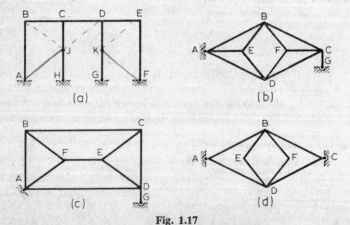

Fig. 1.17

more bars are used than are necessary state which bars could be removed. Justify your conclusions and choice.

All joints lettered are formed by frictionless pins.   (*London*)

SOLUTION (*a*)

The frame shown in Fig. 1.17 (*a*) has 6 free joints B, C, D, E, J and K; thus the number of bars for a statically-determinate frame is $2 \times 6 = 12$. The number of bars provided is 9 (AB, BC, CD, DE, EF, HJ, JC, GK and KD) and therefore 3 more bars are required. The problem is to decide where these bars must be placed.

A and H are pinned supports and to brace J a bar is required from each of these pins. JH already exists and one bar which must be added is AJ. Similarly, G and F are pinned supports, but K is not properly braced and a bar KF must be added. This gives J and K adequately braced. There are various alternatives now open.

One is to brace D by using the existing bar KD and adding a new bar JD. This makes D and F joints from which E can be braced. C can be fixed by the existing bars JC and DC. Finally B can be fixed by the existing bars CB and AB.

As an alternative, B could have been braced by the existing bar AB and a new bar BJ; C would then have been braced by the existing bars BC and JC; D by the existing bars CD and KD; and finally E by the existing bars DE and EF.

One solution is, therefore, to add bars AJ and KF, and either JD, BJ or KE.

SOLUTION (*b*)

The frame shown in Fig. 1.17 (*b*) has 5 free joints; thus 10 bars are required for a statically-determinate frame. A count shows that 11 have been provided and therefore one must be cut out. It does not, however, follow that the 11 bars have been placed so that each joint is properly braced, and in deciding on the final configuration a start may be made from the pinned supports A and G.

The joint D is braced by the existing bar AD and a new bar DG. From A and D, E is braced by the existing bars AE and DE. B is braced from A and E by the existing bars AB and EB. F is braced from D and B by the existing bars DF and BF. Joint C can be braced by the existing bars DC and BC, or by the bars GC and BC. Thus one solution is to cut out CF and CD and add the bar DG; there are, however, a number of alternative solutions.

SOLUTION (*c*)

The frame shown in Fig. 1.17 (*c*) has 5 free joints: 10 bars are required for a statically-determinate frame. Although 10 bars are provided it does not necessarily follow that these are placed to give adequate bracing and a check must be made. Joint D is braced to the supports A and G by the bars AD and GD. No other joint is properly braced since none is connected to two fixed points. E could be fixed by the existing bar DE and a new bar AE. C is fixed from D and E by the existing bars EC and DC. B is fixed from A and C by the existing bars AB and CB. Finally F is fixed from A and B by the existing bars AF and BF, and FE must be removed to retain determinacy. Alternatively it could have been braced from B and E. The first solution would cut out the existing bar EF and the second the bar AF. In each case the insertion of a bar AE is required.

SOLUTION (*d*)

The final frame, that shown in Fig. 1.17 (*d*), is the only one where the bracing is adequate and efficient. There are 4 free joints, and hence 8 bars are required for a statically-determinate frame. This is the number provided and a check shows that each joint is fixed to two previously fixed points: B is braced to the supports A and C by the bars AB and CB; D likewise by the bars AD and CD; E and F are then fixed from the joints B and D by the bars BE, ED and BF, DF respectively.

The solutions are shown in Figs. 1.18 (*a*), (*b*) and (*c*) respectively.

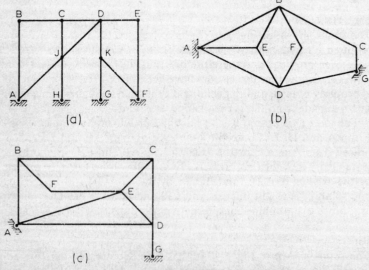

Fig. 1.18

1.8 Examples for Practice

1. Are the frames shown in Fig. 1.19 (*a*)–(*e*)
   (i) stable, (ii) statically determinate? If unstable, what change or addition would make them just stiff? (All joints are pinned.)

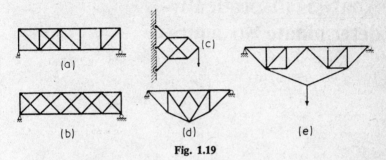

Fig. 1.19

2. How many degrees of redundancy have the structures shown in Figs. 1.20 (*a*)–(*g*)? (Structure (*c*) has two stiff joints and (*d*) has one.)

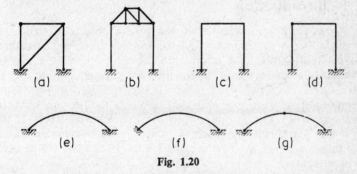

Fig. 1.20

3. The frames shown in Fig. 1.21 (*a*)–(*c*) have stiff joints where indicated.

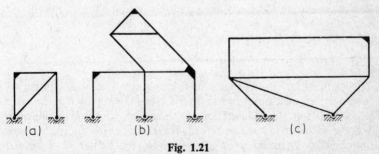

Fig. 1.21

How many redundancies are there in each?

2

Analysis of Statically-determinate Structures

2.1 Introduction

It has already been stated that one of the objects of structural analysis is the determination of the force actions in a structure. Statically-determinate structures are those in which these can be obtained by application of the equations of static equilibrium. Such structures are of two main types: trusses arranged to satisfy the relationship between bars and joints given in the previous chapter, and beams provided with not more than three reactive forces. In the first type the forces are generally direct tensions or compressions; in the second they are bending and shear. Each type will be dealt with separately.

In the analysis of trusses it is generally assumed: (1) that the joints are all pinned, and (2) that external loads are applied at the joints. Various methods which will be separately described can be used, but each is based on the equations of static equilibrium.

2.2 The Force Polygon

The principle of the polygon of forces states that if a number of forces acting at a point are in equilibrium the vector diagram is a closed polygon. In a plane frame or truss each joint is in equilibrium and the vector diagram can be drawn if there are not more than two unknowns. These may be unknown forces acting in a known direction or one force which is unknown in both magnitude and direction.

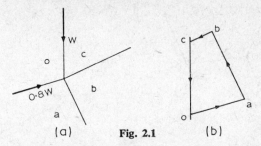

(a)   **Fig. 2.1**   (b)

As an example of the first case consider the joint shown in Fig. 2.1 (*a*), where there is an external load *W* and where the force in the member *oa* has been found to be 0·8*W*. The unknowns are the magnitudes of the forces in *ab* and *bc* and these can be determined by the vector polygon shown in Fig. 2.1 (*b*).

As an example of the second case consider the pinned support shown in Fig. 2.2 (*a*), where the forces in the members *ab*, *bc* and *cd* have already been determined and have the values 1·5*W*, 3·0*W* and 1·2*W*, respectively. The unknown force is the reaction *ad* and the two unknown quantities relating to it, namely its magnitude and direction, can be obtained from the vector polygon shown in Fig. 2.2 (*b*).

The vector polygon is drawn in each case for the joint and the directions shown for the forces are the directions in which they act on the joint in order to maintain its equilibrium. The force which the joint exerts on the bar is equal and opposite. It is the force in the bar itself which is of importance in structural analysis.

The first step in the analysis of a statically-determinate frame is normally to determine the reactions. This can be done either by using the equations of static equilibrium or by the funicular polygon. The latter can be used to determine the two reactive forces to any given system provided that the line of action of one resultant is known and that a point on the other is also known.

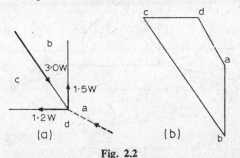

(a)   (b)

Fig. 2.2

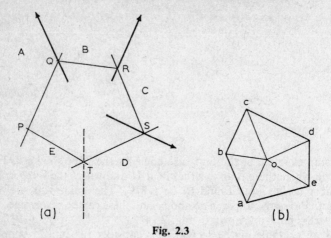

Fig. 2.3

The method is as follows. Consider the system of forces AB, BC and CD shown in Fig. 2.3 (*a*). These are balanced by two reactions one of which, T, acts in the line shown and the other acts through the point P. Draw the incomplete polygon *abcd* as in Fig. 2.3 (*b*) and select any pole *o*. Start the funicular polygon from the known point P by drawing PQ parallel to *ao* then draw QR parallel to *ob*, RS parallel to *oc*, ST parallel to *od* and join TP. From the pole *o* draw *oe* parallel to TP to cut a line through *d*, parallel to the known line of action of T in *e*. Then the reactions are given by *de* and *ae*.

Example 2.1
As an example of the determination of forces by this method consider the roof truss shown in Fig. 2.4, which is supported on a roller

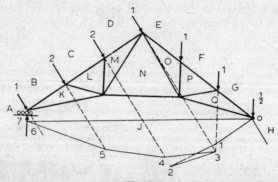

Fig. 2.4

bearing at the left-hand end and a pin at the right-hand support. The loads on the left-hand side are inclined and act perpendicular to the rafter, whilst those on the right act vertically.

SOLUTION

The force diagram *ab . . . gh* is drawn as shown in Fig. 2.5. A pole *o* is selected and the rays *oa, ob . . . oh* drawn. A start is made at the pinned support and the line 01 drawn parallel to *og*, 12 is then

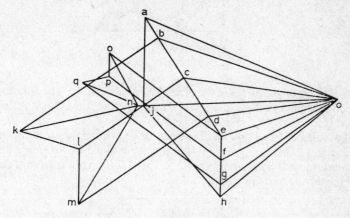

Fig. 2.5

drawn parallel to *fo* and the construction continued and completed by drawing 67 parallel to *oa*. The points 0 and 7 are joined and a ray *oj* drawn from the pole to cut the vertical through *a* in *j*. Then *aj* is the reaction at the left-hand support and *jh* gives the inclined reaction at the pinned support.

    The force polygon for the whole truss can then be drawn by considering any joint at which there are not more than two unknowns. A start can be made at the left-hand support where the known forces are *ja* and *ab*. The unknowns *bk* and *kj* are determined from the force polygon *jabk*; next the forces *kl* and *lc* are determined from the polygon *kbcl*; then *dm* and *ml* from *lcdm*; after which *mn* and *nj* are found from the polygon *klmnj*. The process is carried on until the point *q* on the force polygon is found. The work is then checked by joining *gq* on the force polygon. If the drawing is correct then *gq* in the force polygon will be parallel to GQ in the configuration diagram. If this is not the case then the draughtsmanship must be checked until the error has been traced.

2.3 The Free-body Diagram

Since a structure as a whole is a body in equilibrium, any part of it must also constitute a body in equilibrium. A portion of a structure can (in imagination) be cut free from the whole, and this portion will be in equilibrium under the action of any applied loads acting on it

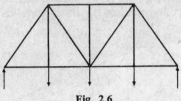

Fig. 2.6

and of the internal forces in the members which are cut. The diagram showing such a portion of a structure together with the forces acting on it is known as a *free-body* diagram.

Two typical free-body diagrams can be drawn for the truss shown in Fig. 2.6. One is Fig. 2.7 (*a*) and this is obtained by making a cut around a joint. The other, shown in Fig. 2.7 (*b*), results from a cut completely across the structure.

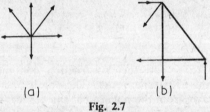

(a)   (b)

Fig. 2.7

The first type of diagram is used in the method of analysis known as the *Method of Joints* and the second is that known as the *Method of Sections*.

2.4 The Method of Joints

The joint shown in Fig. 2.7 (*a*) is in equilibrium, hence

$$\sum P_x = 0$$
$$\sum P_y = 0$$

Thus if only two unknown forces exist at this joint these can be determined by the application of the two equations given.

In the graphical method the force polygon was drawn for any joint at which not more than two unknown force vectors existed. In the method of resolution at the joints free-body diagrams are drawn for each joint in turn, starting with a joint at which there are not more than two unknowns.

If the angles of the members of the frame have trigonometrical properties which are well known, it will be possible for the student (after some practice) to form the equations and solve them mentally, so that this method is sometimes known as the *Method of Inspection*.

Example 2.2
Determine the forces in the members of the frame shown in Fig. 2.8 when it carries the loads shown.

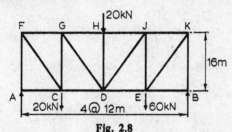

Fig. 2.8

SOLUTION

The first step is to find the reactions. Taking moments about B gives

$$R_A \times 48 = 20 \times 36 + 20 \times 24 + 60 \times 12$$

giving $\qquad R_A = 40 \text{ kN}$

hence $\qquad R_B = 100 - 40 = 60 \text{ kN}$

The free-body diagram is drawn for joint A. It is shown in Fig. 2.9 (*a*).

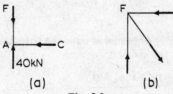

(a)   (b)

Fig. 2.9

Resolving horizontally gives $\quad AC = 0$

Resolving vertically gives $\qquad AF = R_A = 40 \text{ kN}$

The diagram shown in Fig. 2.9 (*b*) is drawn for joint F. Resolving vertically,

$$\tfrac{4}{5} \times FC = AF = 40$$

therefore $\qquad FC = 50$

Free-body diagrams are not drawn for the remaining joints but the analysis is as follows—

Joint C

Resolving vertically, $GC = \frac{4}{5} \times FC - 20$
                          $= 40 - 20 = 20$

Resolving horizontally, $CD = \frac{3}{5} \times FC = 30$

Joint G

Resolving vertically, $\frac{4}{5} \times GD = 20$
                          $GD = 25$

Resolving horizontally, $GH = \frac{3}{5} \times GD + FG$
                          $= 15 + 30 = 45$

Joint H

Resolving vertically, $HD = 20$
Resolving horizontally, $HJ = HG = 45$

Joint D

Resolving vertically, $\frac{4}{5} \times DJ = 20 - \frac{4}{5} \times GD$
                          $= 20 - 20 = 0$

Resolving horizontally, $DE = DC + \frac{3}{5} \times GD$
                          $= 30 + \frac{3}{5} \times 25 = 45$

Joint J

Resolving vertically, $JE = \frac{4}{5} \times DJ = 0$
Resolving horizontally, $JK = JH = 45$

Joint E

Resolving vertically, $\frac{4}{5} \times KE = 60$
                          $KE = 75$

Resolving horizontally, $BE = ED - \frac{3}{5} \times EK$
                          $= 45 - \frac{3}{5} \times 75 = 0$

Joint K

Resolving vertically, $KB = \frac{4}{5} \times KE = 60$

Joint B

The force KB already obtained is equal to the vertical reaction at B, thus checking that the work is correct.

The forces are shown in Fig. 2.10.

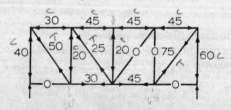

Fig. 2.10

2.5 The Method of Sections

This method consists of applying the equations of static equilibrium to a free-body diagram similar to Fig. 2.7.

If, for example, it is required to find the forces in the members X, Y and Z of the frame shown in Fig. 2.11 (*a*), a section is cut through the panel and the equilibrium of the left-hand portion of the truss considered. This is shown in Fig. 2.11 (*b*). The forces acting are the external loads R_A and W_1 and the internal forces X, Y and Z. These can be determined as follows.

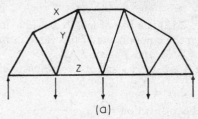 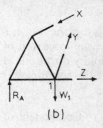

(a)   (b)

Fig. 2.11

Take moments about point 1; this gives a simple equation to determine X. Resolve vertically to obtain Y after X has been determined. Finally, knowing X and Y, resolve horizontally to determine Z.

This method is very long if used to determine all the forces in the members of a frame (the frame shown in Fig. 2.11, for example, would require consideration of eight or nine sections), but is useful if the forces in only a few members are required.

No worked example of this method is given, but it is used in the section on influence lines for framed structures which is given later in this chapter.

2.6 The Method of Tension Coefficients

For a space frame three equations of static equilibrium have to be satisfied at each joint and application of these equations to each joint in turn enables the forces in the members to be found.

The method best adapted for this purpose is similar to that used in the method of resolution at the joints but is somewhat simplified by using the tension coefficient of the member.

The fundamental principles of this method will be given for a plane frame, followed by an example showing the application to a space frame.

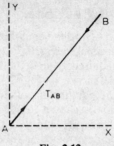

Fig. 2.12

In a plane frame, if AB (Fig. 2.12) is a bar of length L_{AB}, having a tensile force in it of T_{AB}, then the components of this force in the x- and y-directions are $T_{AB} \cos$ BAX and $T_{AB} \sin$ BAX.

If the coordinates of A and B are x_A, y_A and x_B, y_B respectively then

$$\text{Component of } T_{AB} \text{ in the x-direction} = T_{AB} \frac{x_B - x_A}{L_{AB}}$$

$$= t_{AB}(x_B - x_A)$$

where $t_{AB} = T_{AB}/L_{AB}$ and is known as the tension coefficient of the bar AB.

Similarly the component in the y-direction $= t_{AB}(y_B - y_A)$.

If at the joint A in the frame there are a number of bars AB, AC, . . . , AN and external loads X_A, Y_A acting in the x- and y-directions, then, since the joint is in equilibrium the sum of the components of the external and internal forces must be zero in each of these directions.

Expressing these relationships symbolically gives the equations

$$t_{AB}(x_B - x_A) + t_{AC}(x_C - x_A) \ldots + t_{AN}(x_N - x_A) + X_A = 0 \tag{2.1}$$

$$t_{AB}(y_B - y_A) + t_{AC}(y_C - y_A) \ldots + t_{AN}(y_N - y_A) + Y_A = 0 \tag{2.2}$$

A similar pair of equations can be formed for each joint in the frame, giving in all $2j$ equations in the case of a frame having j joints. These equations will contain the tension coefficients as unknowns and if the frame has n members then there are n unknown tension coefficients. But for a plane frame $n = 2j - 3$, hence there are three superfluous equations. These can be used either to determine the reactions or to check the values of the tension coefficients obtained from the previous equations.

In a space frame each joint has three coordinates and the forces have components in three directions, x, y, and z. Thus if there are j joints in a space frame the consideration of the equilibrium in the three directions produces $3j$ equations containing n unknown tension coefficients. But $n = 3j - 6$; hence there are six superfluous equations which can be used either to determine the reactions or to check the values of the tension coefficients.

Having found the tension coefficients t_{AB}, the force in the bar is the product $t_{AB}L_{AB}$.

This is known as the *Method of Tension Coefficients* and the equations of equilibrium are built up by assuming that the bars are in tension. A bar that is in compression has a negative tension coefficient.

The procedure in using the method is as follows—

(1) Take positive directions for x, y and z.

(2) Assume all members in tension.

(3) Write down equations for each joint in the frame. The terms are positive or negative as the tensile forces tend to move the joints in the positive or negative directions of x, y or z. The student should note particularly that the whole build-up of equations contains terms such as $t_{AB}(x_B - x_A)$ and $t_{AB}(x_A - x_B)$. Thus a simple check on the accuracy of the build-up is to note that a positive coefficient to one of the unknowns must be accompanied by an equal negative one in the equation for a connected joint.

(4) Solve equations for t_{AB}, etc.

(5) Check values for t_{AB} from equations of static equilibrium.

(6) Calculate $T_{AB} = L_{AB}t_{AB}$.

Example 2.3

The space frame shown in plan in Fig. 2.13 has the pinned supports A, B and C at the same level. DE is horizontal and at a height of 10 m

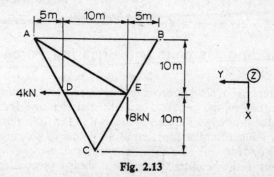

Fig. 2.13

above the plane of the supports. Calculate the forces in the members when the frame carries loads of 8 kN and 4 kN acting in a horizontal plane at E and D respectively.

SOLUTION

In the case of a space frame having pinned supports the number of bars for a statically-determinate frame is $3j_f$, where $j_f =$ number of free joints. The number of equilibrium equations will be $3j_f$, hence sufficient equations are obtained to solve for the unknown tension coefficients.

Table 2.1(a) **Equilibrium Equations**

Joint	Direction	Equations	Tension coefficients
D	x y z	$-10AD + 10CD = 0$ $+5AD - 10DE - 5DC + 4 = 0$ $10AD + 10CD = 0$	$AD = CD = 0$ $DE = 0.4$
E	x y z	$10CE - 10AE - 10BE + 8 = 0$ $10DE + 15AE + 5CE - 5BE = 0$ $10CE + 10BE + 10AE = 0$	$CE = -0.4$ $AE = 0$ $BE = 0.4$

The tension coefficient equations are built up as shown in Table 2.1 (*a*) using the positive directions indicated in the figure. The forces in the members are given in Table 2.1 (*b*).

Table 2.1(b) **Forces in Members**

Member	Length (m)	Tension coefficients	Load
AD	15	0	0
DC	15	0	0
DE	10	0.4	4 kN tension
AE	20.6	0	0
CE	15	-0.4	6 kN compression
BE	15	+0.4	6 kN tension

2.7 Bending Moment and Shearing Force Diagrams

In the beam type of structure, analysis involves the determination of the magnitude of the bending moment and shearing force at all points of the beam. The results are shown in the form of diagrams giving the value of either function at all points in a beam under a given load system.

The signs of the bending moment and shearing force are defined

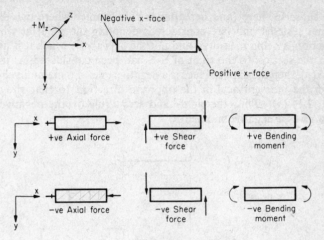

Fig. 2.14

relative to a right-handed set of axes as shown in Fig. 2.14, which also gives the positive and negative faces of a beam along the x-axis. Positive moments are defined as those acting in a right-hand screw direction on the positive face.

Consider the beam ACB shown in Fig. 2.15, in which ACB is the

Fig. 2.15

positive direction of x. Taking the free-body diagram Fig. 2.16 (a) for the section to the left of S–S, the direction of the moment acting on the positive face to maintain equilibrium is as shown. This acts in an opposite direction to the positive direction of M_z shown in Fig. 2.14, and is therefore negative in sign. Hence

$$M_{SS} = -R_A x$$

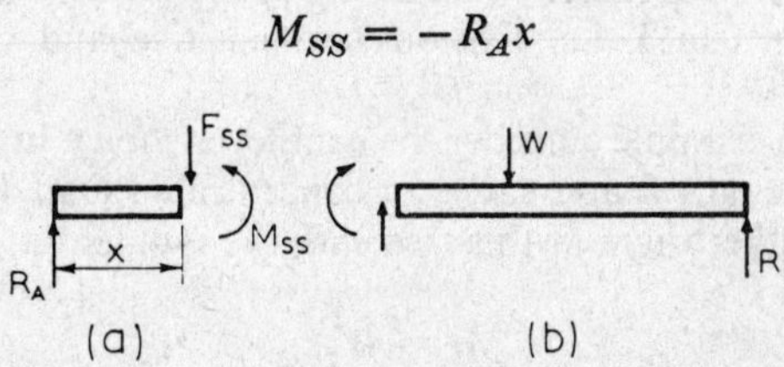

Fig. 2.16

the subscript here (and hereafter in this chapter where two-dimensional systems only are considered) denoting the point at which a function is being measured and not the axis about which it is acting.

If the section to the right of S–S had been considered, as in Fig. 2.16 (*b*), then the exposed face is a negative one. To maintain equilibrium the moment acts in the opposite direction to that shown in Fig. 2.16 (*a*). Thus the right-hand screw rule on the negative face also gives a negative moment.

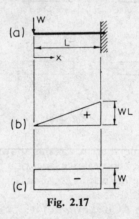

Fig. 2.17

In the common case where the system of coordinates is such that *x* is horizontal and positive from left to right and *y* is positive downwards in the plane of the paper, then it can be said that, considering the free-body diagram to the left of the section shown in Fig. 2.16 (*a*), a clockwise moment acting on the positive face is positive and an anticlockwise one negative.

The sign of the shearing force is such that forces acting in a positive direction of *y* on a positive face of *x* are positive. Thus, for example, in the free-body diagram shown in Fig. 2.16 (*a*) the shearing force on the positive face is acting downwards, i.e. in the positive direction of *y* and is positive.

Axial forces are positive when acting in the positive direction of *x* on a positive *x*-face, i.e. tensions are positive and compressions negative.

As a simple example consider the cantilever shown in Fig. 2.17 (*a*), which has a length *L* and carries a concentrated load *W* at the free end. Taking the origin at this free end, the moment at a distance *x* is given by

$$M = Wx$$

This is a straight-line relationship between the two quantities, the maximum value of M being obtained when $x = L$, and the diagram being as shown in Fig. 2.17 (*b*).

At a distance x from the free end the shearing force F is given by

$$F = -W$$

This is independent of x, and the shearing force diagram for the beam is shown in Fig. 2.17 (*c*).

As a further example consider the simply-supported beam shown in Fig. 2.18 (*a*), which carries a load which varies in intensity from

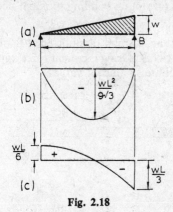

Fig. 2.18

zero at one end to w at the other. The first problem is to find the reactions. To do this take moments about one support. If moments are taken about B, then

$$R_A \times L = w \times \tfrac{1}{2}L \times \tfrac{1}{3}L = \tfrac{1}{6}wL^2$$

Hence $\qquad R_A = \tfrac{1}{6}wL \quad$ and $\quad R_B = \tfrac{1}{3}wL$

At any section at a distance x from the left-hand support the moment M is given by

$$M = -R_A \times x + \tfrac{1}{2}w\,\frac{x^2}{L} \cdot \tfrac{1}{3}x$$

$$= -\tfrac{1}{6}w\,\frac{x}{L}(L^2 - x^2)$$

This is a cubic parabola relationship between M and x, the former having zero value when $x = 0$ and $x = L$. For a maximum numerical value, $dM/dx = 0$, giving

$$L^2 - 3x^2 = 0 \quad \text{or} \quad x = L/\sqrt{3}$$

$$\text{Maximum } M = \frac{w}{6\sqrt{3}}(L^2 - \tfrac{1}{3}L^2) = \frac{wL^2}{9\sqrt{3}}$$

The complete diagram is given in Fig. 2.18 (*b*).

The shearing force F at a point distant x from the left-hand support is given by

$$F = +R_A - \tfrac{1}{2}wx^2/L = +w(\tfrac{1}{6}L - \tfrac{1}{2}x^2/L)$$

This gives a parabolic relationship which is shown in Fig. 2.18 (c). There is no mathematical maximum to the expression for F but the actual maximum occurs at the supports and is $\tfrac{1}{6}wL$ at A and $\tfrac{1}{3}wL$ at B of opposite sign.

The zero shearing force occurs when $x = L/\sqrt{3}$ which is also the point at which the maximum bending moment occurs.

Example 2.4

A beam AF, 20 m long, is simply supported at B and D and is loaded as shown in Fig. 2.19 (a). Draw the shear force and bending moment diagrams stating all the significant values.

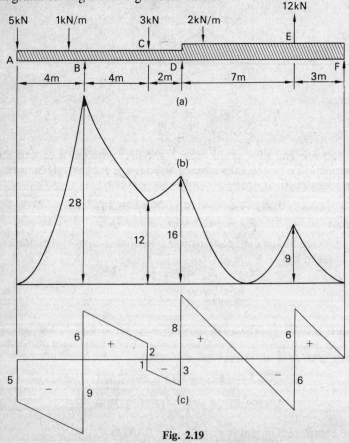

Fig. 2.19

SOLUTION
The first step in the solution is to obtain the reactions. Taking moments about D to determine R_B gives

$$R_B \times 6 = 5 \times 10 + 10 \times 5 + 3 \times 2 + 12 \times 7 - 20 \times 5$$

giving
$$R_B = \tfrac{90}{6} = 15 \, \text{kN}$$

Taking moments about B to determine R_D gives

$$R_D \times 6 = 20 \times 11 + 3 \times 4 + 10 \times 1 - 12 \times 13 - 5 \times 4$$

giving
$$R_D = \tfrac{66}{6} = 11 \, \text{kN}$$

The total load applied to the beam is $(5 + 10 + 3 + 20 - 12) = 26$ kN and it is seen that the calculated value of $R_B + R_D = 15 + 11$ checks with this.

To draw the bending moment diagram consider first the portion of the beam AD and measure x from A, then the moment M_x at x is given by

$$M_x = 5x + \tfrac{1}{2}x^2 - [15(x - 4)] + [3(x - 8)]$$

The use of the square brackets [ ] denotes that the term is only considered when the sign is positive. The relationship is parabolic with discontinuities occurring when $x = 4$ and $x = 8$. The critical values are

$$M_B = 5 \times 4 + \tfrac{16}{2} = 28 \, \text{kN-m}$$
$$M_C = 5 \times 8 + \tfrac{64}{2} - 15 \times 4 = 12 \, \text{kN-m}$$
$$M_D = 5 \times 10 + \tfrac{100}{2} - 15 \times 6 + 3 \times 2 = 16 \, \text{kN-m}$$

If the portion of the beam DF is now considered with x measured from F the moment M_x at x is given by

$$M_x = x^2 - [12(x - 3)]$$

the square-bracket term having the same significance as before. The relationship is parabolic and the critical values are

$$M_E = 9 \, \text{kN-m}$$
$$M_D = 100 - 12 \times 7 = 16 \, \text{kN-m}$$
$$M = 0 \text{ at } x = 6$$

The fact that the same value of M_D is obtained from each section of the beam provides a partial check on the work.

The complete diagram is shown in Fig. 2.19 (*b*). This is drawn with the positive moment above the horizontal zero (and a negative below), and it is noted that the moment is always drawn on the tension side of the beam.

To draw the shear force diagram proceed in a similar way. The shear force F at a point X distant x from A is given by

$$F = -5 - x + [15]_{x \geqslant 4} - [3]_{x \geqslant 8}$$

The critical values are

$$F_A = -5 \text{ kN}$$
$$F_B = -9 \text{ kN} \quad \text{or} \quad +6 \text{ kN}$$
$$F_C = +2 \text{ kN} \quad \text{or} \quad -1 \text{ kN}$$
$$F_D = -3 \text{ kN}$$

Considering the portion FD with x measured from F,

$$F_x = +2x - [12]_{x \geqslant 3}$$

The critical values are

$$F_E = +6 \text{ kN} \quad \text{or} \quad -6 \text{ kN}$$
$$F_D = +8 \text{ kN}$$

The total change of shear at D is $3 + 8 = 11$ kN, which is equal to the calculated value of R_D, thus providing a check on the work.

The shear force diagram is shown in Fig. 2.19 (*c*).

2.8   Rolling Loads and Influence Lines

In the previous sections diagrams have been drawn which show the bending moment at any point on a beam when the load occupies a given position.

Many structures, however, have to be able to support loads which move, and a method must be evolved for dealing with them. The one usually employed is the influence line. An influence line can be defined as a diagram which shows the variation of a function *at a given point* in a structure as a unit load travels from one support to the other. The italicized words are important, for the influence line must be drawn for a given point in the span. A bending moment diagram is a diagram which shows how the bending moment at all points in the span varies with the load in a given position. An influence line for bending moment at a given point is, however, a diagram which shows how the bending moment *at that point* varies as a unit load is at different points in the span.

2.9   Influence Lines for Bending Moment and Shearing Force

Before these diagrams can be drawn the point must be selected. Take the point X in Fig. 2.20 (*a*); the problem then is to draw a diagram showing how the bending moment at X varies as unit load travels across the span. This diagram will be the influence line for bending moment at X.

At any given instant let the load be at a distance a from A. Then considering the forces to the right of X gives the bending moment M_X at X as

$$M_X = -R_B(L - x) + [(a - x)]$$

By moments about A,

$$R_B = a/L$$

Thus M_X is a linear function in a and reaches its maximum value when $a = x$. When $a = 0$ and $a = L$ the values of M_X are zero and the influence line for bending moment is shown in Fig. 2.20 (b).

The shearing force F_X at X is given by

$$F_X = -R_B + [1]_{a>x}$$
$$= -a/L + [1]$$

This is also a linear relationship, but the diagram consists of two parallel lines as shown in Fig. 2.20 (c).

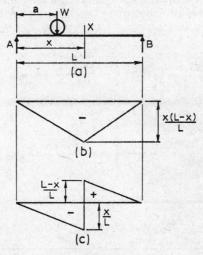

Fig. 2.20

Influence lines are drawn for unit loads because the effect of any given load W is obtained simply by multiplying the effect of a unit load by W.

Thus if at any given instant loads W_1, W_2 and W_3 occupy positions on the span such that the ordinates to the bending moment influence line are m_1, m_2 and m_3 respectively, then the moment at X when the loads are in this position is given by

$$M_X = W_1 m_1 + W_2 m_2 + W_3 m_3$$

If W_1, W_2 and W_3 are a given train of loads which can cross the beam, then the bending moment at X has its maximum value when the function $W_1 m_1 + W_2 m_2 + W_3 m_3$ has its maximum value.

Although an influence line is drawn for a point load, it can also be used for a distributed load.

Consider the beam AXB of Fig. 2.21 (*a*), where the influence line for shear at X is as shown in (*b*); the shearing force at this point due to a uniformly-distributed load of intensity *w* acting over the length *c* is obtained as follows.

Consider an element of load at D of length δc; the load due to this element is $w\,\delta c$ and the corresponding ordinate to the influence line is *h*; hence the shear at X due to the element is $wh\,\delta c$. Therefore

$$\text{Total shear at X due to load } cw = w \int h \, dc$$

But $\int h \, dc$ is the area under the influence line bounded by the length *c*. Thus with a distributed load the area under the influence line is

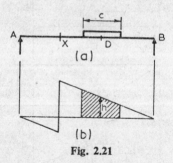

(a)

(b)

Fig. 2.21

multiplied by the intensity of loading to give the effect of the load on the value of the function.

The maximum value of the bending moment at X due to a given load system will occur when the function $\Sigma\,Wm$ has its maximum value. This will depend on the load system. The following systems will be examined in turn—

(1) Single concentrated load W.
(2) Uniformly-distributed load of intensity *w* and length $\geqslant L$.
(3) Uniformly-distributed load of length *c*, where $c < L$.
(4) Any system of point loads.

2.10 Maximum Bending Moment and Shearing Force Due to a Single Moving Load

The influence line for the bending moment at X has its maximum ordinate when the load is at X, the magnitude being $x(L - x)/L$. Thus the maximum bending moment at X due to a single moving load W is $Wx(L - x)/L$. The diagram showing how this function varies with x is known as a *maximum bending moment diagram* and for a single rolling load is the parabola shown in Fig. 2.22. The maximum value of this function occurs when $x = \frac{1}{2}L$, in which case M_X has the value $\frac{1}{4}WL$.

The influence line for shearing force given in Fig. 2.20 (*c*) shows that the maximum values of both positive and negative shear occur when the load is immediately to the right or left of X, their values being $(L - x)/L$ and x/L respectively. Thus the maximum negative shear at X due to a single load W is Wx/L and the maximum positive shear is $W(L - x)/L$. The maximum shearing force diagram is

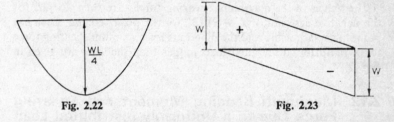

Fig. 2.22   Fig. 2.23

given in Fig. 2.23. An important fact which should be noted from this diagram is that there are both positive and negative values for the maximum shearing force at any point.

2.11 Maximum Bending Moment and Shearing Force Due to a Uniformly-distributed Load Longer than the Span

In dealing with a uniformly-distributed load it is the area under that part of the influence line covered by the load which is critical. Examination of Fig. 2.20 (*b*) shows that this area has its maximum value when the whole span is covered by the load, in which case

$$M_X = \tfrac{1}{2} \times w \times L \times \frac{x(L - x)}{L}$$

$$= \tfrac{1}{2}wx(L - x)$$

This function is parabolic, giving the maximum bending moment diagram for a uniformly-distributed load longer than the span similar to Fig. 2.22, but with the maximum value of $\frac{1}{8}wL^2$ when $x = \frac{1}{2}L$.

Inspection of the influence line for shearing force shows that the maximum negative area corresponds to the case where the load covers the portion AX of the span, and the maximum positive area when XB is covered.

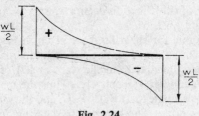

Fig. 2.24

The values of the maximum shearing forces are then $wx^2/2L$ for the negative case and $w(L - x)^2/2L$ for the positive case. Each is a parabolic relationship and the diagrams for maximum shearing force for a uniformly-distributed load longer than the span are given in Fig. 2.24.

2.12 Maximum Bending Moment and Shearing Force Due to a Uniformly-distributed Load Shorter than the Span

The load is of length c and the maximum value of the bending moment will occur when the load is placed so that the area covered by the length c has its maximum value. Let this be when the load is in the position FG as shown in Fig. 2.25. Then

$$\text{Maximum ordinate EX} = x(L - x)/L$$

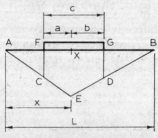

Fig. 2.25

Let XF $= a$ and XG $= b$, with $c = a + b$. Then

$$CF = (x - a)(L - x)/L$$

and $$\text{Area } CFXE = a(2x - a)(L - x)/2L$$

$$DG = x(L - x - b)/L$$

$$\text{Area } DGXE = bx(2L - 2x - b)/2L$$

Thus the area under the influence line A is given by

$$A = \frac{1}{2L}(2x - a)(L - x)a + \frac{1}{2L}(2L - 2x - b)bx$$

Substituting $b = (c - a)$ gives A in terms of the one variable a as

$$A = \frac{1}{2L}(2x - a)(L - x)a + \frac{1}{2L}(2L - 2x - c + a)x(c - a)$$

$$= \frac{1}{2L}\{(2xL - 2x^2)c - a^2(L - x) - (c - a)^2x\}$$

Therefore

$$M_X = \frac{cwx}{L}\left\{(L - x) - \frac{a^2}{2cx}(L - x) - \frac{(c - a)^2}{2c}\right\} \quad (2.1)$$

For A to be a maximum, $dA/da = 0$ or $-2a(L - x) + 2(c - a)x = 0$

i.e. $$\frac{x}{L - x} = \frac{a}{c - a}$$

or $$\frac{x}{L} = \frac{a}{c} \quad \text{i.e. } CF = GD$$

Thus the maximum bending moment at any given point occurs when that point divides the load and span in the same ratio or when

$$\frac{\text{Load to the left}}{\text{Total load}} = \frac{\text{Span to the left}}{\text{Total span}}$$

The actual value of the maximum bending moment due to a uniformly-distributed load of intensity w and length c is obtained by substituting $a = cx/L$ in Eq. (2.1), giving

$$M_X = \frac{cwx}{L}\left\{(L - x) - \frac{cx}{2L^2}(L - x) - \frac{c}{2}\left(1 - \frac{x}{L}\right)^2\right\}$$

$$= \frac{cwx}{L}(L - x)\left(1 - \frac{c}{2L}\right)$$

The maximum negative shearing force will occur when the length c occupies the portion of the span immediately to the left of X. In this case the area under the influence line is

$$\frac{c}{2}\left(\frac{x}{L} + \frac{(x-c)}{x} \cdot \frac{x}{L}\right)$$

or $(c/2L)(2x - c)$. Thus the maximum value of the negative shearing force is

$$\frac{cw}{2L}(2x - c)$$

The maximum positive shearing will occur when the length c is placed immediately to the right of X then the area under the influence line is

$$\frac{c}{2}\left(\frac{L-x}{L} + \frac{L-x-c}{L}\right) = \frac{c}{2L}(2L - 2x - c)$$

giving the maximum positive shearing force due to a uniformly-distributed load of length c and intensity w as

$$\frac{wc}{2L}(2L - 2x - c)$$

2.13 Maximum Bending Moment and Shearing Force Due to a System of Point Loads

Consider in the first instance the two loads W_1 and W_2 separated by a distance a as shown in Fig. 2.26, which also gives the influence line

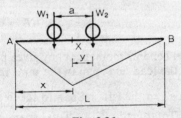

Fig. 2.26

for moment at X. When the load W_2 is at X the moment M_2 is given by

$$M_2 = W_2 x(L - x)\frac{1}{L} + W_1(x - a)(L - x)\frac{1}{L}$$

When the load W_1 is at X the moment M_1 is given by

$$M_1 = W_2 x(L - x - a)\frac{1}{L} + W_1 x(L - x)\frac{1}{L}$$

When neither load is at X but the load W_2 is at a distance y to the right of X, the moment M_Y is given by

$$M_Y = W_2x(L - x - y)\frac{1}{L} + W_1(x - a + y)(L - x)\frac{1}{L}$$

$$= W_2x(L - x)\frac{1}{L} - W_2yx\frac{1}{L} + W_1(x - a)(L - x)\frac{1}{L}$$

$$+ W_1y(L - x)\frac{1}{L}$$

$$= M_2 - \frac{y}{L}\{W_2x - W_1(L - x)\}$$

But M_Y can also be re-written as

$$M_Y = W_2x(L - x - a)\frac{1}{L} + W_2x(a - y)\frac{1}{L} + W_1x(L - x)\frac{1}{L}$$

$$- W_1(L - x)(a - y)\frac{1}{L}$$

$$= M_1 + \frac{(a - y)}{L}\{W_2x - W_1(L - x)\}$$

It is seen that the term in the curled brackets is common to both expressions for M_Y. If it is positive, i.e. $W_2x > W_1(L - x)$, then $M_2 > M_Y > M_1$. If it is negative, i.e. $W_2x < W_1(L - x)$, then $M_1 > M_Y > M_2$.

Thus the maximum bending moment is either M_1 or M_2, or, in other words, the maximum bending moment at a point occurs when a load is at that point.

The next step is to find which load has to be placed at the point in order to give the maximum bending moment at it.

Let the train of loads have a resultant P which at any given time is distant y from the left-hand support. Let P_L be the resultant of the loads to the left of X, with the distance between the two resultants being a. The position is then as shown in Fig. 2.27 and the moment

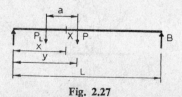

Fig. 2.27

M_X at X is obtained by taking moments of the forces to the left and is given by

$$M_X = -P(L - y)x\frac{1}{L} + P_L(x - y + a)$$

The variable is y, and

$$\frac{dM_X}{dy} = \frac{Px}{L} - P_L$$

For a maximum value of M_X, dM_X/dy passes from positive through zero to a negative value.

If

$$\frac{dM_X}{dy}$$

is positive,

$$\frac{P}{L} > \frac{P_L}{x}, \qquad \text{or} \qquad \frac{x}{L} > \frac{P_L}{P}$$

If it is negative,

$$\frac{P}{L} < \frac{P_L}{x} \qquad \text{or} \qquad \frac{x}{L} < \frac{P_L}{P}$$

Hence the critical load is the one which causes the expression

$$\frac{x}{L} - \frac{P_L}{P}$$

to change sign.

A problem which is similar to that of finding the load position to give the maximum bending moment at a given point is that of finding the load position to give the maximum bending moment under a given load.

This is dealt with in the following way. W_1 is a given load in a system, P being the resultant of all the loads in the system and P_L the resultant of the loads to the left of W_1 (Fig. 2.28). Let W_1 be at a distance x from the left-hand support when the maximum moment occurs under it. Let P and P_L be distant a_1 and a_2, respectively, from

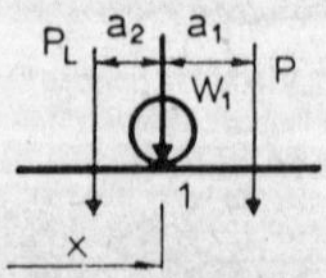

Fig. 2.28

W_1. Then the moment M_1 under W_1 is given by

$$M_1 = - R_A x + P_L a_2$$

$$= - \frac{Px}{L}(L - x - a_1) + P_L a_2$$

For this to be a maximum

$$\frac{dM_1}{dx} = 0, \quad \text{or} \quad L - 2x - a = 0$$

$$x = \tfrac{1}{2}L - \tfrac{1}{2}a_1$$

Thus the maximum bending moment under a given load occurs when the centre of the span bisects the distance between that load and the resultant of the system.

It is not convenient to give any definite rule for the load position for maximum shearing force as this depends on the load system. Normally the maximum negative shear occurs when the right-hand load of the system is at the point and the load wholly to the left, and the maximum positive shear when the left-hand load is at the point and the load wholly to the right. In some cases, however, where the end loads of a system are small compared with the other loads, this rule will not apply. Trial and error methods must be used, but very few trials are necessary.

Example 2.5
The span of a girder is 10 m. The live load system shown in Fig. 2.29(a) may cross the span in either direction. Determine the

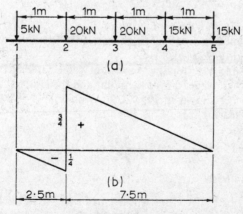

Fig. 2.29

maximum bending moment in the girder and obtain the equivalent uniform loading. Draw the influence line for shear at the left-hand quarter point and calculate the maximum value.

SOLUTION

To solve the first part of the question it is necessary to find the position of the centre of gravity of the load system. Let this be at a distance $\bar{x}$ from load 1. Taking moments about the line of action of this load gives

$$(5 + 20 + 20 + 15 + 15)\bar{x}$$
$$= 20 \times 1 + 20 \times 2 + 15 \times 3 + 15 \times 4$$

leading to

$$\bar{x} = 2 \cdot 2 \text{ m}$$

The maximum bending moment will occur when wheel 3 is at 0·1 m from the centre or when wheel 4 is at 0·4 m from the centre. It should be clear that the former gives a bigger value than the latter. With the load in this position the reaction at the left-hand end is $(4 \cdot 9/10) \times 75$ kN and the moment M_3 under the load is given by

$$M_3 = -\left(\frac{4 \cdot 9}{10} \times 75\right) \times 4 \cdot 9 + 5 \times 2 + 20 \times 1$$
$$= -150 \text{ kN-m}$$

The equivalent uniformly-distributed load w_e to give the same maximum bending moment is given by

$$w_e = \frac{150 \times 8}{10 \times 10} = 12 \text{ kN/m run}$$

The influence line for shearing force at the left-hand quarter-point is shown in Fig. 2.29(*b*).

One possible case for the maximum value under the given load system is when wheel 5 is at the point and the load system (reversed from the order shown in Fig. 2.29(*a*)) placed wholly to the right. Then, the shearing force F_5 is given by

$$F_5 = +\frac{3}{4}\left(15 + \frac{6 \cdot 5}{7 \cdot 5} \times 15 + \frac{5 \cdot 5}{7 \cdot 5} \times 20 + \frac{4 \cdot 5}{7 \cdot 5} \times 20 + \frac{3 \cdot 5}{7 \cdot 5} \times 5\right)$$
$$= +44 \cdot 6 \text{ kN}$$

Another possibility is for wheel 1 to be to the left, wheel 2 at the quarter-point and the remaining loads to the right. In this case the shearing force F_2 is given by

$$F_2 = +\frac{3}{4}\left(20 + \frac{6\cdot5}{7\cdot5} \times 20 + \frac{5\cdot5}{7\cdot5} \times 15 + \frac{4\cdot5}{7\cdot5} \times 15\right) - \frac{1}{4} \times 5 \times \frac{1\cdot5}{2\cdot5}$$

$$= +42\cdot4 \text{ kN}$$

This is less than F_5, hence the maximum shearing force is 44·6 kN.

2.14 Influence Lines for Braced Structures

A trussed girder is frequently used in bridge design where the load is generally moving, so that the student will realize the value of the influence line for this type of structure. In a beam it has been emphasized that the influence line can be drawn only for one point, but in the braced structure one member must be selected and the diagram drawn for it.

Let the problem be to draw the influence lines for the forces in the members DE and EF of the truss shown in Fig. 2.30(*a*).

Let us deal first with the force in the member DE. There are three possibilities: $x < L_1$, $x > L_2$ and $L_2 > x > L_1$.

In the first, the shear across the panel is x/L and if a section is cut through the panel at P–P then the force in DE must have a vertical component of x/L acting downwards to balance this shear, i.e.

$$\text{Force in DE is} \quad F_{DE} = -\frac{x}{L} \text{cosec } \theta$$

the negative sign denoting compression and $\theta = \angle$ CDE.

In the second, the shear is $(x/L - 1)$ and

$$\text{Force } F_{DE} = \left(\frac{L - x}{L}\right) \text{cosec } \theta$$

The third case is that of the load on the panel CD. This member then acts at a simply-supported beam transferring part of the load to each of the supports C and D. The portion transferred to D is

$$\frac{x - L_1}{L_2 - L_1}$$

Hence

$$\text{Shear across the panel} = \frac{x}{L} - \left(\frac{x - L_1}{L_2 - L_1}\right)$$

and

$$\text{Force } F_{DE} = -\left\{\frac{x}{L} - \frac{(x - L_1)}{(L_2 - L_1)}\right\} \text{cosec } \theta$$

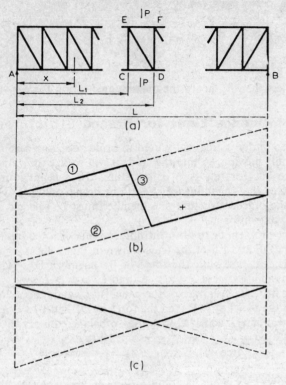

Fig. 2.30

The three equations are each straight lines and the form of the influence line is shown by the full lines in Fig. 2.30(*b*). The dotted lines will form a part of the influence line for any member sloping in the same direction as DE; the change of sign from line 1 to line 2 occurs at the panel of which the actual member forms a part.

In order to get the influence line for the force in the member EF consider the section formed at P–P, and take moments of the forces about D. The lines of action of the forces in the members DE and CD pass through D and will have no moment about this point. Considering the equilibrium of the forces to the right gives the force F_{EF} in EF as

$$F_{EF} = -\frac{R_B(L - L_2)}{d} = -\frac{x(L - L_2)}{Ld} W$$

when $x < L_2$.

In the case where $x > L_2$,

$$F_{EF} = -\frac{x(L - L_2)}{Ld} W + \frac{(x - L_2)}{d} W$$

$$= -\frac{WL_2}{Ld}(L - x)$$

The minus sign denotes compression. The two equations are straight lines which meet vertically below D and give the influence line for the force in EF as the full lines in Fig. 2.30(c).

Member EF is a typical top or bottom chord member and one feature of the influence line is that it is of the same sign for all values of x which is similar to the influence line for bending moment in a simply-supported beam. On the other hand, the influence line for the force in DE, which has a change of sign as the load crosses the panel, is typical of an internal bracing member and is similar to the influence line for shear at a given point in a simply-supported beam.

Example 2.6
A non-parallel boom braced girder which is simply supported, spans 60 metres and has a maximum depth of 12 metres at the centre of the span as shown in Fig. 2.31(a). The form of the top boom is parabolic, the depths at the panel points being $\frac{32}{3}$ and $\frac{20}{3}$ m, but all members may be considered straight between panel points. The load system shown is supported at the bottom boom panel points. Draw the influence line diagrams for the forces in the members X and Z, and determine the maximum possible forces in these members as the load system crosses the span.

SOLUTION
To find the force in member X take a section P–P through the panel and study the equilibrium of the free body to the right of P–P when the unit load is to be left of P–P. Resolve X into two components at D and take moments about C (Fig. 2.31(b)), i.e.

$$12H_X + 30R_B = 0$$

and
$$H_X = X(30/\sqrt{916})$$

or
$$X = (\sqrt{916})H_X/30$$

$$R_B = x/L$$

Thus X is a linear function of x and, for $x = 30$,

$$X = -\frac{\sqrt{916}}{30} \cdot \frac{30}{12} \cdot \frac{30}{60} = -1 \cdot 26 \quad \text{(Compression)}$$

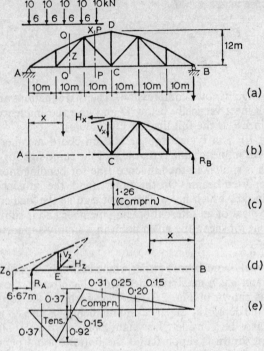

Fig. 2.31

Similarly for the unit load to the right of P–P and using the left-hand free body, X is again seen to be a linear function of x, as shown in Fig. 2.31(c).

The maximum value for the force in this member X_{max} occurs when the central 10-kN load is at the centre of the span. The ordinates to the influence line for the outer and inner wheels are then 0·75 and 1·01 respectively, and

$$X_{max} = 10(1·26 + 2 \times 1·01 + 2 \times 0·75) = 47·8 \text{ kN}$$

To get the influence line diagram for the force in Z, draw a section Q–Q and examine the equilibrium of the free body to the left of Q–Q (Fig. 2.31(d)). Resolve the force in Z (assumed tensile) into its horizontal and vertical components at E, then

$$V_Z = Z(16/21·9)$$

Produce the line of the top boom at section Q–Q to meet BA

produced in Z_0, then the distance Z_0A

$$(10/4)(10\tfrac{2}{3}) - 20 = 6\cdot67 \text{ m}$$

For the free body to the left of Q–Q when the unit load is to the right of the panel, taking moments about Z_0 gives

$$16\cdot67V_Z + 6\cdot67R_A = 0$$

and

$$R_A = \frac{x}{L}$$

where x is measured from B to the load and V_Z is a linear function of x. Thus

$$V_Z = -\frac{6\cdot67x}{16\cdot67 \times 60} = -\frac{x}{150}$$

and when $x = 40$,

$$V_Z = -40/150$$

and

$$Z = \frac{21\cdot9}{16} V_Z = -0\cdot37 \quad \text{(Compression)}$$

In a similar way when the unit load is to the left of the panel, the equilibrium of the free body to the right of Q–Q, again taking moments about Z_0, gives

$$16\cdot67V_Z - 66\cdot67R_B = 0$$

and $R_B = (L - x)/L$, so that V_Z is again linear.
So that, for $x = 50$,

$$V_Z = \frac{10}{60} \times \frac{66\cdot67}{16\cdot67} = 0\cdot67$$

$$Z = \frac{21\cdot9}{16} V_Z = 0\cdot92 \quad \text{(Tension)}$$

The influence line diagram is given in Fig. 2.31(e).
For the maximum compression in Z place the wheels as shown in Fig. 2.32(a), which gives

$$Z = 10(0\cdot37 + 0\cdot31 + 0\cdot25 + 0\cdot20 + 0\cdot15) = 12\cdot8 \text{ kN}$$

For the maximum tension in Z the wheels are placed as shown in Fig. 2.32(b), which gives

$$Z = 10(0\cdot37 + 0\cdot92 + 0\cdot15) = 14\cdot4 \text{ kN}$$

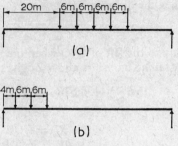

Fig. 2.32

2.15 Structures with Combined Bending and Direct Forces

It has already been said that a stiff joint is capable of transmitting a bending moment. Stiff-jointed structures are generally redundant. There is, however, a simple structure of this type and that is the three-pinned rigid-jointed frame.

This structure is also of interest since in addition to bending moments and shearing forces the members are subject to axial loads.

Figure 2.33 shows a portal which is pinned to supports A and B and has a third pin at C in the middle of the beam DE. The joints at D and E are stiff. As a structure it consists of two members ADC and BEC pinned to supports to brace a point C, and is statically determinate. The four reactive forces H_A, V_A, H_B and V_B can be obtained from the four equations

$$W = V_A + V_B$$
$$P + H_A = H_B$$
$$V_B \times L = W \times a + P \times b$$
$$H_B \times h = V_B \times L/2$$

Having determined the unknown reactions, the bending moment, shear force and thrust diagrams can be drawn.

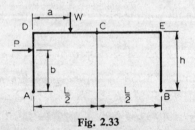

Fig. 2.33

Example 2.7
Calculate the reactive forces and draw the bending moment diagram for the three-pinned portal shown in Fig. 2.34(*a*).

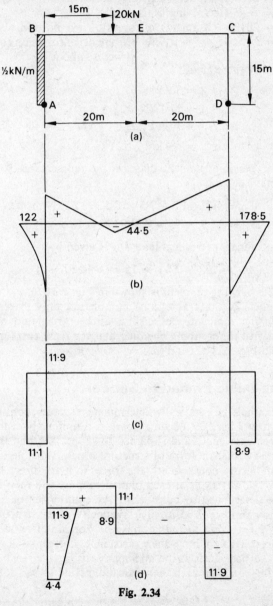

Fig. 2.34

SOLUTION

The equations of equilibrium and their derivations are as follows—

$V_A + V_D = 20$ (Resolving vertically)
$H_A + 7.5 = H_D$ (Resolving horizontally)
   $15H_D = 20V_D$ (Moments of forces on r-h portion about E)
   $40V_D = 20 \times 15 + 7.5 \times 7.5$ (Moments of forces on complete structure about A)

The solution of these gives

$$V_D = 8.9 \text{ kN}$$
$$V_A = 11.1 \text{ kN}$$
$$H_D = \tfrac{4}{3} \times V_D = 11.9 \text{ kN}$$
$$H_A = 4.4 \text{ kN}$$

To draw the bending moment diagram the moments at B and C must be obtained.

$$M_B = 4.4 \times 15 + 7.5 \times 7.5 = 122 \text{ kN-m}$$
$$M_C = 11.9 \times 15 = 178.5 \text{ kN-m}$$

The moment under the vertical load M_L is given by

$$M_L = 122 - 11.1 \times 15 = -44.5 \text{ kN-m}$$

The complete bending moment is shown in Fig. 2.34(*b*).

The members of this frame must be able not only to resist the applied bending moment, but also a direct thrust and a shear perpendicular to the section. The diagrams for these two functions are shown in Figs. 2.34(*c*) and (*d*).

2.16   The Cable Pinned to Supports

As a final example of a statically-determinate structure, consider the cable shown in Fig. 2.35(*a*), which carries a uniformly-distributed load of *w* per unit length and is pinned to supports A and B at the same level. A cable is a form of structure which is unable to resist a bending moment; consequently the forces in it are direct.

There are two reactions at each pinned support and their magnitude can be determined by considering the equilibrium of half the cable, as shown in Fig. 2.35(*b*). The symmetry of the structure and the loading means that the cable must be horizontal at mid-span and the direct tension in it there acts in a horizontal direction. There is no vertical shear at the mid-span section.

Considering equilibrium in the vertical direction gives

$$V_B = \tfrac{1}{2}wL$$

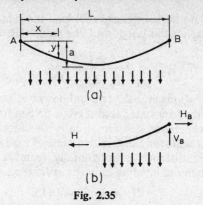

(a)

(b)

Fig. 2.35

Taking moments about B gives

$$H \times a = \tfrac{1}{2}wL \times \tfrac{1}{4}L$$

Thus

$$H = \frac{wL^2}{8a} = H_B$$

The maximum tension T_{max} in the cable occurs at B, and is given by

$$T_{max} = \sqrt{(V_B{}^2 + H_B{}^2)}$$
$$= \sqrt{\left(\frac{w^2L^2}{4} + \frac{w^2L^4}{64a^2}\right)} = \tfrac{1}{2}wL\left(1 + \frac{L^2}{16a^2}\right)^{1/2}$$

If the supports A and B are at different levels, then the method of solution is to examine each part of the cable separately. Figure 2.36 shows a cable in which support B is at a distance h above the other support A and with the lowest point on the cable at a distance a below A. The horizontal distance of O, the lowest point on the cable, from A is an unknown. Let it be x_1. Then considering the portion of the cable AO, the horizontal force H is given by

$$H = \frac{wx_1{}^2}{2a}$$

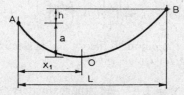

Fig. 2.36

Since there is no horizontal external load the same horizontal force must be acting on the portion of the cable OB. Consideration of this gives

$$H = \frac{w(L - x_1)^2}{2(a + h)}$$

These two equations give the two unknowns H and x_1.

It has already been stated that there is no bending moment at any point in the cable, and consideration of this fact enables the cable shape under a uniformly-distributed vertical load to be determined.

If a section distance x horizontally from A (Fig. 2.35(a)) is considered, then the bending moment at this section is given by

$$Hy + \tfrac{1}{2}wx^2 - \tfrac{1}{2}wLx$$

But this is equal to zero, and hence

$$y = \frac{wx}{2H}(L - x)$$

Substituting for H gives

$$y = \frac{4ax}{L^2}(L - x)$$

which is the equation of a *parabola*. Thus the form of the cable is a parabola.

The length of a cable in terms of its span and depth can be obtained from the knowledge that its form is parabolic.

If δs is the length of a small segment of the cable and s is the total length of the cable, then

$$\delta s^2 = \delta x^2 + \delta y^2$$
$$= \delta x^2 \left\{ 1 + \left(\frac{\delta y}{\delta x}\right)^2 \right\}$$

But

$$y = \frac{4ax}{L^2}(L - x)$$

and

$$\frac{dy}{dx} = \frac{4a}{L} - \frac{8ax}{L^2}$$

$$s = \int_0^L \left\{ 1 + \left(\frac{dy}{dx}\right)^2 \right\}^{1/2} . \, dx$$

Expanding this by the binomial theory and neglecting all terms after the second, since a/L is small, gives

$$s = L + \frac{8}{3}\frac{a^2}{L} \quad \text{approximately}$$

A case which is nearly the same as this is that of a cable hanging under its own weight. In this the load per unit length horizontally is not uniform but equal to $w \sec \alpha$, where α is the inclination of the cable to the horizontal. The curve of the cable under this condition is a *catenary*, which is close to a parabola.

2.17 Examples for Practice

1. Find the forces in the members a, b, c and d of the frame shown in Fig. 2.37.

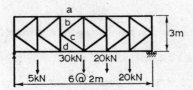

Fig. 2.37

(*Ans.* $F_a = -35{\cdot}5$; $F_b = -20{\cdot}1$; $F_c = +20{\cdot}1$; $F_d = +35{\cdot}5$ kN.)

2. Find the forces in the members of the frame shown in Fig. 2.38 and calculate the reactions at the pinned supports D, E, F and G. ABC is a horizontal plane.

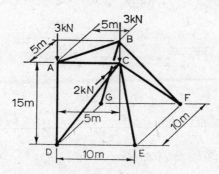

Fig. 2.38

(*Ans.* AD = $-3{\cdot}00$;   CF = $-7{\cdot}22$;   CE = $+4{\cdot}75$;   CD = $-1{\cdot}58$ kN; all other members unloaded; $R_D = 3{\cdot}39$; $R_E = -4{\cdot}75$; $R_F = 7{\cdot}22$ $R_G = 0$ kN.)

3. The frame shown in Fig. 2.39 is pinned to supports A, B and C; F is in the vertical plane ABDE. Find the forces in the members.
   (*Ans.* FE = +5·6; FD = +5·6; FC = −11·2; BE = +5·0; DE = −2·5; AD = +5·0 kN; CD = CE = 0.)

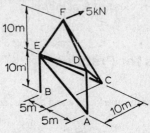

Fig. 2.39

4. Draw the bending moment and shearing force diagrams for the beam shown in Fig. 2.40.
   (*Ans.* R_L = 5·9 kN; R_R = 4·1 kN; M_{max} = 41 kN-m.)

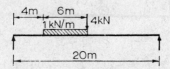

Fig. 2.40

5. The beam system shown in Fig. 2.41 is arranged so that the supports between beams 2 and 4 between 3 and 4 are capable of exerting upward as well as downward reactions. Draw the bending moment and shearing force diagrams.
   (*Ans.* R_{1L} = 5·0; R_{4L} = 0·5; R_{4R} = 19·5 kN.)

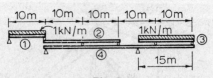

Fig. 2.41

6. Draw the bending moment and shearing force diagrams for the beam shown in Fig. 2.42, which is built-in at A, simply-supported at C, and has a pin inserted at B.
   (*Ans.* R_A = 50 kN; M_A 900 kN-m; R_σ = 10 kN.)

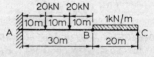

Fig. 2.42

7. The span of a girder is 100 m. The live load system given below may cross the span in either direction. Determine the maximum bending moment in the girder and obtain the equivalent uniform load. Draw the influence line for shear at the left-hand quarter-point and calculate the maximum value.

Vertical point loads (kN)   5 + 20 + 20 + 15 + 15
at horizontal spacing (m)   10 + 10 + 10 + 10

(*Ans.* 1,500 kN-m; 1·2 kN/m; 42·75 kN.)

8. A lattice girder (Fig. 2.43) spans 120 m. It has eight 15-m panels and a curved lower chord. Assuming that one bearing is supported on rollers, draw the influence lines for the forces in BD and DC, and find the maximum force in DC due to the given live loading.
(*Ans.* 58·3 kN.)

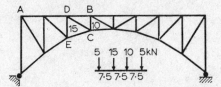

Fig. 2.43

9. A Warren girder having a span of 80 m consists of four equal panels as shown in Fig. 2.44. Draw the influence lines for the forces in members *a* and *b*, and find the maximum force in the member *a* due to a uniformly-distributed live load of 2 kN/m run and 30 m long.
(*Ans.* 46·9 kN.)

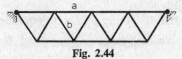

Fig. 2.44

10. Draw the bending moment, shearing force and thrust diagrams for the 3-pinned portal shown in Fig. 2.45.
(*Ans.* $V_L = 6·25$; $H_L = 1·75$; $V_R = 13·75$; $H_R = 8·25$ kN.)

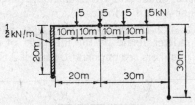

Fig. 2.45

11. A cable carrying a uniformly distributed vertical load of 30 N/m run is placed between two supports 200 m apart, one being 20 m higher than the other. The lowest point of the cable is to be 10 m below the lower support.
  Calculate (*a*) the length of cable required, and (*b*) the horizontal pull at the supports.
(*Ans.* 222·6 m; 8·05 kN.)

3

Simple Stress Systems

3.1 Introduction

The forces in and the deformations of a structure must satisfy two
conditions
  (1) Equilibrium,
  (2) Compatibility.
The first ensures that the forces satisfy the equilibrium requirements;
the second that the deformations satisfy the geometrical requirements.

In analyzing statically-determinate structures no consideration
was given to compatibility because the members of such structures
will always fit together without any straining and a slight change in
length only produces small alterations in the geometry, with corresponding small changes in the forces.

If however the statically-determinate frame ABCD shown in
Fig. 3.1 is made indeterminate by the inclusion of a member AC,
the lengths of the members are important, since if AC is not an exact

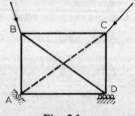

Fig. 3.1

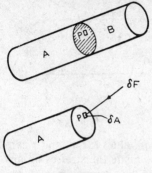

Fig. 3.2

fit the remainder of the frame will have to be strained to get it into place. Thus in statically-indeterminate structures compatibility is important and both conditions have to be taken into account.

Member deformations depend on the characteristics of the material of which it is made, in particular the load-deformation or stress–strain relationship.

Stress is defined as the transmission of force across a boundary. Consider the solid body shown in Fig. 3.2, separated by a plane into two parts A and B. Each part exerts a force on the other, and over a small area δA at a point P let this force on part A be δF in an arbitrary direction. The intensity of this force is $\delta F/\delta A$ and the *stress* at P is the limit of $\delta F/\delta A$ as $\delta A \to 0$.

$$\text{Stress } \sigma = \text{Force}/\text{Area}$$

in, for example, N/mm². The force action has magnitude and direction, and in addition the plane through P has an orientation. These three quantities must be specified for the stress σ to be completely known.

3.2 Stress Components

In the general case σ can be resolved into two components, a normal stress σ_n and a tangential or shear stress τ lying in the plane. The normal σ_n is a tension or compression; and the planar force τ is a traction across the plane. Each has a different effect on the deformation of the body.

If the normal to the plane is taken as the x-axis, then the traction or shear force τ can be resolved into two components τ_y and τ_z acting in the y- and z-directions respectively (*see* Fig. 3.3).

There are an infinite number of possible planes through P, and

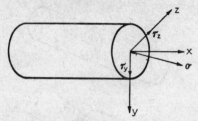

Fig. 3.3

hence an infinite set of values for σ_n and τ can define the stress at a point. In order to calculate the stresses on any arbitrary plane it is necessary, in the three dimensional case, to specify six quantities.

Consider an elementary cube around the point P as shown in Fig. 3.4. On the yz-plane there is a normal stress σ_x and shear stresses τ_{xy} and τ_{xz}. On the xz- and xy-planes the corresponding forces are σ_y, τ_{yx} and τ_{yz} and σ_z, τ_{zx} and τ_{zy}.

Since the block is in equilibrium, the sum of the components of the forces along the three axes must vanish and also the moments about any three axes. If axes through the centre of the cube are taken it is found that

$$\left. \begin{aligned} \tau_{yz} &= \tau_{zy} \\ \tau_{yx} &= \tau_{xy} \\ \tau_{xz} &= \tau_{zx} \end{aligned} \right\} \quad (3.1)$$

These are known as *complementary* shear stresses.

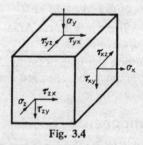

Fig. 3.4

3.3 Stresses on a Plane: Three-dimensional Case

If the three normal and shear stresses acting on any one plane are known then it is possible to calculate the stresses on any other plane.

The plane ABC shown in Fig. 3.5 has a unit area and the normal to the plane has direction cosines l, m and n with the three axes. This means also that the areas OCB, OCA and OAB are l, m and n respectively.

With the stresses on the three normal planes as shown in the figure, the total forces on the plane ABC f_x, f_y and f_z in the x-, y- and z-directions respectively are given by

$$\left.\begin{aligned} f_x &= l\sigma_x + m\tau_{yx} + n\tau_{zx} \\ f_y &= m\sigma_y + n\tau_{zy} + l\tau_{xy} \\ f_z &= n\sigma_z + l\tau_{xz} + m\tau_{yz} \end{aligned}\right\} \quad (3.2)$$

Since the body is in equilibrium, the normal force σ_n on the plane ABC (which since the area of this plane is unity is also the stress) is

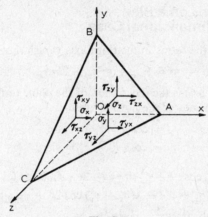

Fig. 3.5

therefore given by

$$\begin{aligned} \sigma_n &= lf_x + mf_y + nf_z \\ &= l^2\sigma_x + m^2\sigma_y + n^2\sigma_z + 2lm\tau_{xy} \\ &\quad + 2ln\tau_{xz} + 2mn\tau_{yz} \end{aligned} \quad (3.3)$$

The resultant force σ_r on the section is given by

$$\sigma_r = \sqrt{(f_x^2 + f_y^2 + f_z^2)} \quad (3.4)$$

The general expression for this is lengthy and any particular case can best be worked out arithmetically.

The shear stress τ on the plane ABC can be worked out from the fact that

$$\sigma_r^2 = \tau^2 + \sigma_n^2$$

or $$\tau^2 = \sigma_r^2 - \sigma_n^2 \quad (3.5)$$

As with σ_r, the general expression is complicated, but any particular case can be solved arithmetically.

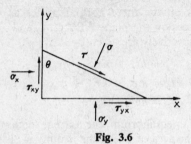

Fig. 3.6

3.4 Stresses on a Plane: Two-dimensional Case

In this case all the terms in the third axis vanish, giving

$$\sigma = l^2\sigma_x + m^2\sigma_y + 2lm\tau_{xy}$$

If θ (*see* Fig. 3.6) is the inclination of the plane to the y-axis, then the direction cosines of the normal are

$$l = \cos\theta$$

and

$$m = \cos(90 - \theta) = \sin\theta$$

Thus

$$\sigma = \sigma_x \cos^2\theta + \sigma_y \sin^2\theta + 2\tau_{xy}\sin\theta\cos\theta$$
$$= \tfrac{1}{2}(\sigma_x + \sigma_y) + \tfrac{1}{2}(\sigma_x - \sigma_y)\cos 2\theta + \tau_{xy}\sin 2\theta \qquad (3.6)$$

and the shearing stress τ is given by

$$\tau = -\sigma_x \sin\theta\cos\theta + \sigma_y \sin\theta\cos\theta + \tau_{yx}(\cos^2\theta - \sin^2\theta)$$
$$= -\tfrac{1}{2}(\sigma_x - \sigma_y)\sin 2\theta + \tau_{yx}\cos 2\theta \qquad (3.7)$$

3.5 Stress Resultants

Figure 3.7 shows an arbitrary plane cross-section of a body under stress. At every point in the plane the stress can be resolved into its three components σ_x, τ_{xy} and τ_{xz}. If the coordinates of an element

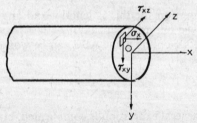

Fig. 3.7

of area δA are y and z, then the resultant forces on the section are

Along the x-axis $\qquad \sum \sigma_x \, \delta A = F_x$

Along the y-axis $\qquad \sum \tau_{xy} \, \delta A = F_y$

Along the z-axis $\qquad \sum \tau_{xz} \, \delta A = F_z$

The moments of the forces about the different axes are—

About the x-axis $\qquad \sum (\tau_{xy}z - \tau_{xz}y) \, \delta A = -M_x$

About the y-axis $\qquad \sum \sigma_x z \, \delta A = M_y$

About the z-axis $\qquad \sum \sigma_x y \, \delta A = -M_z$

Thus the internal stresses on the plane cross-section are equivalent, statically, to forces parallel to the axes acting at O together with moments about the three axes.

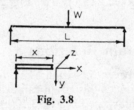

Fig. 3.8

This set of internal stress resultants must be in equilibrium with the external forces. For example, in the case of a simply-supported beam (Fig. 3.8) of span L, carrying a concentrated load W at midspan, at a section distant x from the left-hand support the external forces are

$$M_x = M_y = 0: \quad M_z = -\tfrac{1}{2}Wx$$
$$F_x = F_z = 0: \quad F_y = \tfrac{1}{2}W$$

Hence at this section the internal stress distribution must be such as to give a resistance moment of $\tfrac{1}{2}Wx$ and a shear of $\tfrac{1}{2}W$.

3.6 Normal and Shear Strain

If a prism of uniform material is subject to a uniform normal stress, it will elongate by an amount e as shown in Fig. 3.9(a). The normal strain ϵ is defined as the extension (or compression) per unit length, i.e.

$$\epsilon = \frac{e}{L}$$

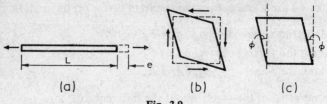

Fig. 3.9

This is a dimensionless quantity. A plane element (as the dotted square in Fig. 3.9(*b*)) if subjected to equal shears on all four faces will deform as shown to a parallelogram. The angle between two adjacent sides has changed from 90° to $(90 \pm \phi)°$. ϕ is defined as the shear strain and, like normal strain, is a dimensionless quantity. Usually in engineering applications an axis parallel to one of the deformed sides is taken as shown in Fig. 3.9(*c*), and ϕ is then the shear strain.

3.7 Stress–Strain Relations

The relationship between the strain as shown in Fig. 3.9(*a*) and the stress applied to produce this strain can exhibit one of four characteristic behaviours illustrated in Fig. 3.10(*a*)–(*d*).

In each of these the slope of the curve relating stress to strain is the vital factor. This slope is symbolized by E, where $E = d\sigma/d\epsilon$.

In Fig. 3.10(*a*), $E = \infty$, i.e. there is no strain and the material is said to be *rigid*.

In (*b*), $E =$ constant, and the stress–strain relationship is linear.

In (*c*), $E = f(\sigma)$.

In (*d*), $E = 0$: there is no definite strain for any value of the stress and the material is said to be *plastic*.

The behaviour on unloading can either follow the full or dotted curves shown in Fig. 3.10 (*b*) or (*c*). Whenever the material returns to its original dimensions along the loading curve it is said to be

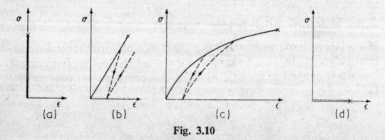

Fig. 3.10

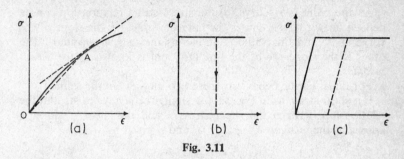

Fig. 3.11

elastic, but if it behaves as indicated by the dotted lines, i.e. if it departs from the original loading curve, it is *inelastic*.

Materials which fracture when the strains are small are known as *brittle*, whilst materials which have an appreciable deformation before failure are said to be *ductile*.

Real materials generally have complex stress–strain relationships, but for purposes of analysis behaviour is simplified into either—

(1) That exemplified by the single curve of Fig. 3.11 (*a*), or
(2) Behaviour as rigid material up to some yield stress σ_y, after which it is plastic (Fig. 3.11 (*b*)), or
(3) A combination of linear elastic and plastic behaviour (Fig. 3.11 (*c*)).

A uniform rod of length L and cross-sectional area A (Fig. 3.12), which is subjected to a uniform state of stress σ, i.e. a force $F = \sigma A$, and with a value of E which is constant for the stress range used, has

$$\text{Strain} = \frac{e}{L}$$

$$\text{Stress} = \frac{F}{A}$$

Hence
$$E = \frac{FL}{Ae} \qquad (3.8)$$

This constant value of E is called *Young's Modulus* or the *Modulus of Elasticity*.

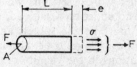

Fig. 3.12

At any point A in curve (a) the stress–strain relationship can be defined in two ways. One is the tangent $d\sigma/d\epsilon$, i.e. the slope of the curve at A, and this value is known as the *tangent modulus*. The other is the slope σ/ϵ of the line OA, and is known as the *secant modulus*.

For linear elastic behaviour these two moduli are the same.

For shear strain as in Fig. 3.9 (c), similar behaviour exists, but the relationship between the shear stress τ and the shear strain γ is known as the *modulus of rigidity G*, and is given by

$$G = \frac{\tau}{\gamma} \tag{3.9}$$

3.8 Simple Tension Member

The expression for the deformation FL/AE given above assumes that the stress resulting from the load is uniform, (a) at all points in a given cross-section, and (b) throughout the length of the bar.

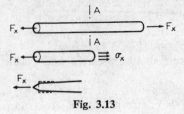

Fig. 3.13

If the load is applied axially, then for equilibrium between the external and internal forces the resultant of the latter must act along the axis. This means a symmetrical but not necessarily uniform distribution of the internal forces.

If a simple experiment is carried out on a piece of very low modulus material the application of an axial load to the end causes a deformation in the vicinity of the load as shown in Fig. 3.13. With such a deformation, sections which were originally plane do not remain so and it is not possible to define in a simple manner the stress distribution at sections in the load vicinity.

At some distance away from the end, however, there is no evidence of distortion, and a series of adjacent cross-sections have similar properties, or, in other words, sections which were plane remain plane. For this to be so under an axial load, the strain at all points in any cross-section must be the same; and if the stress–strain relationship is linear the stress must be uniform throughout.

Neglecting deformations in the vicinity of the load is a standard procedure known as the *Principle of St. Venant*.

Example 3.1

The rigid block shown in Fig. 3.14 exerts a force of 2,000 N and is supported by three wires. The outer two are of steel and have an area of 100 mm². The middle one is of aluminium and has an area of 200 mm². If the elastic moduli of the two materials are 225 kN/mm² and 75 kN/mm² respectively, calculate the stresses in the wires.

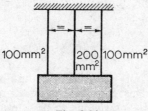

100mm²   200 mm²   100mm²

Fig. 3.14

SOLUTION

From the symmetry of the arrangement, the stress in each steel wire is the same. Let σ_S be this stress and σ_A the stress in the aluminium. Equilibrium gives

$$200\sigma_S + 200\sigma_A = 2,000$$
$$\sigma_S + \sigma_A = 10$$

Compatibility gives the same strain in the two materials, i.e.

$$\frac{\sigma_S}{E_S} = \frac{\sigma_A}{E_A}$$
$$\sigma_S = \frac{225}{75}\sigma_A = 3\sigma_A$$

Substituting gives
$$4\sigma_A = 10$$

or
$$\sigma_A = 2 \cdot 5 \text{ N/mm}^2$$
$$\sigma_S = 7 \cdot 5 \text{ N/mm}^2$$

3.9 Poisson's Ratio

The element of Fig. 3.15, which initially has the shape indicated by the full lines, will deform under a tension as shown to the shape

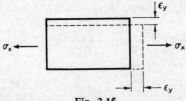

ϵ_y

σ_x σ_x

ϵ_y

Fig. 3.15

indicated by the dotted lines. The longitudinal strain ϵ_x is accompanied by a lateral strain ϵ_y and the ratio ϵ_y/ϵ_x is known as *Poisson's Ratio* and generally symbolized by v.

If, in addition to the force producing the stress σ_x, there is a force on the perpendicular plane producing a stress σ_y then the strains ϵ_x and ϵ_y due to the combined stresses are given by

$$\epsilon_x = \frac{1}{E}(\sigma_x - v\sigma_y)$$

$$\epsilon_y = \frac{1}{E}(\sigma_y - v\sigma_x)$$

Poisson's ratio not only gives the relationship between the longitudinal and lateral strains, but also enables the relationship between the moduli of elasticity and rigidity to be determined.

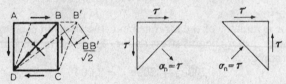

Fig. 3.16

The element of area ABCD, of unit length sides, shown in Fig. 3.16 is subjected to shear stresses τ on each face as defined in the engineering form (p.60). This produces a direct tension and compression of magnitude τ on the two diagonals (*see* eqn. (3.6)).

The strain on the diagonal BD is therefore equal to

$$\frac{1}{E}(\tau + v\tau) = \frac{\tau}{E}(1 + v)$$

If the effect of the shear is to cause the deformation shown by the dotted lines, with B moving to B′, then BB′ is the shear strain and is approximately equal to τ/G.

The increase in length of the diagonal DB is BB′$/\sqrt{2}$, i.e. $\tau/G\sqrt{2}$, and since the original length was $\sqrt{2}$, the strain is $\tau/2G$. But this strain is also equal to $(\tau/E)(1 + v)$. Hence

$$\frac{1 + v}{E} = \frac{1}{2G}$$

or

$$G = \frac{E}{2(1 + v)} \qquad (3.10)$$

3.10 General Assumptions made in the Engineer's Theory

The findings of this section of the chapter can be summarized by stating that if a force F is applied axially to a uniform straight member of homogeneous material having a constant modulus of elasticity E, then apart from local stresses in the vicinity of the load there is a uniform state of stress of intensity F/A, where A is the cross-sectional area; and that the elongation is FL/AE, where L is the original length; and that sections plane before the load is applied remain so afterwards.

The direct axial load is the simplest that can be dealt with. It is now proposed to determine the stresses resulting from other types of loading using the same assumptions, namely—

(1) Material homogeneous.
(2) Sections plane before application of load remain so after loading (*Navier's hypothesis*).
(3) The stress–strain relationship is linear for the loading considered.

3.11 Torsion of Circular Shafts

Figure 3.17 (*a*) shows a shaft which is firmly fixed at one end and to which a torque T is applied at the free end.

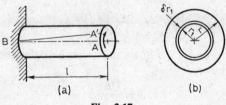

Fig. 3.17

The effect of the application of this torque is to cause the line AB to be deformed into A′B.

If the length of the shaft is l and its radius r, then if AB remains straight,

$$AA' = r\theta$$

where θ = angle of twist. If plane sections remain plane, the only deformation is a shearing deformation. The shear strain is

$$\gamma = \frac{AA'}{l} = \frac{r\theta}{l}$$

But

$$\gamma = \frac{\tau}{G}$$

where τ is the shear stress at the outermost fibre. Equating the values of γ gives

$$\frac{\tau}{G} = \frac{r\theta}{l} \qquad (3.11)$$

It is seen that τ is proportional to the radius. Hence, at a distance r_1 from the centre, the shear stress is given by $r_1 G\theta/l$.

Considering the element of thickness δr_1 at radius r_1, the complementary nature of the shear stresses shows that τ acts along an elementary ring of cross-section $2\pi r_1 \, \delta r_1$ (*see* Fig. 3.17 (*b*)).

The total force δF on this ring is given by

$$\delta F = \tau \, . \, 2\pi r_1 \, \delta r_1$$

The moment of this force about the axis is

$$\delta T = r_1 \, \delta F = \tau \, . \, 2\pi r_1{}^2 \, . \, \delta r_1$$

The total torque T is given by

$$T = \int_0^r \tau \, . \, 2\pi r_1{}^2 \, \delta r_1$$

$$= \frac{G\theta}{l} \int_0^r 2\pi r_1{}^3 \, \delta r_1$$

The expression under the integral sign is the polar second moment of area I_p. Hence

$$T = \frac{G\theta}{l} I_p$$

Combining the two relationships gives

$$\frac{\tau}{r} = \frac{G\theta}{l} = \frac{T}{I_p} \qquad (3.12)$$

For a solid shaft of diameter d,

$$I_p = \pi d^4/32$$

whilst for a hollow shaft of external and internal diameters d_0 and d_i respectively,

$$I_p = \pi(d_0{}^4 - d_i{}^4)/32$$

Example 3.2

What is the angle of twist (in degrees) in a 3 m length of a hollow shaft 150 mm external and 75 mm internal diameter when it is subjected to a torque which produces a maximum stress of 75 N/mm²?

Find also the shear stress at the inside edge. Take $G = 75$ kN/mm².

SOLUTION
From eqn. (3.11),

$$\theta = \frac{\tau l}{rG}$$

Substituting,

$$\theta = \frac{75 \times 3 \times 10^3}{75 \times 75 \times 10^3} = 0 \cdot 040 \text{ radians}$$

$$= 2 \cdot 39°$$

From eqn. (3.11), τ is proportional to r, hence at the inside edge

$$\tau = \frac{37 \cdot 5}{75} \times 75 = 37 \cdot 5 \text{ N/mm}^2$$

3.12 Pure Bending of Straight Uniform Beams

Figure 3.18 (a) shows the cross-section of a beam which has an axis of symmetry. A length of beam δx is considered, as shown in (b) and moments M are applied at each end, the plane of the moment coinciding with the axis of symmetry. In addition to the three assumptions previously given (p. 65) it will also be assumed that E has the same value in compression as it has in tension.

The deformed shape of the beam is shown in Fig. 3.18 (b) and it is clear from this that the fibres on the convex side extend, i.e. they

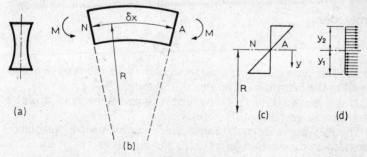

(a)   (c)   (d)

(b)

Fig. 3.18

are in tension, whilst those on the concave side shorten, i.e. they are in compression. Hence if the material is continuous from top to bottom there must be one fibre which neither extends nor shortens, i.e. remains at its original length. This fibre is known as the *neutral axis*.

If the deformed shape is such that the neutral axis is bent in the form of a circular arc of radius R, then the angle $\delta\phi$ subtended at the centre of this arc by the length δx is given by $\delta\phi = \delta x/R$ or $\delta x = R\,\delta\phi$.

Since plane sections remain plane, the strain diagram across the section is linear, as shown in Fig. 3.18 (c) and at a distance y from the neutral axis the new length of the fibre is $(R - y)\,\delta\phi$. Thus the change in length is $-y\,\delta\phi$ and the strain is $-y(\delta\phi/\delta x)$ or $-y/R$.

Hence the longitudinal stress σ in a fibre distance y from the neutral axis is given by

$$\sigma = -\frac{E}{R}\,y$$

where tensile stresses are taken as positive, or

$$\frac{\sigma}{y} = \frac{-E}{R} \tag{3.13}$$

If δA is the area of a strip of a beam distance y from the neutral axis, then

$$\text{Total compression} = \int_0^{y_1} \sigma\, dA = \frac{E}{R}\int_0^{y_1} y\, dA$$

and

$$\text{Total tension} = \frac{E}{R}\int_0^{-y_2} y\, dA$$

If the beam is subject to bending forces only, there is no direct load on the beam and the total tension must equal the total compression, i.e.

$$\int_0^{y_1} y\, dA = \int_0^{-y_2} y\, dA$$

or the first moment of area of the section above the neutral axis is equal to the first moment of area of the section below.

This is the condition that the neutral axis passes through the centre of gravity or *centroid* of the section.

The member is in equilibrium, hence the moment of resistance provided by the internal forces about the neutral axis must equal the externally applied moment.

The moment of the internal forces about the neutral axis is given by

$$\frac{E}{R}\int_0^{y_1} y^2\,dA + \frac{E}{R}\int_0^{-y_2} y^2\,dA = \frac{E}{R}\int_{-y_2}^{y_1} y^2\,dA$$

The applied moment is M, and hence the relationship is

$$M = \frac{E}{R}\int_{-y_2}^{y_1} y^2\,dA$$

The expression in the integral sign is the second moment of area of the section about the neutral axis. If this is taken as I then the equation can be combined with the previous E/R relationship to give

$$+\frac{M}{I} = +\frac{E}{R} = \frac{-\sigma}{y} \qquad (3.14)$$

It is clear from this equation that the maximum stress due to bending occurs at the outermost fibres and the normal problem in design is to calculate the dimensions of the section such that the maximum stress does not exceed a given amount. The value of the maximum compression stress σ_{y1} is given by

$$\sigma_{y1} = \frac{My_1}{I}$$

This assumes $y_1 > y_2$; if the opposite is the case then the maximum stress is $\sigma_{y_2} = M_{y_2}/I$.)

The expression I/y_1 is known as the *section modulus in compression* and I/y_2 is the *modulus in tension*. It is generally symbolized by Z. Using this the equation becomes

$$\sigma = \frac{M}{Z} \quad \text{or} \quad M = \sigma Z \qquad (3.15)$$

If the section has two axes of symmetry then $y_1 = y_2$ and the moduli of section in compression and tension are the same.

The simplest beam section is the rectangle. If this has a width b and depth d, the second moment of area is $bd^3/12$ and the section modulus $bd^2/6$. The distribution of stress is as given in Fig. 3.19 (*a*) and it is seen that the material near the neutral axis has very little stress and could be removed from the rectangular section without any appreciable loss of strength. This would leave the I-section shown in Fig. 3.19 (*b*). If t_f is the thickness of the flange and t_w that

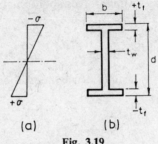

(a)   (b)

Fig. 3.19

of the web, then

$$I = \tfrac{1}{12}bd^3 - (b - t_w)(d - 2t_f)^3 \times \tfrac{1}{12}$$

The concentration of area at those sections where it is most required to carry the stress produces an economical section, thus accounting for the universal use of the I-section beam.

These sections are rolled up to 1 metre deep, but above this the plate-web girder-construction is used.

The fact that the web is thin and the stress is small means that it contributes little to the resistance to bending and it is generally assumed that this is provided entirely by the flanges.

Since the flange depth is small there is little variation in stress across its depth and the stress is usually assumed uniform. With this assumption the moment of resistance of the section is given by

$$M = \sigma b t_f (d - t_f) \tag{3.16}$$

Example 3.3
Calculate the moment of resistance of the beam section shown in Fig. 3.20 if the stresses in the upper and lower flanges are limited to 25 N/mm² and 40 N/mm², respectively.

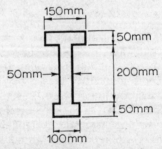

Fig. 3.20

SOLUTION

The position of the neutral axis must first be found. Taking first moments of areas about the top flange gives

$$150 \times 50 = 7{,}500 \times 25 = 1{,}875 \times 10^2$$
$$200 \times 50 = 10{,}000 \times 150 = 15{,}000 \times 10^2$$
$$100 \times 50 = 5{,}000 \times 275 = 13{,}750 \times 10^2$$

$$\overline{22{,}500 \times \bar{y} = 30{,}625 \times 10^2}$$

Hence $\bar{y} = 30{,}625/225 = 136 \cdot 1$ mm.

It is now necessary to find the second moment of area of the section about the neutral axis. This can be done by considering each rectangle separately as follows—

Top flange $(150 \times 50^3)/12 = 15 \cdot 6 \times 10^5$
$7{,}500 \times 111 \cdot 1^2 = 930 \cdot 00 \times 10^5$
Web $(50 \times 200^3)/12 = 334 \cdot 00 \times 10^5$
$10{,}000 \times 13 \cdot 9^2 = 19 \cdot 40 \times 10^5$
Bottom flange $(100 \times 50^3)/12 = 10 \cdot 00 \times 10^5$
$5{,}000 \times 138 \cdot 9^2 = 964 \cdot 00 \times 10^5$

$$\overline{2{,}273 \times 10^5}$$

Based on the top flange, the moment of resistance M_t, is given by

$$M_t = (25 \times 2{,}273 \times 10^5)/(136 \cdot 1 \times 10^6) = 41 \cdot 6 \text{ kN-m}$$

The moment M_b based on the bottom flange is

$$M_b = (40 \times 2{,}273 \times 10^5)/(163 \cdot 9 \times 10^6) = 55 \cdot 5 \text{ kN-m}$$

The former is the lesser of the two and will therefore govern the carrying capacity. Hence

$$\text{Moment of resistance} = 41 \cdot 6 \text{ kN-m}$$

3.13   Reinforced Concrete Beams

The theory just developed has been based on the assumption that the beam is composed of uniform material throughout. This is frequently the case, but a very common example of a non-uniform material is reinforced concrete where, since concrete is weak in tension, steel bars have been added.

A simple theory for these beams is the *straight-line no-tension theory*; in addition to the assumptions inherent in the standard theory of bending, this theory also assumes that concrete can take no tension.

The material generally used as reinforcement is steel, which will satisfy the assumption of linear elasticity within working stresses. Concrete does not, and the tendency therefore is to use methods based on the ultimate moment of resistance analogous to the full plastic moment of steel beams as discussed on p. 103.

The straight-line no-tension theory is developed as follows. Consider the beam cross-section shown in Fig. 3.21 (*a*). Following from the usual assumption that plane sections remain plane, a cross-section will rotate about the neutral axis to give the linear strain

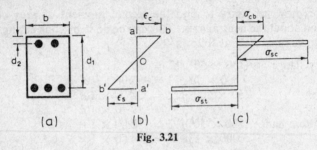

(a)   (b)   (c)

Fig. 3.21

distribution of Fig. 3.21 (*b*). Since concrete can take no tension the concrete stress distribution will be as shown in Fig. 3.21 (*c*), varying from a maximum compression σ_{cb} to zero stress at and below the neutral axis. The steel stresses σ_{sc} and σ_{st} are assumed to be uniform over the small depth taken up by the reinforcing bars and are shown with the concrete stress in Fig. 3.21 (*c*). The discontinuity in stress distribution results from the differing E-values for steel and concrete.

The compatibility condition gives

$$\text{Strain in steel} = \text{strain in surrounding concrete}$$

From Fig. 3.21 (*b*)

$$\frac{a'b'}{Oa'} = \frac{ab}{Oa}$$

$$\frac{\sigma_{st}}{E_s(d_1 - d_n)} = \frac{\sigma_{cb}}{E_c \cdot d_n}$$

where E_s and E_c are Young's Moduli for steel and concrete respectively and d_n is depth of the neutral axis. The ratio of these moduli E_s/E_c is generally denoted by m and substituting gives

$$\sigma_{st} = m\sigma_{cb}\left(\frac{d_1 - d_n}{d_n}\right) \tag{3.17}$$

or

Steel stress = m(concrete stress at same depth if concrete took tension)

Similarly

$$\sigma_{sc} = m\sigma_{cb}\left(\frac{d_n - d_2}{d_n}\right)$$

Since in the case of pure bending the total compression equals the total tension, the equilibrium condition gives

$$\sigma_{cb} \cdot \tfrac{1}{2} d_n b + \sigma_{sc} A_{sc} - \sigma_{cb} A_{sc}\left(\frac{d_n - d_2}{d_n}\right) = A_{st}\sigma_{sc}$$

where the third term allows for the concrete displaced by A_{sc}. Substituting for σ_{sc} and σ_{st} gives

$$\sigma_{cb} \cdot \tfrac{1}{2} bd_n + (m - 1)A_{sc}\sigma_{cb}\left(\frac{d_n - d_2}{d_n}\right) = m\sigma_{cb}\left(\frac{d_1 - d_n}{d_n}\right)A_{st}$$

or

$$\frac{bd_n}{2} + (m - 1)A_{sc}\left(\frac{d_n - d_2}{d_n}\right) = mA_{st}\left(\frac{d_1 - d_n}{d_1}\right) \quad (3.18)$$

which can be solved to get the neutral axis depth d_n.

If

$(m - 1)A_{sc}$ = equivalent concrete area of compression steel

mA_{st} = equivalent concrete area of tension steel

then eqn. (3.18) states that the first moment of area about the neutral axis of the equivalent sections above and below the neutral axis are equal, i.e. the beam behaves as a concrete beam with the steel replaced by an equivalent concrete area. This section is known as the *transformed section* and the standard beam analysis developed in Section 3.12 can be applied to it, i.e.

$$\text{Equivalent } I = \tfrac{1}{3}bd_n{}^3 + (m - 1)A_{sc}(d_n - d_2)^2 + mA_{st}(d_1 - d_n)^2$$

and

$$\sigma_{cb} = \frac{Md_n}{I}$$

The moment of resistance M_R of the section can also be expressed in terms of the steel areas and stresses as

$$M_R = \tfrac{1}{3}\sigma_{cb}bd_n{}^2 + (m - 1)A_{sc}\sigma_{cb}\frac{(d_n - d_2)^2}{d_n} + A_s \sigma_{st}(d_1 - d_n)$$

$$(3.19)$$

One important difference between a composite beam of this type and a uniform beam is that in the former there are two permissible stresses to be taken into consideration as against one in the latter.

It is seen that the ratio of stresses in a reinforced concrete beam depends on the neutral axis depth, which in turn depends on the amount of steel present. Examination of eqns. (3.17) and (3.18) shows that there is for a given beam only one value of the steel area which will allow both materials to reach their permissible stress values. The beam is then said to be in a state of *balanced design*.

Example 3.4

A reinforced concrete beam 400 mm wide has to carry a bending moment of 320 kN-m, the permissible stresses being 8 N/mm² in the concrete and 135 N/mm² in the steel. Calculate the relative economy in steel and concrete as between

    (a) a singly-reinforced section to balanced design, and
    (b) a doubly-reinforced section with effective depth restricted to 560 mm, both materials working at their full stress. Take $m = 15$.

SOLUTION
Let d_1 be the effective depth of the singly-reinforced section and d_{nb} the depth to the neutral axis at balanced design.

The strain relationship gives

$$\frac{15 \times 8}{135} = \frac{d_{nb}}{d_1 - d_{nb}}$$

leading to
$$d_{nb} = 0.47 d_1$$

The moment of resistance of the singly-reinforced section at balanced design M_{rb} is given by

$$M_{rb} = \tfrac{1}{2} \times 8 \times 0.47 d_1 \times 400 (d_1 - 0.157 d_1) = 632 d_1{}^2$$

This must be equal to the applied bending moment of 320×10^6 N-mm. Hence
$$d_1 = 10^3 \sqrt{(320/632)} = 711 \text{ mm}$$

From the equilibrium condition,

$$A_{st} = (\tfrac{1}{2} \times 8 \times 711 \times 0.47 \times 400)/135$$
$$= 3{,}940 \text{ mm}^2$$

In the second case the beam section is as shown in Fig. 3.22. Both concrete and steel will develop their full stresses in compression and tension respectively. Thus

$$d_n = 0.47d_1$$
$$= 263 \text{ mm}$$
$$d_2 = 40 \text{ mm}$$
$$d_n - d_2 = 223 \text{ mm}$$

Total compression $= \frac{1}{2} \times 8 \times 400 \times 263$
$$+ (223/263) \times 14 \times 8 \times A_{sc}$$

Total tension $= 135A_{st}$

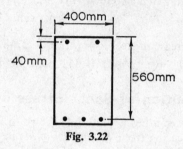

Fig. 3.22

These are equal, hence

$$\tfrac{1}{2}(8 \times 400 \times 263) + \frac{223}{263} \times 14 \times 8 \times A_{sc} = 135A_{st}$$

which on expanding and simplifying gives

$$4,440 + A_{sc} = 1.42A_{st}$$

The moment of resistance of the section about the neutral axis is given by the sum of the following three quantities—

Due to concrete
$$8 \times 263^2 \times 400 \times \tfrac{1}{3} = 74 \times 10^6$$

Due to comp. steel
$$(223^2/263) \times 14 \times 8 \times A_{sc} = 2.12 \times 10^4 A_{sc}$$

Due to tension steel
$$135A_{st} \times 297 = 4.00 \times 10^4 A_{sc}$$

This sum must equal the applied bending moment of 320×10^6 N-mm, i.e.

$$74 \times 10^6 + 2.12 \times 10^4 A_{sc} + 4.00 \times 10^4 A_{st} = 320 \times 10^6$$

or
$$A_{sc} + 1.89A_{st} = 11,580$$

but
$$A_{sc} = 1{\cdot}42A_{st} - 4{,}440$$
Substituting gives

$$3{\cdot}31A_{st} = 16{,}020$$
leading to
$$A_{st} = 4{,}850 \text{ mm}^2$$
and
$$A_{sc} = 2{,}430 \text{ mm}^2$$

$$\text{The saving in concrete} = (711 - 560)/711$$
$$= 21{\cdot}2 \text{ per cent}$$

$$\text{The increase in steel} = (7{,}280 - 3{,}940)/3{,}940$$
$$= 84{\cdot}8 \text{ per cent}$$

3.14 Distribution of Shear Stress in a Beam

A short length δx of a beam width b is shown in Fig. 3.23 (*a*) with the moment increasing from M to $(M + \delta M)$ in this length. The axial stress distribution is shown in Figs. (*b*) and (*c*). Any slice

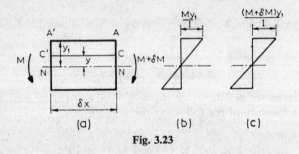

Fig. 3.23

A'ACC' (Fig. 3.23 (*a*)) is in equilibrium under the following forces—
(1) The force F on the face A'C'.
(2) The force $F + \delta F$ on the face AC.
(3) A shearing force along CC'.
If τ_y is the intensity of shear on CC', then for horizontal equilibrium

$$(F + \delta F) - F - \tau_y b \, \delta x = 0$$

The force F is given by

$$-\int_{+y}^{+y_1} \frac{Myb}{I} \, dy$$

(since the stress σ_y due to bending $= -My/I$), and

$$F + \delta F = -\int_{+v}^{+v_1} \frac{(M + \delta M)_y b}{I} \, dy$$

The equilibrium equation then gives

$$-\frac{\delta M}{I} \int_{+v}^{+v_1} by \, dy - \tau_y b \, \delta x = 0$$

(The width b is not necessarily constant, so it is kept within the integral sign.) Putting

$$\int_{+v}^{+v_1} by \, dy = Q \qquad \text{and} \qquad \frac{dM}{dx} = -V \quad (\textit{see} \text{ p. 110})$$

gives

$$\tau_y = +\frac{VQ}{Ib} \tag{3.20}$$

τ_y is the horizontal shear stress across the section, which is equal to the intensity of vertical shear. (*See* eqn. 3.1.)

The integral $Q = \int_y^{v_1} by \, dy =$ first moment of area of the cross-section above y about the neutral axis. $V =$ total shear force at section.

It is seen that at the top of the cross-section, i.e. along the line AA', the shear stress is zero and that the maximum value occurs at the maximum value of Q, i.e. at the neutral axis.

In the rectangular section shown in Fig. 3.24 (*a*) the shear stress at the neutral axis τ_{max} is given by

$$\tau_{max} = \frac{V(\tfrac{1}{2}bd)(\tfrac{1}{4}d)}{\tfrac{1}{12}bd^3 \cdot b} = \frac{3V}{2bd}$$

since

$$Q = b\int_0^{d/2} y \, dy = \tfrac{1}{2}bd \cdot \tfrac{1}{4}d$$

The average shear stress $\tau_{av} = V/bd$ and the maximum value is 1·5 times this. Since the denominator is constant and Q varies parabolically with y, the distribution of shear stress across the section is parabolic, as shown in Fig. 3.24 (*b*).

In the I-section shown in Fig. 3.25 (*a*), at a section just inside the bottom of the web the stress is given by

$$\tau_w = \frac{V}{It_w} bt_f \tfrac{1}{2}(d - t_f)$$

since $Q = bt_f \tfrac{1}{2}(d - t_f)$.

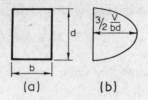

Fig. 3.24

At the same section but just inside the flange the shear stress nominally is

$$\tau_f = \frac{V}{Ib} bt_f \tfrac{1}{2}(d - t_f)$$

The distribution across the section is shown in Fig. 3.25 (b) and this indicates that there is a sudden large discontinuity of shear; though the variation of shear stress across the flange is not determined it is clear that most of the shear force is carried by the web. In designing plate web girders it is therefore assumed that all the shear at the section is carried by the web plate.

The thickness of the web is small and the shear stress is constant over this small width. It is convenient to introduce a new term q, defined as shear flow and given by

$$q = \tau t_w \quad \text{(force per unit length)}$$

$$= \frac{VQ}{I} \tag{3.21}$$

There is however in the I-section the problem of the shear distribution in the flanges due to the fact that these overhang the web and that τ_{xy} must be zero at their upper and lower surfaces.

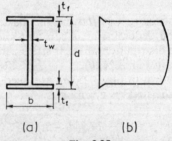

Fig. 3.25

Assuming that the axial stress σ due to bending is constant over the small depth of flange thickness,

$$\sigma = \frac{My}{I} = \frac{Md'}{2I} = \text{a constant}$$

with $d' = d - t_f$. Considering, as before, the equilibrium of the small element of the flange shown in Fig. 3.26 cut off to expose a positive z-face, then for equilibrium

$$\sum X = 0$$

i.e. $$\tau_{zx} t_f \, \delta x + \int_0^{+z} [(\sigma + \delta\sigma) - \sigma] t_f \, . \, dz = 0$$

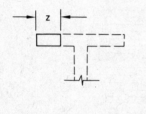 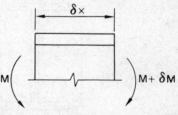

Fig. 3.26

Substituting for σ and $\delta\sigma$ in terms of M and δM and using the shear flow,

$$q_{zx} = \tau_{zx} t_f$$

$$q_{zx} = \frac{\delta M}{\delta x} \int_0^{+z} \left(\frac{t_f y}{I}\right) dz$$

$$= -\frac{VQ}{I}$$

(as before but with changed sign),

$$q_{zx} = q_{xz} \quad (\textit{see} \text{ eqn. (3.1)})$$

where

$$Q = \int_0^{+z} t_f y \, dy$$

$$= \text{1st moment about the neutral axis of the flange cross-sectional area outside } z$$

and, since $y = \text{constant} = d'/2$,

$$Q = [t_f \tfrac{1}{2} d' z]_0^z$$

i.e. a linear function in z altering in sign as the signs of y and z alter.

Thus the shear flow over the complete cross-section under positive shear force is as shown in Fig. 3.27. If it is assumed that the flanges carry all the moment (i.e. $\sigma = 0$ in the web) and that σ is constant over the flange thickness, then the value of the stress σ is given by

$$\sigma = \frac{M}{t_f b d'}$$

but

$$\sigma = \frac{My}{I} = \frac{Md'}{2I}$$

Therefore

$$I \simeq \tfrac{1}{2} t_f b d'^2$$

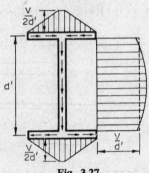

Fig. 3.27

and for the maximum shear flow in the flange at the flange–web junction

$$Q = (t_f \tfrac{1}{2} b) \tfrac{1}{2} d'$$

$$q_{xz} = \frac{VQ}{I} = \frac{V}{2d'}$$

Similarly if q_{xy} is assumed constant down the web,

$$V = q_{xy} d' \quad \text{or} \quad q_{xy} = \frac{V}{d'}$$

as shown in Fig. 3.27.

3.15   Shear Connectors

A type of beam which is being frequently used is that shown in Fig. 3.28 (*a*), which consists of a concrete deck connected to a steel box girder by means of shear connectors at the interface.

The bending stresses in such a section can be calculated by the *transformed section* method previously given for reinforced concrete, provided that at the interface the strain at the bottom of the concrete slab is the same as that on the top of the steel box. One condition which must be fulfilled for this to be so is adequate shear connexion at the interface. The connectors used are generally headed studs welded on to the steel girder (*see* Fig. 3.28 (*b*)). The load carried by such a stud will depend on its length and diameter and the stress in the surrounding concrete. Although much research has been carried out on the subject, the value of the load-carrying capacity based on these three variables has not yet been determined, and an

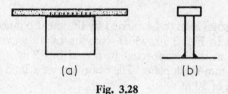

(a)   (b)

Fig. 3.28

experimental approach must be used. Nevertheless, when once the load-carrying capacity of a particular stud has been determined, the number of studs required can be found as follows.

The shear stress at the interface τ_i is, by eqn. (3.20),

$$\tau_i = \frac{VQ}{Ib}$$

where the symbols have the meaning given on p. 77.

The shear per unit length of beam is $\tau_i b$, and so

$$\tau_i b = \frac{VQ}{I}$$

If the load carried is a uniformly-distributed load of intensity w per unit length and the span of the beam is l, then the shearing force diagram is as shown in Fig. 3.29, and the shear V, at a distance x from the centre, is wx. Thus the total shear to be carried by the connectors is

$$\frac{Q}{I} \int_0^{l/2} wx \, dx = \frac{Qwl^2}{8I}$$

for each half of the span.

This analysis assumes rigid connectors, i.e. automatic strain compatibility at interface. In practice the connectors are not rigid and design details require to meet additional criteria.

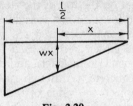

Fig. 3.29

Example 3.5

Calculate the total load to be carried by the shear connectors for the
section shown in Fig. 3.30, which consists of a concrete deck 8 m
wide by 200 mm thick connected to a steel box 3 m × 4 m overall,
made from 15-mm-thick plate. The beam carries a load of 150 kN/m
run on a span of 50 m.

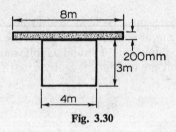

Fig. 3.30

SOLUTION

The first problem is to determine the position of the neutral axis.
This will be done using the transformed section with a modular
ratio of 15.

Concrete Deck	8 × 0·2	= 1·6 × 0·1	= 0·16
Top Steel	15 × 3·970 × 0·015	= 0·90 × 0·207	= 0·184
Side Steel	15 × 3·0 × 0·030	= 1·35 × 1·7	= 2·29
Bottom Steel	15 × 3·970 × 0·015	= 0·90 × 3·193	= 2·87
		$4·75 \times \bar{y}$	= 5·504

giving

$$\bar{y} = \frac{5 \cdot 504}{4 \cdot 75} = 1 \cdot 16 \text{ m}$$

I and Q for the transformed section will be obtained in feet units.

Calculation of I_{NN}   m⁴

Concrete $8 \times 0.2^3 \times \frac{1}{12} = 0.005$

$8 \times 0.2 \times 1.06^2 = 1.790$

Top steel $0.90 \times 0.953^2 = 0.815$

Bottom steel $0.90 \times 2.033^2 = 3.730$

Side steel $15 \times 0.030 \times 3.0^3 \times \frac{1}{12} = 1.014$

$1.35 \times 0.54^2 = 0.394$

$\overline{}$

7.748

$Q_{interface} = 8 \times 0.2 \times 1.060 = 1.696 \text{ m}^3$

Hence total load to be carried by shear connectors on half-span is given by

$$\frac{1.696 \times 150 \times 50 \times 50}{8 \times 7.748} = 10{,}260 \text{ kN}$$

3.16 Principal Stresses

In a previous section (p. 58), expressions have been given for the normal and shear stresses at a point on an inclined plane in a two-dimensional stress system (eqns. (3.6) and (3.7)). It is seen on examining these that the shearing stress τ has zero value when

$$(\sigma_x - \sigma_y) \sin 2\theta = 2\tau_{xy} \cos 2\theta$$

i.e. when

$$\tan 2\theta = \frac{2\tau_{xy}}{\sigma_x - \sigma_y}$$

The values of θ satisfying this equation are

$$\left. \begin{array}{c} \theta = \frac{1}{2} \tan^{-1} \dfrac{2\tau_{xy}}{\sigma_x - \sigma_y} \\[2em] \text{or} \\[1em] \theta = \frac{1}{2} \tan^{-1} \dfrac{2\tau_{xy}}{\sigma_x - \sigma_y} + \frac{1}{2}\pi \end{array} \right\} \quad (3.22)$$

There are thus two planes at right angles to one another on which there is no shearing stress. These planes are known as *principal planes* and the normal forces acting on these planes are called *principal stresses*.

The direct stress σ on any plane has a maximum value when $d\sigma/d\theta$ is equal to zero, i.e. when

$$-(\sigma_x - \sigma_y)\sin 2\theta + 2\tau_{xy}\cos 2\theta = 0$$

This is the same condition that the value of τ is zero. Hence the principal stresses are the maximum and minimum direct stresses which can occur.

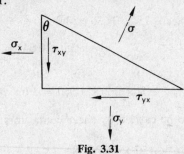

Fig. 3.31

The directions of the principal planes are known when θ is calculated, and knowing the directions we can find the stresses by substitution in the expressions for stress.

The principal stress can also be found by consideration of the two-dimensional stress system shown in Fig. 3.31, where σ is a principal stress acting on a plane of unit area inclined at θ to a given stress system.

Resolving in the x-direction gives

$$\sigma\cos\theta - \sigma_x\cos\theta = \tau_{xy}\sin\theta$$

thus

$$(\sigma - \sigma_x) = \tau_{xy}\tan\theta$$

Equilibrium in the y-direction gives

$$\sigma\sin\theta - \sigma_y\sin\theta = \tau_{xy}\cos\theta$$

thus

$$(\sigma - \sigma_y) = \tau_{xy}\cot\theta$$

Eliminating θ,

$$(\sigma - \sigma_x)(\sigma - \sigma_y) = \tau_{xy}^2$$

Solving this quadratic for σ gives the two principal stresses σ_1 and σ_2 in a more convenient form as

$$\sigma_1 = \tfrac{1}{2}(\sigma_x + \sigma_y) + \tfrac{1}{2}\sqrt{\{(\sigma_x - \sigma_y)^2 + 4\tau_{xy}^2\}} \qquad (3.23)$$

$$\sigma_2 = \tfrac{1}{2}(\sigma_x + \sigma_y) - \tfrac{1}{2}\sqrt{\{(\sigma_x - \sigma_y)^2 + 4\tau_{xy}^2\}} \qquad (3.24)$$

From which

$$\sigma_1 + \sigma_2 = \sigma_x + \sigma_y \tag{3.25}$$

The principal planes are planes of zero shearing stress; the magnitude of the latter will vary with the inclination of the plane. For a maximum

$$\frac{d\tau}{d\theta} = -(\sigma_x - \sigma_y) \cos 2\theta - 2\tau_{xy} \sin 2\theta = 0$$

i.e. the maximum occurs when

$$\cot 2\theta = -\frac{2\tau_{xy}}{(\sigma_x - \sigma_y)} \tag{3.26}$$

Comparison of eqns. (3.22) and (3.26) shows that the planes of maximum shearing stress are inclined at 45° to the planes of principal stress.

Substituting the value of θ in the equation for the intensity of shear stress,

$$\tau_{max} = \sqrt{[\{\tfrac{1}{2}(\sigma_x - \sigma_y)\}^2 + \tau_{xy}^2]}$$

But the right-hand side can be expressed in terms of the principal stresses σ_1 and σ_2 as

$$\tau_{max} = \sigma_1 - \tfrac{1}{2}(\sigma_x + \sigma_y)$$

or

$$\tau_{max} = -\sigma_2 + \tfrac{1}{2}(\sigma_x + \sigma_y)$$

Adding,

$$2\tau_{max} = (\sigma_1 - \sigma_2)$$

or

$$\tau_{max} = \tfrac{1}{2}(\sigma_1 - \sigma_2) \tag{3.27}$$

Expressed in words, the maximum shearing stress is one-half the difference of the principal stresses and acts on planes at 45° to the principal planes.

3.17. Mohr's Circle of Stress

The normal stress σ_θ and the shear stress τ_θ on any plane inclined at an angle θ to the plane on which the major principal stress acts are given from eqns. (3.6) and (3.7) (p. 58) by

$$\sigma_\theta = \tfrac{1}{2}(\sigma_1 + \sigma_2) + \tfrac{1}{2}(\sigma_1 - \sigma_2) \cos 2\theta$$

and

$$\tau_\theta = -\tfrac{1}{2}(\sigma_1 - \sigma_2) \sin 2\theta$$

These expressions can be obtained graphically by a method due to Mohr.

Take a pair of orthogonal axes, one to represent the direct and the other the shear stress. With the origin at O, mark off abscissae OA = σ_2 and OB = σ_1. Let C be the mid-point of AB. Then

$$OC = OA + \tfrac{1}{2}AB$$
$$= \sigma_2 + \tfrac{1}{2}(\sigma_1 - \sigma_2)$$
$$= \tfrac{1}{2}(\sigma_1 + \sigma_2)$$

Draw a circle (Fig. 3.32) with centre C and diameter AB. Let P

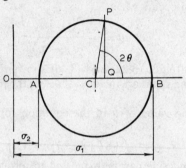

Fig. 3.32

be any point on this circle such that $\angle$ PCB = 2θ. Then if PQ is perpendicular to OB,

$$PQ = \tfrac{1}{2}AB \times \sin 2\theta = -\tfrac{1}{2}(\sigma_1 - \sigma_2) \sin 2\theta$$
$$OQ = OC + CQ = \tfrac{1}{2}(\sigma_1 + \sigma_2) + \tfrac{1}{2}(\sigma_1 - \sigma_2) \cos 2\theta$$

Thus PQ represents the shear stress on a plane inclined at θ to the major principle stress and OQ the normal stress.

This graphical construction enables the shear and normal stresses to be obtained from the principal stresses and conversely enables the principal stresses to be obtained if the normal and shear stresses on a given plane are known.

Example 3.6
Across a plane A in a stressed material there is transmitted a normal tensile stress of 50 N/mm² accompanied by an unknown shear stress. The maximum principal stress in the material is 120 N/mm² tension on a plane making 55° with plane A. Determine the second principal stress and calculate the strain normal to plane A. (E = 200 kN/mm²; ν = 0·3)

SOLUTION
The forces on an element are given in Fig. 3.33 (a).

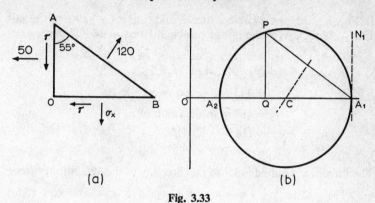

Fig. 3.33

Resolving in the x-direction gives

$$\sigma_x \times OB + \tau \times OA = 120 \times AB \sin 55°$$
$$(120 - \sigma_x) \sin 55° = \tau \cos 55°$$

In the y-direction,

$$50 \times OA + \tau \times OB = 120 \times AB \cos 55°$$
$$50 \cos 55° + \tau \sin 55° = 120 \cos 55°$$
$$70 \cos 55° = \tau \sin 55°$$

Hence $\tau = 48.9$ N/mm², and

$$120 - \sigma_x = 48.9 \cot 55°$$
$$= 35.1$$

Hence

$$\sigma_x = 84.9 \text{ N/mm}^2$$

Applying eqn. (3.25),

$$120 + \sigma_2 = 84.9 + 50$$
$$\sigma_2 = 14.9 \text{ N/mm}^2$$

The Mohr circle construction is given in Fig. 3.33 (*b*), where

$$OA_1 = 120 = \text{major principal stress}$$
$$OQ = 50 \ = \text{normal stress on OA}$$

Draw A_1P at 55° to A_1N_1; PQ is vertical and equal to $\tau = 48.9$ N/mm². Bisect PA_1 at right angles, cutting OA_1 at C. It can be shown that, if C is the centre of a circle passing through A_1 and P, then $\angle PCA_1 = 2 \times \angle PA_1N_1$. Thus $\angle PCA_1$ corresponds to 2θ

in Fig. 3.32. Draw a circle centre C and radius CA_1 cutting the axis in A_2. Now OA_2 is the minor principal stress = 14·9 N/mm².

$$\epsilon_1 = \text{strain in direction of } \sigma_1$$
$$= (120 - 14\cdot9v)/E$$
$$= 115\cdot5/200{,}000 = 5\cdot77 \times 10^{-4}$$
$$\epsilon_2 = \text{strain in direction of } \sigma_2$$
$$= (14\cdot9 - 120v)/E$$
$$= -0\cdot105 \times 10^{-4}$$

The direction required is 35° to the direction of the 120 N/mm² stress.

$$\epsilon = \epsilon_1 \cos^2 \theta + \epsilon_2 \sin^2 \theta \qquad (\textit{see} \text{ eqn. (3.6)})$$
$$= 0\cdot329\epsilon_1 + 0\cdot670\epsilon_2$$
$$= 1\cdot790 \times 10^{-4}$$

3.18 The Strain Gauge Rosette

In the laboratory the state of stress at a given point in a member is determined by the use of strain gauges. These can be either electrical or mechanical, the electric resistance gauge being one of the most convenient to use.

In the two dimensional case the state of strain is fully defined by three quantities, ϵ_x, ϵ_y, γ_{xy}. The shear strain γ_{xy} is very difficult to measure. If, however, the strains in three different directions can be measured, the results give three quantities from which three unknowns can be determined. These three strains are measured by a set of gauges known as a strain gauge rosette.

In Fig. 3.34(a) let ϵ_I and ϵ_{II} be the principal strains and let the strains measured by the three gauges be ϵ_1, ϵ_2 and ϵ_3. The angles between the gauges α and β are known but the direction θ between gauge 1 and the major principal stress is unknown. The relationship between the measured and the principal strains is given) (from eqns (3.6) and(3.7) by the following equations

$$\epsilon_1 = \tfrac{1}{2}(\epsilon_I + \epsilon_{II}) + \tfrac{1}{2}(\epsilon_I - \epsilon_{II}) \cos 2\theta$$
$$\epsilon_2 = \tfrac{1}{2}(\epsilon_I + \epsilon_{II}) + \tfrac{1}{2}(\epsilon_I - \epsilon_{II}) \cos 2(\theta + \alpha)$$
$$\epsilon_3 = \tfrac{1}{2}(\epsilon_I + \epsilon_{II}) + \tfrac{1}{2}(\epsilon_I - \epsilon_{II}) \cos 2(\theta + \alpha + \beta)$$

from the three values of the measured strains the principal strains ϵ_I and ϵ_{II} and the direction θ can be calculated.

Consider as an example the 45° rosette shown in Fig. 3.34(b). In this case $\alpha = \beta = 45°$ and the equations become

$$\epsilon_1 = \tfrac{1}{2}(\epsilon_I + \epsilon_{II}) + \tfrac{1}{2}(\epsilon_I - \epsilon_{II}) \cos 2\theta$$
$$\epsilon_2 = \tfrac{1}{2}(\epsilon_I + \epsilon_{II}) - \tfrac{1}{2}(\epsilon_I - \epsilon_{II}) \sin 2\theta$$
$$\epsilon_3 = \tfrac{1}{2}(\epsilon_I + \epsilon_{II}) - \tfrac{1}{2}(\epsilon_I - \epsilon_{II}) \cos 2\theta$$

Eliminating θ gives ϵ_I and ϵ_{II} as the two roots of the equation

$$\epsilon^2 - (\epsilon_1 + \epsilon_3)\epsilon + \epsilon_1\epsilon_3 - \tfrac{1}{4}(\epsilon_1 + \epsilon_3 - 2\epsilon_2)^2 = 0$$

and
$$\tan \theta = \frac{2\epsilon_2 - \epsilon_1 - \epsilon_3}{\epsilon_3 - \epsilon_1}$$

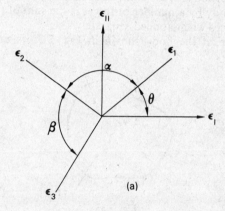

(a)

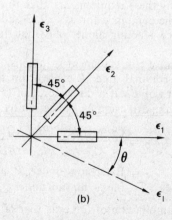

(b)

Fig. 3.34

3.19 Combined Force Actions: Principle of Superposition

The stresses caused by a single force action, i.e. tension, bending, etc., can be calculated by the methods previously given. Often, however, the member is not subject to a single force alone but to a combination of forces, and the problem is to find what stresses are caused by the combined forces.

When the members under stress are made from elastic materials having a linear load-displacement diagram, combined forces can be dealt with by means of the principle of superposition.

This principle states that, if the displacements at all points in a body are proportional to the loads causing them, the effect produced upon such a body by a number of forces is the sum of the effects of the several forces when applied separately.

For example, if the body shown in Fig. 3.35 is made from a

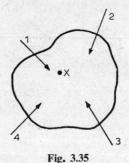

Fig. 3.35

material satisfying these requirements, then the displacement of the point X under the combined forces 1, 2, 3 and 4 is the sum of the effect under force 1 acting alone, plus that due to force 2 acting alone, etc.

A simple proof of this statement is to consider a uniform rod of length L cross-sectional area A made from a material having a Young's Modulus of E. If a direct force of $(P + Q)$ is applied to such a rod then the extension δ_{P+Q} is given by

$$\delta_{P+Q} = \frac{(P + Q)L}{AE} = \frac{PL}{AE} + \frac{QL}{AE}$$

= Extension under P alone
plus extension under Q alone

A common combination of forces is that of bending with direct force. As an example of this consider the column shown in Fig. 3.36

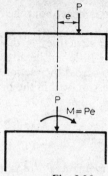

Fig. 3.36

which carries a direct load P at a distance e from the centroidal axis of the section. This is equivalent to loading the member with a direct load P plus a bending moment Pe. Thus the stress σ at any section distant y from the centroidal axis is given by

$$\sigma = \frac{-P}{A} - \frac{Pey}{I} \qquad (3.28)$$

where A is the cross-sectional area and I the second moment of area of the cross-section. If Ak^2 (where $k = \sqrt{(I/A)}$ = radius of gyration) is substituted for I, the expression for the stress becomes

$$\sigma = \frac{-P}{A}\left(1 + \frac{ey}{k^2}\right) \qquad (3.29)$$

This combination of loading is very common in structures. One example is the gravity dam shown in Fig. 3.37, where the base is subjected to a direct load due to the weight of the structure, together with an overturning moment and a shear due to the water pressure. Another example is the chimney shown in Fig. 3.38, where the wind

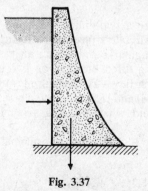

Fig. 3.37

Fig. 3.38

pressure produces a bending moment about the base which combines
with the weight of the stack.

One characteristic of such structures is that they are normally
built of a material which has little tensile strength; as a result no
tension is allowed to develop across the section.

For this to be so, the expression within the brackets in eqn. (3.29)
must always be positive in sign, i.e.

$$1 \geqslant \frac{ey}{k^2} \quad \text{or} \quad \frac{My}{Pk^2}$$

where M is the overturning moment, P the vertical load, and e the
eccentricity with which the resultant load acts.

In a rectangular section, of width b and depth d the maximum
tensile stress will occur at the extreme fibre, i.e. where $y = \frac{1}{2}d$. Thus
the condition for tension not to be developed is

$$1 \geqslant \frac{e \times \frac{1}{2}d}{\frac{1}{12}d^2}, \quad \text{i.e.} \quad e \leqslant \frac{1}{6}d$$

For this to hold, the resultant thrust must always lie within the
middle third of the section.

Another very common combination of loadings is that of bending
plus shear. This occurs in all beams subject to non-uniform bending.

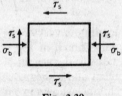

Fig. 3.39

The problem can be reduced to a two-dimensional one by considering an element cut from a stressless face, i.e. a vertical plane
parallel to the x-axis in a rectangular beam. Such an element is
shown in Fig. 3.39 as subject to a direct stress σ_b on two opposite
faces and shear stresses τ_s on each face. The magnitudes of σ_b and
τ_s can be obtained from eqns. (3.14) and (3.20) respectively, assuming
that the simple theories still hold when there is a combination
of stresses.

The combination shown in Fig. 3.38 gives rise to principal stresses
the greater of which is

$$\sigma_1 = \frac{1}{2}\sigma_b + \frac{1}{2}\sqrt{(\sigma_b{}^2 + 4\tau_s{}^2)}$$

The inclination of the plane on which this force acts is
$\frac{1}{2}\tan^{-1}(\tau_s/\sigma_b)$, measured relative to the longitudinal axis of the
beam. It depends on the ratio τ_s/σ_b which in turn depends on the
position of the element relative to the neutral axis. At this line
$\sigma_b = 0$, hence the inclination is 45°. At the outermost fibre $\tau_s =$
0, hence the principal stress acts parallel to the axis.

Lines drawn on a member to show the directions of the principal
stresses due to the loading to which it is being subjected are known
as *stress trajectories*. The stress trajectories for a rectangular cantilever carrying a point load at the end are shown in Fig. 3.40.

It is not possible to generalize where torsion is combined with
non-uniform bending, as the resultant stress distribution and the
stress trajectories depend on the relative magnitudes of the two stress
systems.

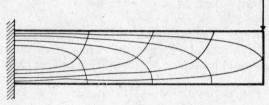

Fig. 3.40

3.20 Bending of Bars of Unsymmetrical Section

A straight bar of arbitrary cross-section with a right-hand set of
axes through the centroid is shown in Fig. 3.41. If the beam bends

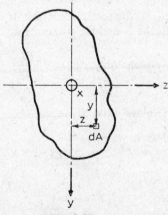

Fig. 3.41

in the x,y-plane only such that the neutral axis is the z-axis, then the stress on any element dA is proportional to the y-coordinate of that element. Let R_y be the radius of curvature of the deflected beam in the xy-plane. Then by eqn. (3.13) (p. 68)

$$\frac{\sigma}{y} = -\frac{E}{R_y}$$

and the moment of resistance about the z-axis is M_z where

$$M_z = -\int \sigma\, y\, dA = \frac{E}{R_y} \int y^2\, dA = \frac{EI_{zz}}{R_y}$$

Also the moment of resistance about the y-axis is

$$M_y = \int \sigma z\, dA = -\frac{E}{R_y} \int yz\, dA = -\frac{EI_{yz}}{R_y}$$

Bending, however, may take place in the x,z-plane only such that the y-axis is the neutral axis and the radius of curvature is R_z. For the moment of resistance M_y about the y-axis we have

$$M_y = \int \sigma z\, dA = -\frac{E}{R_z} \int z^2\, dA = -\frac{EI_{yy}}{R_z}$$

Also

$$M_z = -\int \sigma y\, dA = \frac{E}{R_z} \int yz\, dA = \frac{EI_{zy}}{R_z}$$

The axes about which I_{zy} (the product of inertia) is zero are known as the *principal axes*. In a study of the stresses due to asymmetrical bending a knowledge of the position of the principal axes for a given section is necessary.

The usual practical case is that for which M_y is zero and bending occurs about both the z- and the y-axes. But,

For bending in x,y-plane only, $M_y = -\dfrac{EI_{yz}}{R_y}$

and

For bending in x,z-plane only, $M_y = -\dfrac{EI_y}{R_z}$

Then for $M_y = 0$ we have

$$-\frac{EI_{yz}}{R_y} - \frac{EI_y}{R_z} = 0$$

or

$$\frac{1}{R_z} = -\frac{1}{R_y} \frac{I_{yz}}{I_y}$$

also

$$M_z = \frac{EI_z}{R_y} + \frac{EI_{yz}}{R_z}$$

i.e.

$$M_z = \frac{E}{R_y} I_z \left\{ \frac{I_z I_y - I_{yz}^2}{I_y I_z} \right\}$$

From this the curvatures and stresses can be calculated. Thus

$$\sigma = -\frac{Ey}{R_y} - \frac{Ez}{R_z}$$

$$= \frac{-M_z(yI_y - zI_{yz})}{(I_y I_z - I_{yz}^2)}$$

For a symmetrical section, or when the y- and the z-axes are principal axes, I_{yz} is shown in the following section to be zero and the above reduces to the orthodox case, namely

$$\sigma = -\frac{M_z y}{I_y}$$

The alternative case of a bending moment about the y-axis only, i.e. $M_z = 0$, is dealt with in the same way to give

$$\sigma = \frac{M_y(zI_z - yI_{yz})}{(I_y I_z - I_{yz}^2)}$$

3.21 Principal Axes of a Section

Consider any section such as that shown in Fig. 3.42 and choose any pair of axes Gy, Gz through the centroid. Take a second pair of axes GY, GZ inclined at θ to the first pair.

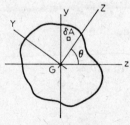

Fig. 3.42

The coordinates of an element δA are z, y under the first system and z', y' under the second. The relationship between the two systems is

$$\left.\begin{array}{c} z' = z \cos \theta + y \sin \theta \\ y' = -z \sin \theta + y \cos \theta \end{array}\right\}$$

The second moments of area about the axes GZ and GY are $I_{z'z'}$ and $I_{y'y'}$.

$$\begin{aligned} I_{z'z'} &= \sum \delta A \times y'^2 \\ &= \sum \delta A \, (-z \sin \theta + y \cos \theta)^2 \\ &= \sum \delta A \, (z^2 \sin^2 \theta + y^2 \cos^2 \theta - 2zy \sin \theta \cos \theta) \\ &= \sin^2 \theta \sum \delta A \, z^2 + \cos^2 \theta \sum \delta A \, y^2 - \sin 2\theta \sum \delta A \, zy \end{aligned}$$

The expression $\sum \delta A \, zy$ is called the *product* moment of area and is denoted by I_{zy}. Thus

$$I_{z'z'} = I_{yy} \sin^2 \theta + I_{zz} \cos^2 \theta - I_{zy} \sin 2\theta \qquad (3.30)$$

In a similar way it can be shown that

$$I_{y'y'} = I_{yy} \cos^2 \theta + I_{zz} \sin^2 \theta + I_{zy} \sin 2\theta \qquad (3.31)$$

Adding,

$$\begin{aligned} I_{z'z'} + I_{y'y'} &= I_{yy}(\sin^2 \theta + \cos \theta^2) + I_{zz}(\sin^2 \theta + \cos^2 \theta) \\ &= I_{yy} + I_{zz} \end{aligned} \qquad (3.32)$$

Hence it is seen that the sum of the second moments of area about any pair of rectangular axes is a constant.

The variation of $I_{z'z'}$ with θ is

$$\frac{dI_{z'}}{d\theta} = I_{yy}2 \sin \theta \cos \theta + I_{zz}2 \cos \theta(-\sin \theta) - I_{zy}2 \cos 2\theta$$

$$= \sin 2\theta(I_{yy} - I_{zz}) - 2I_{zy} \cos 2\theta$$

For zero value of this function

$$\tan 2\theta = \frac{2I_{zy}}{I_{yy} - I_{zz}}$$

Therefore

$$2\theta = \tan^{-1}\left(\frac{2I_{zy}}{I_{yy} - I_{zz}}\right) \pm 180°$$

There are therefore only two positions where the second moment of area takes a turning value: one is a maximum and the other is a minimum. The axes having these two values are known as the *principal axes* and the second moments of area are the *principal moments*, one the *major* and the other the *minor*.

If Gu and Gv are the principal axes inclined at θ to Gz and Gy, then the product moment of area $I_{z'y'}$ or I_{uv} is

$$\sum \delta A \times z'y' = \sum \delta A \,(z\cos\theta + y\sin\theta)(-z\sin\theta + y\cos\theta)$$

$$= \sum \delta A\, zy(\cos^2\theta - \sin^2\theta)$$

$$- \sum \delta A\, z^2\sin\theta\cos\theta + \sum \delta A\, y^2\sin\theta\cos\theta$$

$$= (I_{zz} - I_{yy})\tfrac{1}{2}\sin 2\theta + I_{zy}\cos 2\theta$$

which is equal to zero when

$$\tan 2\theta = -\frac{2I_{zy}}{I_{zz} - I_{yy}}$$

Hence the product moment of area about principal axes is zero. This explains why axes of symmetry are principal axes, for in this case each positive product has a corresponding negative one and the two cancel out. The values of I_{uu} and I_{vv} are obtained as follows—

$$I_{uu} = I_{zz}\cos^2\alpha + I_{yy}\sin^2\alpha - I_{zy}\sin 2\alpha$$

$$= I_{zz}\tfrac{1}{2}(1 + \cos 2\alpha) + I_{yy}\tfrac{1}{2}(1 - \cos 2\alpha)$$

$$+ \sin 2\alpha\tfrac{1}{2}(I_{zz} - I_{yy})\tan 2\alpha$$

$$= \tfrac{1}{2}(I_{zz} + I_{yy}) + \tfrac{1}{2}(I_{zz} - I_{yy})\left(\cos 2\alpha + \frac{\sin^2 2\alpha}{\cos 2\alpha}\right)$$

$$= \tfrac{1}{2}(I_{zz} + I_{yy}) + \tfrac{1}{2}(I_{zz} - I_{yy})\sec 2\alpha \qquad (3.33)$$

Similarly,

$$I_{vv} = \tfrac{1}{2}(I_{zz} + I_{yy}) - \tfrac{1}{2}(I_{zz} - I_{yy})\sec 2\alpha$$

since

$$I_{uu} + I_{vv} = I_{zz} + I_{yy} \qquad (3.34)$$

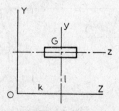

Fig. 3.43

It is useful to study the problem of the product of inertia about parallel axes (*see* Fig. 3.43)—

$$I_{ZY} = \sum \delta A \times ZY = \sum \delta A \, (z + k)(y + l)$$
$$= \sum \delta A \, zy + \sum \delta A \, lz + \sum \delta A \, yk + \sum \delta A \, kl$$
$$= I_{zy} + 0 \text{ (since G is centroid)} + 0 + Akl$$

If z and y are principal axes, $I_{zy} = 0$ and

$$I_{ZY} = Akl$$

3.22 Stresses Due to Asymmetrical Bending

A beam of a section without an axis of symmetry is shown in Fig. 3.44 (*a*), where a moment M is applied in the x,y-plane.

In order to calculate the stress at points in the cross-section due to the applied moment it is resolved into two perpendicular components in the direction of the principal axes. The cross-section with these axes is shown in Fig. 3.44 (*b*) with M resolved into two components $M \cos \theta$ about Gv and $M \sin \theta$ about Gu.

The stress σ at any point u, v is

$$\sigma = + \frac{M \cos \theta \, u}{I_{vv}} + \frac{M \sin \theta \, v}{I_{uu}}$$

In order to find the position of the neutral axis the expression for stress must be equated to zero, giving

$$\frac{u \cos \theta}{I_{vv}} = - \frac{v \sin \theta}{I_{uu}}$$

or

$$\frac{v}{u} = - \frac{I_{uu}}{I_{vv}} \cot \theta = \tan \beta$$

It is clear from the minus sign that β is always in the adjacent quadrant to θ.

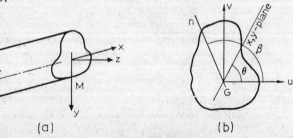

(a)   (b)

Fig. 3.44

The assumption that plane sections remain plane and that the neutral axis is one of zero strain means that the position of maximum strain is at that point on the section which is furthest from the neutral axis.

Since stress is assumed to be proportional to strain then the maximum stress will occur at the point furthest from the neutral axis. Determination of the position of this axis as shown enables in turn the point most distant from the axis to be found.

Since bending takes place about both axes deformations take place about both axes also. The deflexions are thus not only in the x–y plane as in the standard symmetrical section but also in the x–z plane.

Example 3.7

A 150 mm × 100 mm × 10 mm angle carries a load on the shorter leg, the longer leg being vertically upwards. Find the permissible bending moment if the maximum stress is 150 N/mm² [$I_{zz} = 5\cdot49 \times 10^6$ mm⁴; $I_{vv} = 2\cdot03 \times 10^6$ mm⁴; $I_{min} = 1\cdot42 \times 10^6$ mm⁴; $\theta = 21\cdot0°$; $h = 23\cdot8$ mm; $k = 48\cdot7$ mm].

SOLUTION
The section is shown in Fig. 3.44.

$$I_{max} + I_{min} = 5\cdot49 + 2\cdot03 = 7\cdot52 \times 10^6$$

Therefore

$$I_{max} = 6\cdot10 \times 10^6 \text{ mm}^4 \text{ units}$$

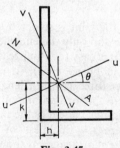

Fig. 3.45

The applied moment must be resolved into two components $M \cos 21\cdot0°$ acting about uu and $M \sin 21\cdot0°$ acting about vv.

The neutral axis will be approximately in the position shown and it is clear that the point farthest from the axis is the inner top point of the vertical leg, the coordinates of which in the zy-system are

—13·8 and 101·3. Transposing these to the *uv*-system,

$$u = -13\cdot8 \cos 21 + 101\cdot3 \sin 21$$
$$= -12\cdot9 + 36\cdot3 = 24\cdot6$$
$$v = 13\cdot8 \sin 21 + 101\cdot3 \cos 21$$
$$= 4\cdot95 + 94\cdot5 = 99\cdot8$$

Hence the maximum stress is

$$\frac{M \sin 21 \times 24\cdot6}{1\cdot42 \times 10^6} + \frac{M \cos 21 \times 99\cdot8}{6\cdot10 \times 10^6}$$

But the maximum stress = 150 N/mm², leading to

$$M = 6\cdot88 \text{ kN-m}$$

3.23 Torsion of Thin-walled Tubes

If the simple rectangular plate shown in Fig. 3.46 (*a*) is subjected to pure shear it distorts to the parallelogram shown dotted, the lengths of the sides being equal to those of the original rectangle. In this shape it can be formed into a thin-walled circular cylindrical tube

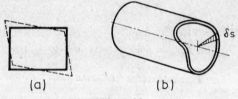

(a)   (b)

Fig. 3.46

and will fit together. If the joint is soldered the shears on the connected edges will balance and only the end shears acting around the circumferences remain. It is easy to visualize that any other shape of tube can be formed as in Fig. 3.46 (*b*). It is therefore reasonable to assume that the shear force per unit length of the periphery is constant and tangential to the median line of the plate.

This force per unit length is known as the *shear flow*. If $q =$ shear flow, then $q = \tau t$, where $t =$ thickness of tube. The torsional moment δT resisted by the short length δs shown in the figure is given by

$$\delta T = q \, \delta s \times r$$
$$= q \times 2 \quad \text{(shaded area)}$$

The total torque T resisted by the tube is therefore

$$T = q \times 2 \times \text{area enclosed by median line}$$
$$= 2qA \qquad\qquad\qquad (3.35)$$

In a thin-walled tube of circular section

$$A = (\pi/4) \times d^2$$

and hence

$$T = \tfrac{1}{2}q\pi d^2$$

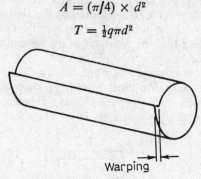

Warping

Fig. 3.47

The torsion of an open section or split tube as shown in Fig. 3.47 has to be dealt with in a different way. There is some shear resistance along the cut and the end cross-sections change shape. The resistance to twist is only that of a simple flat plate and is low. It depends on the resistance which the cross-section can offer to warping. Whilst this is readily dealt with, the solution is lengthy and outwith the scope of this book.

3.24 Shear of Thin-walled Sections: Shear Centre

The application of load to an angle section as in the Example 3.7 causes difficulties in dealing not only with the stress due to bending but also with the stresses due to shear. Consider the angle shown in Fig. 3.48 where the load is applied parallel to the longer leg. The stress can be obtained from eqns. (3.20) and (3.21) and their distribution is shown in the figure.

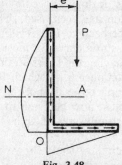

Fig. 3.48

It can be seen that the shear flows have a resultant moment about all points in the plane except O. This point through which the resultant of the shear forces act is known as the *shear centre*. Unless the line of action of the applied load passes through O, the section will twist due to the fact that the resultant of the shear flows will have a moment about the line of action of the load.

In Fig. 3.48 the moment is $P \times e$ and this torque will cause stresses on the section additional to those resulting from pure bending.

Another common section where twisting may occur is the thin-walled open channel shown in Fig. 3.49. The shear centre is at S,

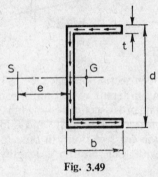

Fig. 3.49

located at e from the centre line of the web. It can be shown that

$$e = \frac{tb^2d^2}{4I_{zz}}$$

I_{zz} is approximately equal to $\frac{1}{2}tbd^2$. Hence $e \simeq \frac{1}{2}b$.

It should be noted that the shear centre is on that side of the web remote from the centroid of the section.

3.25 Plastic Torsion

In the previous sections dealing with combinations of loading the principle of superposition has been used. This is based on elastic behaviour. On reference back to Fig. 3.10 it is seen that this only covers part of the stress-strain curve for most materials. In an ideal elastic-plastic material having the stress-strain relationship given in Fig. 3.50, the linear relationship on which the principle of super-position is based would only cover the portion OA of the stress-strain curve. Consideration must now be given to members subjected to loads which result in strains greater than that corresponding to point A.

Consider firstly the circular shaft shown in Fig. 3.51 which is subjected to a gradually increasing torsion. Until the strain on the

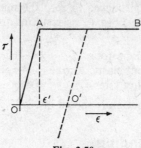

Fig. 3.50

outermost fibre has reached ϵ', the distribution of stress across a radius will vary uniformly from zero at the centre, to a maximum at the circumference (Fig. 3.51(a)). As the torque is increased (assuming that under pure shear $\tau_y = $ constant), the strain at the circumference increases, but the stress remains equal to the yield stress τ_y and the distribution of stress across the section is as shown in Fig. 3.50 (b), where part of section is plastic and the remainder elastic. Further application of load produces further strain, until ultimately the whole section is, for all practical purposes, strained at a value exceeding ϵ'. (Obviously the centre-point of the shaft must be unstrained and points immediate to it may be strained at less than ϵ'.) Under these conditions the stress distribution is as shown in Fig. 3.51 (c) with the whole cross-section stressed at τ_y.

The torsional moment of resistance is then

$$T_p = \tau_y \times 2\pi \int_0^R r^2 \, dr$$

$$= \tau_y 2\pi R^3/3 \quad \text{or} \quad \tau_y \pi d^3/12 \qquad (3.36)$$

The maximum torque resisted under elastic conditions is $\tau_y \pi d^3/16$ and so it can be seen that under fully plastic conditions the torsional moment of resistance is about 33 per cent greater than this.

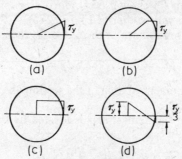

Fig. 3.51

If now the torque is reduced to zero the release of strain occurs elastically. If τ_e is the fibre stress resulting from an elastic stress distribution,

$$\text{Unloading torque} = -\tau_y(\pi d^3/12)$$
$$\text{Equivalent elastic torque} = -\tau_e(\pi d^3/16)$$

Therefore   Reversal of stress $= \frac{4}{3}\tau_y$

The final stress distribution is shown in Fig. 3.50 (d).

3.26 Plastic Bending

A rectangular beam subjected to a gradually increasing bending moment takes up a stress distribution in the early stages as shown in Fig. 3.52 (a) (with the usual assumptions in the standard theory of

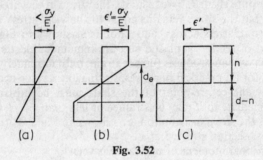

Fig. 3.52

bending). It continues to be of this form until, with a moment $M_y = \sigma_y bd^3/6$, the fibre stress reaches the yield stress σ_y corresponding to a strain ϵ' (assuming yield occurs at $\sigma = \text{constant} = \sigma_y$). Further application of moment causes yield to spread further into the beam, giving a stress distribution as in Fig. 3.52 (b). Ultimately the whole section (with the exception of the neutral axis and immediately adjacent fibres) will be strained at ϵ' or greater, and the stress distribution will be that of Fig. 3.52 (c).

The moment of resistance of the section is then

$$M_p = \sigma_y \int_{-(d-n)}^{n} by\,dy = \sigma_y \times S \qquad (3.37)$$

where S = first moment of area of the section about the new plastic neutral axis.

The position of the neutral axis is located by the solution of the equation

$$\int_0^n b\,dy = \int_0^{d-n} b\,dy$$

It is only the same as for elastic bending for sections symmetrical about the z-axis.

If the section is rectangular, $S = \frac{1}{4}bd^2$, and hence $M_p = \sigma_y \frac{1}{4}bd^2$.

For the general case of bending, where the strain in the beam does not exceed ϵ', the moment of resistance M_R is σZ, where Z is the section modulus. For a rectangular section, $Z = \frac{1}{6}bd^2$. The factor S/Z is known as the *shape factor* of the section, and for a rectangle is 1·5. For most I-section joists it is about 1·15.

If a short section of beam is subjected to uniform bending, as in Fig. 3.53 (*a*), then the relationship between M and the curvature ϕ where $\phi = d\theta/dx$ is plotted in Fig. 3.53 (*b*). OA corresponds to the state of stress shown in Fig. 3.52 (*a*), where the strain on the section nowhere exceeds ϵ'; AB corresponds to the state of stress shown in Fig. 3.52 (*b*); the point B corresponds to Fig. 3.52 (*c*) where all the section has reached yield stress.

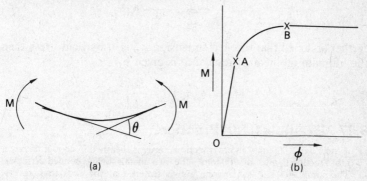

(a)   (b)

Fig. 3.53

Between A and B from the stress system of Fig. 3.52(*b*),

$$\frac{M}{M_y} = \frac{3}{2}\left(1 - \frac{d_e^2}{3d^2}\right)$$

$$\phi = \frac{\sigma_y \times \delta x}{\frac{1}{2}d_e \times E}$$

The ratio between the ordinates to B and A gives the shape factor of the section.

The state of stress shown in Fig. 3.52 (*c*) gives the maximum moment of resistance of the cross-section. This is sometimes known as the *ultimate moment of resistance* or *fully plastic moment* of the section; for uniform beams made from a material having a definite yield point it is $\sigma_y S$; for composite beams it depends on the materials used and the method of connexion.

In the case of the reinforced concrete beam analysed in Section 3.13 where the concrete is assumed to take no tension the stress distribution at collapse is generally assumed to be as in Fig. 3.54. If it is

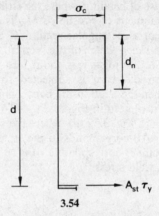

3.54

further assumed that the steel in tension A_{st} is at its yield stress then the ultimate moment of resistance is given by

$$M_p = A_{st}\sigma_y \left(d - \frac{dn}{2}\right)$$

3.27 Examples for Practice

1. A steel bar 1,000 mm² in cross-sectional area is sheathed in copper having a cross-sectional area of 700 mm², the two metals being bonded together. The composite bar is 5 m long and is fastened at the ends to supports. It is free from stress when the temperature is 20°C. Find the total load in the bar and the stress in the steel and in the copper when the temperature falls to 0°C, if the rigidity of the supports permits contraction in length of only 0·5 mm. (E for steel = 230 kN/mm²; for copper = 100 kN/mm²; coefficient of expansion for steel = $1·1 \times 10^{-5}$ per deg C, for copper = $1·8 \times 10^{-5}$ per deg C.)
   (*Ans.* 45·8 kN tension; steel 27·6 N/mm²; copper 26·0 N/mm².)

2. A beam consists of a timber baulk 200 mm deep by 100 mm wide, with a steel plate 100 mm wide by 12 mm thick securely fastened to the lower surface. The beam spans an opening of 4 m. If the maximum tensile stress in the timber is limited to 2·75 N/mm², calculate the value of (*a*) the uniformly-distributed load which may be carried, (*b*) the maximum compressive stress in the timber, and (*c*) the stress in the steel plate, which may be assumed constant across the thickness and equal to the mean value. Take the modular ratio as 15.
   (*Ans.* 4·85 N/m; 8·50 N/mm²; 46·2 N/mm².)

3. Figure 3.55 shows the section of a doubly-reinforced concrete beam. Under a test load on a span of 10 m, strain gauges show zero strain at a depth of 350 mm below the top fibre. Taking this as the depth of the neutral axis,

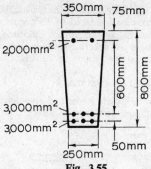

Fig. 3.55

calculate the maximum central load which could be applied so that the permissible stresses of 8 N/mm² in the concrete and 135 N/mm² in the steel are not exceeded. Take $m = 15$ and use the exact lever arm. Obtain the maximum shear stress immediately above the upper layer of tension steel, assuming the bars to be 40 mm diameter.
(*Ans.* 166 kN; 0·435 N/mm².)

4. A tubular member is made from a thin plate of thickness t forming a rectangular section a by b as shown in Fig. 3.56. Find an approximate expression for the maximum shear stress due to shear force S and torque T. Take $a = 2b$.

(*Ans.* $\left(\dfrac{15}{28} \cdot \dfrac{S}{bt} + \dfrac{T}{4b^2t}\right)$.)

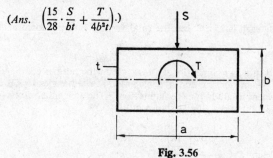

Fig. 3.56

5. The cross-section of a built-up girder is shown in Fig. 3.57. It is to be rein-

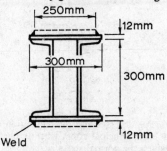

Fig. 3.57

forced by welding 250 mm wide plates to the top and bottom flanges, so that the bending moment may be increased by 75 per cent with the maximum stress remaining approximately the same. Calculate a suitable thickness for the plate and estimate the shearing force in each weld per metre run of beam if the total shear at the section is 700 kN. For each channel $I = 65 \times 10^6$ mm^4 units.

(*Ans.*   18 mm;   470 kN/m for each weld.)

6. Figure 3.58 is a section of a plate girder and gives also the properties of the angles connecting the flanges to the web. Find the pitch of the rivets if the

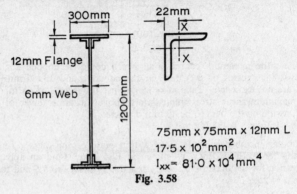

Fig. 3.58

shear carried by each is 25 kN and the total vertical shearing force at the section is 1,300 kN.

(*Ans.*   100 mm.)

7. On a plane AB in a two-dimensional stress system the normal component of stress is 50 N/mm^2 tension and in a plane BC at right angles to AB the normal component is 30 N/mm^2 compression. The tangential stress on these planes is 20 N/mm^2 acting as shown in Fig. 3.59.

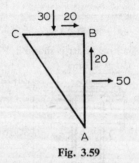

Fig. 3.59

Find (1) the principal stresses and the inclination of the principal planes to AB, identifying each principal stress with its plane and (2) the position of the planes on which the normal stress is zero.

(*Ans.*   54·75 N/mm^2;   −34·75 N/mm^2;   13·4°;   −38°;   −115·2°.)

8. In a two-dimensional stress system the principal stresses are 50 N/mm² and 20 N/mm², both tension. Find the magnitude and obliquity of the resultant stress on a plane inclined at 30° to the plane on which the major principal stress acts.

Find also the inclination of the plane on which the resultant stress has maximum obliquity, the components of stress in this plane and the value of the maximum obliquity. Obliquity is defined as the angle between the force and the normal to the plane on which it acts.
   (*Ans.* 44·4 N/mm², 17°; 57·7°; 28·5 and 13·6 N/mm², 25·4°.)

9. Figure 3.60 shows the plan of a foundation. Vertical loading is carried, having a resultant force acting at a point on the GX-axis. Find its position if the stress at the corner B is zero.
   (*Ans.* $x = +0.73$ m.)

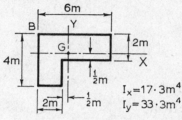

Fig. 3.60

10. The second moment of area about the normal axis Gx of the 100 mm × 100 mm × 12 mm angle shown in Fig. 3.61 is 2.25×10^6 mm⁴. The minimum second moment of area is 0.87×10^6 mm⁴. At a section a bending moment M acts about the Gx-axis. Find the value of M for a maximum stress of 80 N/mm².
    (*Ans.* 2·01 kN-m.)

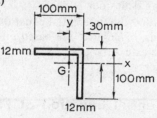

Fig. 3.61

11. Find the shear centre of a semicircular trough section of thin material, mean radius r and thickness t. The second moment of area of the section about the axis of symmetry is $\frac{1}{2}\pi r^3 t$.
    (*Ans.* $4r/\pi$ from the centre.)

12. An I-section beam has a flange 100 mm wide by 12 mm thick and a web 10 mm thick. It is required to be fully plastic when carrying a direct load of 450 kN and a bending moment of 200 kN-m. Calculate the overall depth of section required if the yield stress is 225 N/mm².
    (*Ans.* 250 mm.)

4

Deformation of Beams

4.1 Introduction

A structure will carry loads provided that the load-carrying capacity of the members is not exceeded. It does not necessarily follow, however, that *any* structure carrying the required loads will be satisfactory. It has already been stated that a structure must have adequate stiffness. This normally means that there is an acceptable limit to the deflexion of the structure under load and leads to the problem of calculating deflexions.

We can use here either the differential equation of flexure or one of the energy methods. The latter are based on the principle that the work done by the external loads is stored up in the structure in the form of strain energy.

4.2 The Differential Equation of Flexure

It has already been shown that the relationship between the radius of curvature, the applied bending moment and the flexural rigidity is

$$\frac{1}{R} = \frac{M}{EI} \tag{4.1}$$

The equation for the radius of curvature for a plane curve in the x,y-plane is

$$\frac{1}{R} = \frac{d^2y/dx^2}{\{1 + (dy/dx)^2\}^{3/2}} \tag{4.2}$$

In the majority of the structures used in practice, dy/dx is small and its square can be neglected. Consequently

$$\frac{1}{R} = \frac{d^2y}{dx^2}$$

(The cases where dy/dx is not small require the use of elliptic integrals in their solution and are considered to be outwith the scope of this book.) Thus the differential equation of bending can be written

$$EI\frac{d^2y}{dx^2} = M \qquad (4.3)$$

Integrating this equation twice gives the value of y. The constants of integration are obtained from the known end conditions.

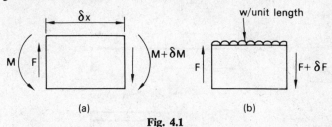

Fig. 4.1

These differential equations can be carried further by considering the equilibrium of the free bodies shown in Fig. 4.1 (*a*) and (*b*).

Figure 4.1 (*a*) leads to the equilibrium equation in the form

$$M + \delta M - M + F\,\delta x = 0$$

or $\qquad\qquad F = -\dfrac{dM}{dx} = -EI\dfrac{d^3y}{dx^3}$ for constant EI $\qquad (4.4)$

Fig. 4.1 (*b*) leads to

$$F + \delta F + w\,\delta x - F = 0$$

or $\qquad\qquad w = -\dfrac{dF}{dx} = EI\dfrac{d^4y}{dx^4} \qquad (4.5)$

These results can be checked by reference back to the example given in Fig. 2.18, where the load varies from zero at $x = 0$ to w at $x = L$.

$$M_x = -\frac{wx}{6L}(L^2 - x^2)$$

$$\frac{dM_x}{dx} = -\frac{wL}{6} + \frac{wx^2}{2L}$$

which is seen to be equal to $-F$.

$$\frac{dF}{dx} = -\frac{wx}{L}$$

i.e. the intensity of loading at a point distant x from the origin.

4.3 Cantilever with Concentrated Load

Consider the cantilever AB shown in Fig. 4.2. End A is built-in and a vertical load W is applied at B. Measuring x from A, the differential equation of flexure is

$$EI\frac{d^2y}{dx^2} = W(l - x)$$

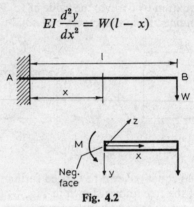

Fig. 4.2

Integrating once,

$$EI\,dy/dx = W(lx - \tfrac{1}{2}x^2) + C_1$$

where C_1 is a constant of integration. When $x = 0$, the slope dy/dx is zero because the cantilever must be securely built-in. Hence $C_1 = 0$. Integrating again,

$$EIy = W(\tfrac{1}{2}lx^2 - \tfrac{1}{6}x^3) + C_2$$

But when $x = 0$, $y = 0$, hence $C_2 = 0$, and the deflected form of the beam is

$$EIy = W(\tfrac{1}{2}lx^2 - \tfrac{1}{6}x^3)$$

At the load, where $x = l$, the deflexion is

$$y = \frac{Wl^3}{3EI} \qquad (4.6)$$

If, instead of being placed at the end, the load is placed as in Fig. 4.3 at a point C distant l_1 from A, the constants of integration C_1

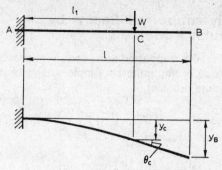

Fig. 4.3

and C_2 will be zero as before, but the equation of the deflected form will only apply from A to C. Since there is no moment from B to C there is no change of curvature and the portion CB is straight. Thus, the deflexion at B is

$$y_B = y_C + \theta_C(l - l_1) = \frac{Wl_1^3}{3EI} + (l - l_1)\frac{Wl_1^2}{2EI} \qquad (4.7)$$

For a cantilever carrying a uniformly-distributed load of intensity w per unit length (Fig. 4.4), the differential equation of flexure is

$$EI\frac{d^2y}{dx^2} = \tfrac{1}{2}w(l - x)^2$$

The constants C_1 and C_2 will be zero as before. The result of integrating twice is to give the equation

$$EIy = \tfrac{1}{2}w(\tfrac{1}{2}l^2x^2 - \tfrac{1}{3}lx^3 + \tfrac{1}{12}x^4)$$

and the deflexion at the end is

$$y_B = \frac{wl^4}{8EI} \qquad (4.8)$$

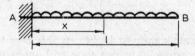

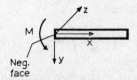

Fig. 4.4

4.4 Simply-supported Beams by Macaulay's Method

The beam in Fig. 4.5, supported at its ends as shown and carrying a single-point load, is not quite so simple as the cantilever. When $x \leqslant a$ the bending moment is

$$M = -\left(\frac{l - a}{l}\right)Wx$$

and when $x > a$ it becomes

$$M = -\left(\frac{l - a}{l}\right)Wx + W(x - a)$$

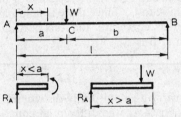

Fig. 4.5

From this it would appear that two differential equations have to be dealt with, one for AC and the other for CB. These two equations would involve four constants of integration which can be determined from the conditions that when $x = 0$ and $x = l$, $y = 0$, and that when $x = a$ the values of dy/dx and y are the same for both sections of the beam.

Macaulay's method, however, simplifies this in the following way. The differential equation is written down as

$$EI \frac{d^2y}{dx^2} = -\left(\frac{l - a}{l}\right)Wx + [W(x - a)] \qquad (4.9)$$

The square brackets signify as before (p. 29) that the term is only evaluated when positive. Integrate thus, without expanding bracketed terms, and we have

$$EI \frac{dy}{dx} = -\left(\frac{l - a}{l}\right)\tfrac{1}{2}Wx^2 + [\tfrac{1}{2}W(x - a)^2] + C_1$$

i.e. the term $(x - a)$ is a new variable only included in the range $x > a$ and has to be retained as a single term. This equation has the correct form when $x < a$, and since the square-bracketted term vanishes when $x = a$, the equation can be used for the whole beam since it gives dy/dx continuous at C. Integrating again gives

$$EIy = -\left(\frac{l-a}{l}\right)\tfrac{1}{6}Wx^3 + [\tfrac{1}{6}W(x-a)^3] + C_1 x + C_2$$

The constants C_1 and C_2 can be determined from the end conditions. When $x = 0$, $y = 0$ the square-bracketted terms are omitted, and hence $C_2 = 0$. When $x = l$, $y = 0$ the square-bracketted terms are included, and hence

$$C_1 = \tfrac{1}{6}(l-a)Wl - \frac{W}{6l}(l-a)^3$$

$$= \frac{Wa}{6l}(l-a)(2l-a)$$

The deflected form is therefore

$$y = \frac{W}{6EI}\frac{x(l-a)}{l}(2al - a^2 - x^2) + \frac{W}{6EI}[x-a]^3 \quad (4.10)$$

The value of this function when $x = a$, i.e. under the load, is

$$y = \frac{W}{6EI}a(l-a)(2al - a^2 - a^2) \cdot \frac{1}{l}$$

$$= \frac{Wa^2b^2}{3EIl}$$

which simplifies when $a = b = \tfrac{1}{2}l$ to

$$y = \frac{Wl^3}{48EI}$$

The rules for applying Macaulay's method are as follows—
1. Select an origin at one end of the beam.
2. Write down the bending moment for a section in the portion of the beam furthest from the origin, taking the free body which includes the origin.
3. Integrate expressions such as $(x - a)$ which only occur when positive as $\tfrac{1}{2}(x-a)^2$, etc.
4. Uniformly-distributed loads must always be made to extend to the remote support, introducing negative loads if necessary.

As an example of the latter, consider the beam shown in Fig. 4.6 (a) which is carrying a uniformly-distributed load on the left-hand half of the span. This is equivalent to the beam shown in Fig. 4.6 (b), which has a uniformly-distributed load acting downwards for the complete length and a load of equal intensity acting upwards on the right-hand side.

The reactions are $R_A = \frac{3}{8}wL$ and $R_B = \frac{1}{8}wL$. The differential equation of flexure is

$$M_x = EI \frac{d^2y}{dx^2} = -\tfrac{3}{8}wLx + \tfrac{1}{2}wx^2 - \tfrac{1}{2}w[(x - \tfrac{1}{2}L)^2]$$

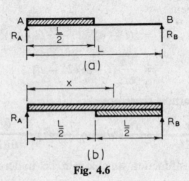

(a)

(b)

Fig. 4.6

Integrating once gives

$$EI \frac{dy}{dx} = -\tfrac{3}{8}wL\tfrac{1}{2}x^2 + \tfrac{1}{6}wx^3 - \tfrac{1}{2}w[\tfrac{1}{3}(x - \tfrac{1}{2}L)^3] + C_1$$

Integrating again gives

$$EIy = -\tfrac{3}{48}wLx^3 + \tfrac{1}{24}wx^4 - \tfrac{1}{2}w[\tfrac{1}{12}(x - \tfrac{1}{2}L)^4] + C_1x + C_2$$

When $x = 0$, $y = 0$, hence $C_2 = 0$. When $x = L$, $y = 0$, hence

$$C_1 = wL^3\left(\frac{3}{48} - \frac{1}{24} + \frac{1}{384}\right) = \tfrac{9}{384}wL^3$$

Thus the deflected form is given by

$$EIy = -\tfrac{3}{48}wLx^3 + \tfrac{1}{24}wx^4 - \tfrac{1}{2}w[\tfrac{1}{12}(x - \tfrac{1}{2}L)^4] + \tfrac{9}{384}wL^3x \quad (4.11)$$

At the mid-point, i.e. where $x = \frac{1}{2}L$, the deflexion becomes

$$y = \frac{5wL^4}{768EI}$$

A similar result would have been obtained using the principle of superposition.

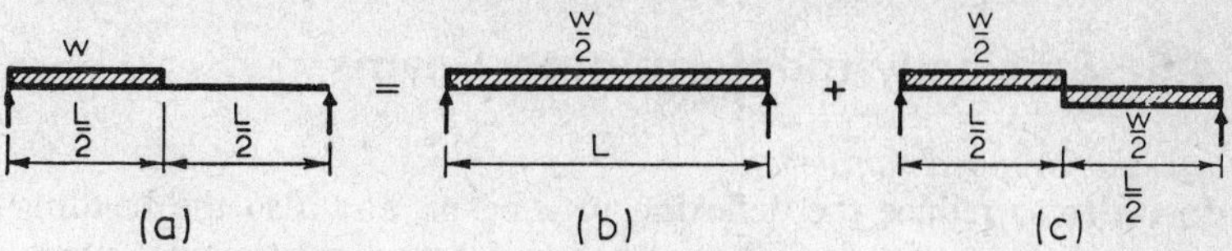

Fig. 4.7

The beam with a load on the left-hand portion (Fig. 4.7 (a)) is equivalent to the combination shown in (b) plus (c). The skew-symmetrical loading system shown in (c) has no central deflexion. Thus the central deflexion is entirely due to the load system (b). This amounts to

$$\frac{5}{384}\left(\tfrac{1}{2}w\right) \cdot \frac{L^4}{EI} = \frac{5}{768}\frac{wL^4}{EI}$$

The beam with a load distributed uniformly over the whole span is common; values for the slope at the ends and the central deflexion are given for it and some other common loadings in Table 4.1.

Table 4.1 **Slope and Deflexion Coefficients**

Type of beam and loading	Max. slope	Max. deflexion	Factors	
			Slope	Deflexion
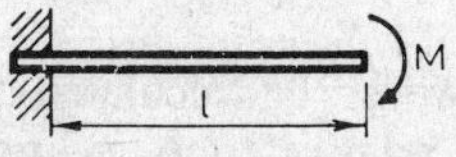	1	$\dfrac{1}{2}$	$\dfrac{Ml}{EI}$	$\dfrac{Ml^2}{EI}$
	$\dfrac{1}{2}$	$\dfrac{1}{8}$		
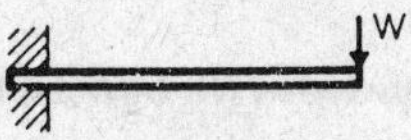	$\dfrac{1}{2}$	$\dfrac{1}{3}$		
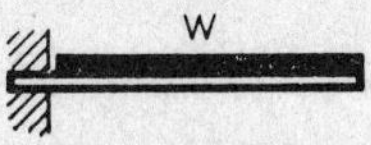	$\dfrac{1}{6}$	$\dfrac{1}{8}$	$\dfrac{Wl^2}{EI}$	$\dfrac{Wl^3}{EI}$
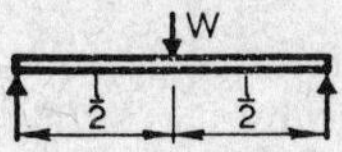	$\dfrac{1}{16}$	$\dfrac{1}{48}$		
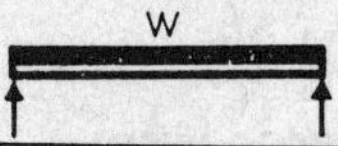	$\dfrac{1}{24}$	$\dfrac{5}{384}$		

4.5 Statically-indeterminate Beams

(*a*) *The propped cantilever*
In order to reduce the deflexion in a beam, and also the bending
moment, a prop is sometimes placed under a beam at a given place.
Such a beam is shown in Fig. 4.8 (*a*), where the cantilever AB, which
is built-in at A, has a prop at the end B. This structure is now statically-indeterminate since the necessary reactions for equilibrium can
all be provided at the built-in end A.

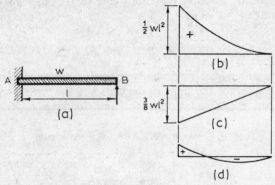

Fig. 4.8

The indeterminate reaction at B can, however, be calculated quite
easily if the final deflected position of B relative to A is known.
If, for example, B and A remain at the same level, then the upward
deflexion from the prop force R_B must be equal to the downward
deflexion without the prop due to the load *wl*, i.e.

$$y_B \text{ due to prop} = \frac{R_B l^3}{3EI} \text{ upwards (see eqn. (4.6))}$$

$$y_B \text{ due to distributed load} = \frac{wl^4}{8EI} \text{ (see eqn. (4.8))}$$

Thus $R_B = \frac{3}{8}wL$.
The introduction of a prop alters the bending moment. The value
of this function at a distance *x* from B is given by

$$M = \tfrac{1}{2}wx^2 - \tfrac{3}{8}wlx = wx(\tfrac{1}{2}x - \tfrac{3}{8}l) \tag{4.12}$$

or   (Cantilever moment) — (moment due to prop)

The diagram for the former is parabolic, as (*b*), whilst that due to the prop is the straight line of (*c*). The combined diagram (*d*) represents eqn. (4.12).

It is seen that this is very different from the simple cantilever. Not only is the maximum bending moment much less, but there is a change in sign over that portion of the beam near to the prop.

(b) The two-span continuous beam

Another common propped beam is shown in Fig. 4.9, where a simply-supported beam AB carrying a uniformly-distributed load

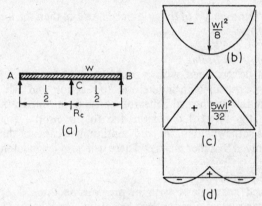

Fig. 4.9

has a prop C at the mid-span. If A, C and B are at the same level, then the upward deflexion due to the prop force R_C must equal the downward deflexion at that point due to the uniformly-distributed loads.

$$y_C \text{ due to prop} = \frac{R_C l^3}{48EI} \quad \text{(from eqn. (4.10))}$$

y_C due to uniformly-distributed load $= \dfrac{5}{384} \dfrac{wl^4}{EI}$ (from Table 4.1)
Thus $R_C = \frac{5}{8}wl$.

The reactions at A and B must, by symmetry, be equal and their total value is $\frac{3}{8}wl$; hence each is $\frac{3}{16}wl$.

The bending moment diagram for this beam is the algebraic sum of the moments due to the distributed load and the moments due to the prop. That due to the former is shown in (*b*), the maximum value being $\frac{1}{8}wl^2$. That due to the prop is shown in (*c*), the maximum value being $5wl^2/32$. The combined bending moment diagram is shown in

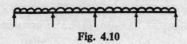

Fig. 4.10

(*d*). This exhibits the same characteristics, namely a reversal of sign and a reduction of maximum moment, as are evident in the diagram for the propped cantilever shown in Fig. 4.8 (*d*).

Figure 4.9 (*d*) could have been obtained directly by consideration of the expression for the moment M_x at a point distance x from A. This is given by

$$M = -\tfrac{3}{16}wlx + \tfrac{1}{2}wx^2 - [\tfrac{5}{8}wl(x - \tfrac{1}{2}l)]$$

If the prop removes $1/n$ of the original deflexion, then the load carried by it is $(1/n)\tfrac{5}{8}wl$.

(*c*) *Multiple-span beams*
Beams can be provided with a number of props as in Fig. 4.10. Each prop constitutes an indeterminate reaction and can be found by equating the downward deflexion at the point due to the applied loads with the upward deflexion due to the prop. The process, however, is somewhat lengthy and continuous beams of this type are generally solved by other means. These will be given in later chapters.

(*d*) *The built-in beam*
The propped beam dealt with in previous sections is one of the simplest examples of a redundant structure. Another is the built-in beam shown in Fig. 4.11. A cantilever, i.e. a beam built-in at one end, is a statically-determinate structure. By building-in the other end three additional reactions (i.e. redundancies) are introduced; if, however, the applied load acts vertically then these are reduced to two as shown. Normally the fixing moment at each end is taken as the redundant force since, if these are known, the vertical reactions can be determined.

Built-in beams can be analysed by applying the differential equation of flexure (*see* p. 110), since the effect of building-in the ends is to stop the slope changing there.

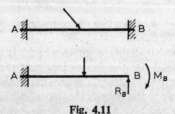

Fig. 4.11

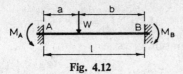

Fig. 4.12

As an example, take the beam shown in Fig. 4.12, which is built-in at A and B and carries a single load W at a distance *a* from A and *b* from B.

The differential equation of flexure is

$$EI \frac{d^2y}{dx^2} = M_A\left(\frac{l-x}{l}\right) + M_B \cdot \frac{x}{l} - \frac{Wb}{l} \cdot x + [W(x-a)]$$

(This equation is derived by superimposing the effects of fixity on the simply-supported beam.) Integrating once gives

$$EI \frac{dy}{dx} = \frac{M_A}{l}\left(lx - \frac{x^2}{2}\right) + M_B \frac{x^2}{2l} - \frac{Wbx^2}{2l} + [\tfrac{1}{2}W(x-a)^2] + C_1$$

When $x = 0$, $dy/dx = 0$, hence $C_1 = 0$. But when $x = l$, $dy/dx = 0$, hence

$$M_A\tfrac{1}{2}l + M_B\tfrac{1}{2}l - \tfrac{1}{2}Wbl + \tfrac{1}{2}Wb^2 = 0$$

Table 4.2 **Fixing-moment Coefficients**

Loading or end deflexion	*at A*	*at B*	*Factor*
	$+\frac{1}{8}$	$+\frac{1}{8}$	
	$+\beta^2\alpha$	$+\alpha^2\beta$	Wl
	$+\frac{1}{12}$	$+\frac{1}{12}$	
	-4	$+2$	$\dfrac{EI\theta}{l}$
	-1	$+1$	$\dfrac{6EI\delta}{l^2}$

or
$$M_A + M_B = Wb - \frac{Wb^2}{l}$$

The second integration gives

$$EIy = \frac{M_A}{l}(\tfrac{1}{2}lx^2 - \tfrac{1}{6}x^3) + \frac{M_B}{l} \cdot \tfrac{1}{6}x^3 - \frac{Wbx^3}{6l} + [W\tfrac{1}{6}(x - a)^3] + C_2$$

When $x = 0$, $y = 0$, hence $C_2 = 0$. Also when $x = l$, $y = 0$, hence

$$M_A\tfrac{1}{3}l^2 + M_B\tfrac{1}{6}l^2 - \tfrac{1}{6}Wbl^2 + \tfrac{1}{6}Wb^3 = 0$$

giving
$$2M_A + M_B = Wb - \frac{Wb^3}{l^2}$$

hence
$$M_A = \frac{Wb^2}{l} - \frac{Wb^3}{l^2}$$

$$= \frac{Wb^2}{l^2}(l - b) = \frac{Wab^2}{l^2}$$

The built-in beam is a common type of redundant structure and the fixing moments introduced by various loadings are given in Table 4.2.

4.6   The Moment-Area Method

The Moment–Area Theorem concerns the relationship between the deflexion and the area of the M/EI diagram. Consider the cantilever shown in Fig. 4.13, which carries a uniformly-distributed load per unit length. At any point B distance x from A, the change of slope $\delta\theta$ between the two ends of an element of length δx is

$$\delta\theta = \frac{M}{EI}\delta x$$

(since $1/R = \delta\theta/\delta x = M/EI$).

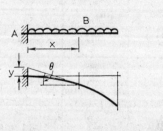

Fig. 4.13

The total change of slope between A and B is

$$\theta = \int_A^B \frac{M}{EI}\,dx$$

$$= \text{Area of } M/EI \text{ diagram between these points} \quad (4.13)$$

The incremental change of slope $\delta\theta$ at B produces at A a deflexion from the tangent at B equal to x multiplied by this change of slope at B.

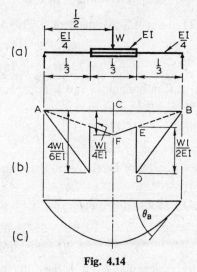

(a)

(b)

(c)

Fig. 4.14

The total deflexion of A from the tangent at B is

$$y = \int_A^B \frac{M}{EI} x\,dx$$

$$= \text{First moment about A of area of } M/EI \text{ diagram} \quad (4.14)$$

As an example of the application of this theorem consider the beam shown in Fig. 4.14 (a). The flexural rigidity of the middle third is four times that of each end and it carries a central point-load W.

The M/EI diagram is shown in Fig. 4.14 (b). The difference in slope between the centre C and the support B is

$$_C\theta_B = \triangle\text{BED} + \triangle\text{CFB}$$

$$= \frac{Wl}{2EI} \cdot \tfrac{1}{3}l \cdot \frac{1}{2} + \frac{Wl}{4EI} \cdot \tfrac{1}{2}l \cdot \frac{1}{2} = \frac{Wl^2}{12EI} + \frac{Wl^2}{16EI} = \frac{7Wl^2}{48EI}$$

Because the loading is symmetrical, the slope at C is zero. Hence

$$\theta_B = \frac{7Wl^2}{48EI}$$

The intercept of the tangent from B at the centre line is

$$\frac{l}{2} \times \frac{7Wl^2}{48EI} = \frac{7Wl^3}{96EI}$$

The deflexion of B relative to the tangent at C is

Moment of △BED about B + Moment of △CFB about B

$$= {}_B\delta_C = \frac{Wl^2}{12EI} \cdot \frac{2l}{9} + \frac{Wl^2}{16EI} \cdot \frac{l}{3} = \frac{17Wl^3}{432EI}$$

Since the tangent at C is horizontal and B remains stationary, this gives the central deflexion as

$$\frac{17Wl^3}{432EI}$$

As a check, the deflexion of C relative to the tangent at B will be calculated. This is given by

Moment of △BED about C + Moment of △CFB about C

$$= \frac{Wl^2}{12EI} \cdot \frac{5l}{18} + \frac{Wl^2}{16EI} \cdot \frac{l}{6} = \frac{29Wl^3}{864EI}$$

The actual central deflexion is given by the difference between this value and the intercept of the tangent at B on the centre line, i.e.

$$\delta_C = \frac{7Wl^3}{96EI} - \frac{29Wl^3}{864EI}$$

$$= \frac{17Wl^3}{432EI}$$

A glance at Fig. 4.14 (c) shows that in each case the deflexion of the point in question relative to the tangent at the other point is upwards. The reason seems clear when looking at the deflexion diagrams but it has a mathematical explanation. The bending moment due to the applied load is negative in sign: so, therefore, are the areas of the diagrams and the deflexions.

One thing that should be noticed about the answer to this example is the effect of the reduction in flexural rigidity on the magnitude of the central deflexion. It is approximately three times as much as when the beam is of uniform flexural rigidity throughout.

This method will now be applied to solve the propped-beam problem shown in Fig. 4.9 (a). This will be divided into the uniformly-distributed load, the bending moment diagram for which is shown in (b), and the prop, the bending moment diagram for which is shown in (c).

Then, considering (b)

$$_C\delta_B = (\tfrac{2}{3} \times \tfrac{1}{8}wl^2 \times \tfrac{1}{2}l) \times \tfrac{5}{16}l = 5wl^4/384$$

Considering (c)

$$_C\delta_B = -(R_C\tfrac{1}{4}l \times \tfrac{1}{4}l)(\tfrac{1}{3}l) = -R_Cl^3/48$$

But since there is no final deflexion at C, these are equal, hence

$$R_C = \tfrac{5}{8}wl$$

i.e. the force on the prop R_C is $\tfrac{5}{8}wl$.

4.7 Deflexion Due to Shear

In the foregoing analysis only the effect of the strains resulting from direct stresses due to bending has been considered when calculating the deflexion. Strains, however, are caused by shearing forces and these strains too will result in a deflexion of one section of the beam relative to its neighbour. An approximate treatment which shows the relationship between the deflexions due to shear and those due to bending is given below. For a more accurate treatment of this the reader is referred to Chap. 5 Section 16.

If τ_N is the intensity of shear stress at the neutral axis, the resulting shear strain γ_N is given by

$$\gamma_N = \frac{\tau_N}{G}$$

where G is the modulus of rigidity. The resulting deflexion between two cross-sections δx apart, as shown in Fig. 4.15 (and assuming that there is no change in τ_N), is

$$\delta y_S = \gamma_N \delta x = \frac{\tau_N}{G} \cdot \delta x$$

Hence

$$\frac{dy_S}{dx} = \frac{\tau_N}{G} \tag{4.15}$$

is the differential equation for the shear deflexion y_S.

In a beam of rectangular cross-section subjected to a shearing force V,

$$\tau_N = \frac{3}{2}\frac{V}{bd}$$

Substituting gives the equation as

$$\frac{dy_S}{dx} = \frac{3V}{2Gbd}$$

If, as an example, a cantilever carrying a concentrated load W at the end is considered, the shearing force V is constant and equal to

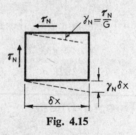

Fig. 4.15

W for all values of x. Integrating,

$$y_S = \frac{3Wx}{2Gbd} + C_1$$

Measuring x from the built-in end gives $y_S = 0$ when $x = 0$, hence $C_1 = 0$. When $x = l$,

$$y_S = \frac{3Wl}{2Gbd} \qquad (4.16)$$

The deflexion resulting from flexural strains is $Wl^3/3EI$, hence the total deflexion δ at $x = l$ is

$$\delta = \frac{Wl^3}{3EI} + \frac{3Wl}{2Gbd}$$

But $I = bd^3/12$, hence

$$\delta = \frac{4Wl^3}{Ebd^3}\left\{1 + \frac{3E}{8G}\left(\frac{d}{l}\right)^2\right\}$$

The second term within the bracket gives that amount of the deflexion which results from shear strains. The expression $3E/8G$ is of the order of unity and so we see that the deflexion resulting from shear strains depends on the depth:span ratio. If this is of the order of

1:10, then the deflexion due to bending should be increased by about 1 per cent to get the total deflexion.

It is clear that shear deflexion is only important in the case of deep beams.

4.8 Strain Energy

It was stated in the opening paragraphs of this chapter that energy methods are used for calculating deflexions. The main principle behind these methods is that the work done by loads is stored up in

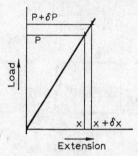

Fig. 4.16

the structure in the form of strain energy. It is now necessary to find expressions for the strain energy caused by different types of loading.

If the strains are within the elastic limit, then it is the strain energy stored in the structure which causes it to return to its unstrained state when the load is removed. In deriving expressions for strain energy it will be assumed that in all cases the strains are within the elastic limit.

(a) Strain energy due to direct load

Figure 4.16 shows the stress-strain or load-displacement relationship for a linear elastic body. Let it be a rod of length l, cross-sectional area A, and with a modulus of elasticity E. Under a load P let the extension be x, increasing to $x + \delta x$ when the load is increased by δP.

Then the work during this increase is the average load multiplied by the distance through which it moves, i.e.

$$(P + \tfrac{1}{2}\delta P)\,\delta x$$

This work is stored in the bar in the form of strain energy. If δU is the increment of strain energy resulting from the increase of load, then (neglecting second-order infinitesimals)

$$\delta U = P \, \delta x$$

But
$$\delta x = \frac{l \, \delta P}{AE}$$

Hence the strain energy U_D stored in the bar as the load increases from 0 to P_1 is given by

$$U_D = \frac{l}{AE} \int_0^{P_1} P \, dP$$

$$= \frac{P_1^2 l}{2AE} \qquad\qquad (4.17)$$

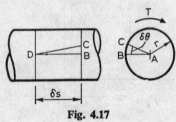

Fig. 4.17

or, if σ_1 is the stress intensity due to load P_1, and v is the volume of rod,

$$U_D = \frac{\sigma_1^2 v}{2E} \qquad\qquad (4.18)$$

(b) Strain energy due to torsion

Consider the circular shaft shown in Fig. 4.17, of radius r, subjected to a torque T, which is constant over a length δs. The line DB is strained into the position DC, the angle BAC being $\delta\theta$. The shear strain

$$\frac{BC}{BD} = \frac{r \, \delta\theta}{\delta s}$$

But the shear strain is

$$\frac{\tau}{G} = \frac{Tr}{GJ}$$

Thus
$$\frac{r \, \delta\theta}{\delta s} = \frac{Tr}{GJ}$$

or
$$\delta\theta = \frac{T\delta s}{GJ}$$

The work done $= \frac{1}{2}T\,\delta\theta =$ strain energy. Thus the strain energy due to torsion δU_T in the length δs is given by

$$\delta U_T = \frac{T^2\,\delta s}{2GJ}$$

and the total strain energy is

$$U_T = \int \frac{T^2\,ds}{2GJ} \tag{4.19}$$

(c) Strain energy due to bending

Figure 4.18 shows a small length of beam δs to which equal and opposite moments M are applied. Their effect is to cause the beam to

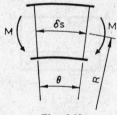

Fig. 4.18

bend into a circular arc. The radius R of the neutral axis is obtained from the relationship

$$\frac{1}{R} = \frac{M}{EI}$$

This bending of the beam means that work is done by the moments; its value is $\frac{1}{2}M\theta$, where θ is the relative angular displacement of the two faces. θ can be found from the expression $\theta = \delta s/R$. Hence

$$\text{Work done} = \frac{1}{2}\frac{M\,\delta s}{R}$$

Substituting for $1/R$ gives the strain energy U_B of the beam from the expression

$$U_B = \int \frac{M^2\,ds}{2EI} \tag{4.20}$$

the integration being carried out along the length of the neutral axis of the beam. It should be noted that $\int \frac{1}{2}M^2\,ds$ is the first moment of area of the bending moment diagram about its base.

130   *Deformation of Beams*

(d) Strain energy due to shear

Figure 4.19 shows two sections AC and BD distance δs apart, which are subjected to a shearing force V. If it is assumed that this produces a uniform shear stress across the section, then the shear strain γ is given by

$$\gamma = \frac{V}{AG}$$

where A is the cross-sectional area and G is the modulus of rigidity.

The work done is $\frac{1}{2} V \, \delta s \times \gamma$; hence the strain energy due to shear U_S is given by

$$U_S = \int \frac{V^2 \, ds}{2AG} \qquad (4.21)$$

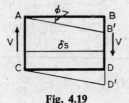

Fig. 4.19

This expression will only hold if the shear stress is uniform. Generally it can be said that

$$U_S = K \int \frac{V^2}{2AG} \, ds$$

where K is a factor depending on the shape of the section.

4.9 Single-load Deflexions by Strain Energy

The increase in strain energy of a body when a load is applied is equal to the work done by the load as it deflects. Thus if the load applied is W and the increase in strain energy U, the deflexion δ is found from the relationship

$$\tfrac{1}{2} W \delta = U$$

or

$$\delta = 2U/W$$

This principle will be applied in the first instance to the beam shown in Fig. 4.20. If the flexural rigidity is constant throughout, then for the strain energy U_B we have

$$U_B = \frac{1}{2EI} \int_0^l M^2 \, dx$$

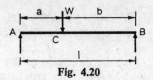

Fig. 4.20

and for the deflexion

$$\delta = \frac{2U}{W}$$

$$R_A = (b/l)W \quad \text{and} \quad R_B = (a/l)W$$

From A to C, $M_x = -(b/l)Wx$, therefore

$$[U_B]_{AC} = \frac{1}{2EI}\int_0^a \frac{b^2}{l^2} W^2 x^2 \, dx = \frac{W^2 a^3 b^2}{6EIl^2}$$

From B to C, $M_x = -(a/l)Wx$, therefore

$$[U_B]_{BC} = \frac{1}{2EI}\int_0^b \frac{a^2 W^2 x^2}{l^2} \, dx = \frac{W^2 a^2 b^3}{6EIl^2}$$

The total strain energy is

$$U_B = \frac{W^2 a^2 b^2}{6EIl}$$

Hence

$$\delta = \frac{2U_B}{W} = \frac{Wa^2 b^2}{3EIl}$$

In this example the strain energy due to shear has been neglected. This is usual in such cases. The deflexion resulting from shear in slender beams has been shown to be, in general, negligible with regard to that due to bending.

It is clear that this method could only be used where a single load is applied to a beam. For example, the cantilever shown in Fig. 4.21, which carries loads W_1 and W_2, will store up strain energy as a result of the work done by each of these loads and the relationship is

$$\tfrac{1}{2}W_1\delta_1 + \tfrac{1}{2}W_2\delta_2 = U_B$$

This cannot be used to solve for either δ_1 or δ_2.

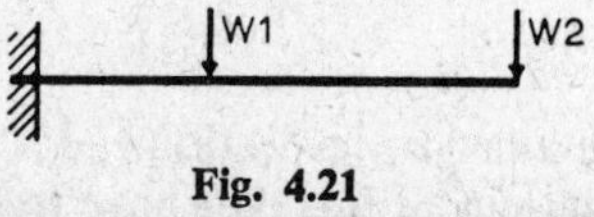

Fig. 4.21

The previous example used the strain energy due to bending. A further example will now be given where the strain energy is due to direct forces.

The problem is to find the deflexion at the load point G in the frame shown in Fig. 4.22 (a). The cross-sectional areas of the members are given in square brackets on the figure and $E = 230$ kN/mm².

If the 5 kN load is symbolized by W and if δ is the deflexion at the load-point, then

$$\tfrac{1}{2}W\delta = U$$

where U is the increase in strain energy.

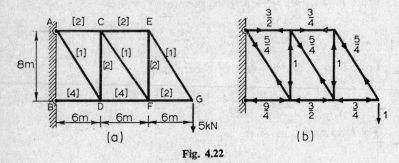

Fig. 4.22

In the pin-jointed frame structure the energy is

$$U_D = \frac{P^2 l}{2AE}$$

where P is the force in any member.

But P depends on W and can be expressed as αW, where α is the force in the member when unit load acts at the load point. Thus

$$U = \sum \frac{\alpha^2 W^2 l}{2AE}$$

$$= W^2 \sum \frac{\alpha^2 l}{2AE}$$

Hence

$$\delta = W \sum \frac{\alpha^2 l}{AE} \qquad (4.22)$$

The value of α is obtained by inspection and is shown in Fig. 4.22 (b). Whenever summations of this type have to be carried out the best plan is to tabulate the data in Table 4.3. From this we have

$$\text{Deflexion} = \frac{5 \times 759 \cdot 9}{230} = 16 \cdot 5 \text{ mm}$$

Table 4.3

Member	Length (m)	α	Area (mm² × 10²)	$\alpha^2 l / A$ (mm⁻¹)
AC	6	$+\frac{3}{2}$	2	67·5
CE	6	$+\frac{3}{4}$	2	16·9
BD	6	$-\frac{3}{4}$	4	76·0
DF	6	$-\frac{3}{2}$	4	33·7
FG	6	$-\frac{3}{4}$	2	16·9
AD	10	$+\frac{5}{4}$	1	156·3
CF	10	$+\frac{5}{4}$	1	156·3
EG	10	$+\frac{5}{4}$	1	156·3
CD	8	-1	2	40·0
EF	8	-1	2	40·0
				$\Sigma = 759\cdot9$

4.10 Examples for Practice

1. A uniform shaft ABC is simply-supported in bearings A and B and overhangs to C. AB $= l$, BC $= a$. When a transverse force P acts at C, show that the maximum deflexion in the portion AB is $Pal^2/9\sqrt{3}$.

2. A uniform beam 10 m long is supported at points 1 m from each end. It carries loads of 30 kN at each end and 120 kN in the centre. Find the deflexion under the loads and the slope at the supports.
   (*Ans.* $(1/EI) \times \{[-350, +1,040] \text{ m}; \ (1/EI) \times 360 \text{ rad}\}$.)

3. A beam 10 m span, freely supported at each end, carries a point load of 100 kN at a distance 2·5 m from one end. The maximum deflexion of the beam is not to exceed 1/500 of the span. Calculate the value of the second moment of area of the section, expressing the result in millimetre-units, and the depth of the beam, if the maximum stress is 90 N/mm². Take $E = 230$ kN/mm².
   (*Ans.* $2 \cdot 52 \times 10^8$ mm units; 242 mm.)

4. A simply-supported beam of span length L is loaded with two equal loads W at points distant $L/4$ from each end. Show that the central deflexion is $(11/384) \cdot (WL^3/EI)$.
   A beam so loaded is to be propped at the centre so that the maximum positive and negative bending moments are numerically equal. Determine the central reaction and the central deflexion in terms of W, L and EI.
   (*Ans.* $\frac{4}{3}W$; $(1/1,152) \cdot (WL^3/EI)$.)

5. A horizontal beam of flexural rigidity EI and length $2L$ is supported at its ends and centre by three springs. Each spring deflects unit distance under a load λ. Two equal loads W are carried at points mid-way between the springs. Show that an end reaction R may be obtained from $R(1 + 9\beta) = W(\frac{5}{16} + 6\beta)$, where $\beta = EI/\lambda L^3$.

6. A beam AB is built-in at A and B and carries a load which varies uniformly from w at A to $2w$ at B. The flexural rigidity of the beam also varies uniformly, its value at B being twice that at A.

   Determine the fixing moments at A and B and the maximum deflexion.
   (*Ans.* $M_A = 0 \cdot 100wl^2$, $M_B = 0 \cdot 150wl^2$, $0 \cdot 00271wl^4/EI$.)

7. An I-section beam, with depth of 400 mm and second moment of area of $3 \cdot 00 \times 10^8$ mm⁴ units, is built-in horizontally at both ends of a span of 5 m, the ends being on the same level. Estimate the uniformly-distributed load which the joist will carry if the maximum stress is 90 N/mm². What will be the change in the maximum stress if the left support yields until the slope there is $1/450$ downwards to the right, the ends remaining on the same level?
   ($E = 230$ kN/mm².)
   (*Ans.* 324 kN; 123 N/mm².)

8. A simple crane pillar is shown in Fig. 4.23. The geometrical properties of

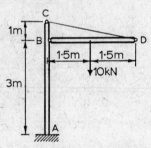

Fig. 4.23

the members are given in the table—

Member	A (mm²)	I (mm⁴)
ABC	40×10^2	$13 \cdot 0 \times 10^6$
BD	25×10^2	$5 \cdot 0 \times 10^6$
CD	3×10^2	—

For the given position of the load calculate the strain energy stored in the crane and the deflexion at the load point. ($E = 200$ kN/mm².)
(*Ans.* 222·2 kN-mm; 44·4 mm.)

9. A simply-supported beam of span l carries a load of W at each of the third points. The central half of the beam has a flexural rigidity EI, whilst each of the end quarters is of flexural rigidity $EI/3$. Determine the central deflexion.
   (*Ans.* $0 \cdot 046 \, Wl^3/EI$.)

5

Deformation of Structures

5.1 Introduction

A structure is, essentially, an assembly of many separate elements. Previous chapters have dealt with the deformations of rod, beam and shaft elements, so that the next step is to determine the overall deformations of an assembly of such elements.

Two procedures are developed in this chapter, based on
(a) Geometrical Principles,
(b) the Principle of Virtual Work.

5.2 Linear Elasticity

A structure is linearly elastic when all displacements are linear functions of all loads, i.e.

$$d = FW$$

where d = any displacement,
W = any load,
F = constant called the flexibility,
= deformation per unit load.

5.3 Superposition

In the particular case of the structure in Fig. 5.1 (a) having a single load W_2 at the point 2, the displacement of point 1 in a defined direction can be written

$$d_1 = f_{12}W_2$$

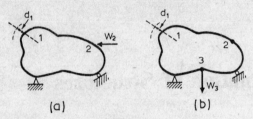

Fig. 5.1

where d_1 = displacement at 1 in a given direction,
$\quad f_{12}$ = constant = flexibility coefficient,
$\quad\quad$ = value of d_1 for unit value of W_2.

Similarly for a single load W_3 at point 3 as in Fig. 5.1 (*b*),

$$d_1 = f_{13}W_3$$

where again f_{13} = constant,
$\quad\quad\quad$ = flexibility coefficient,
$\quad\quad\quad$ = the value of d_1 for unit W_3.

Then, in the case of linear elasticity, that is to say both f_{12} and f_{13} constants, when both loads act together

$$d_1 = f_{12}W_2 + f_{13}W_3 \tag{5.1}$$

This can be re-arranged to

$$d_1 = \left(f_{12}\frac{W_2}{W_3} + f_{13}\right)W_3$$

$$= FW_3 \tag{5.2}$$

where F is a new flexibility coefficient. While the principle of super-position (5.1) always holds, the altered load-deflexion relation (5.2) is only linear if F remains constant, i.e. W_2/W_3 = constant, i.e. all loads must change at a constant rate. This condition is not referred to again but is implicit in all that follows.

5.4 Two-bar Plane Frame

In the simple two-bar plane frame in Fig. 5.2, displacements are readily determined by simple geometry.

$$L_1 = \text{original length of AC},$$
$$L_1 + \delta L_1 = \text{loaded length of AC},$$
$$(x, y) = \text{original coordinates of point C},$$
$$[(x + \delta x), (y + \delta y)] = \text{displaced coordinates of point C}.$$

Then $$L_1{}^2 = x^2 + y^2 \qquad (5.3)$$
and $$(L_1 + \delta L_1)^2 = (x + \delta x)^2 + (y + \delta y)^2 \qquad (5.4)$$
which is a compatibility condition.

Expanding eqn. (5.4) and making use of (5.3),

$$2L_1\, \delta L_1 + (\delta L_1)^2 = 2x \cdot \delta x + (\delta x)^2 + 2y\, \delta y + (\delta y)^2$$

Neglecting the squares of small quantities,

$$L_1\, \delta L_1 = x\, \delta x + y\, \delta y$$

or $$\delta L_1 = \frac{x}{L_1} \delta x + \frac{y}{L_1} \delta y$$

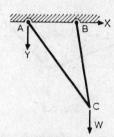

Fig. 5.2

One compatibility equation is then

$$\delta L_1 = l_1\, \delta x + m_1\, \delta y \qquad (5.5)$$

where $l_1 = x/L_1 = \cos$ (angle between AC and the x-axis), $m_1 = y/L_1 = \cos$ (angle between AC and the y-axis) and l_1 and m_1 are called the direction cosines of AC relative to the axes at A.

Similarly, using BC, the other compatibility equation is found—

$$\delta L_2 = l_2\, \delta x + m_2\, \delta y \qquad (5.6)$$

If p_1 and p_2 are the tensile axial *stress resultants* (page 153) in AC and BC, respectively, then

$$\delta L_1 = \frac{p_1 L_1}{A_1 E_1}$$

and $$\delta L_2 = \frac{p_2 L_2}{A_2 E_2}$$

For the equilibrium of joint C, $\Sigma Y = 0$, i.e.

$$-\frac{y}{L_1} p_1 - \frac{y}{L_2} p_2 + W = 0$$

or
$$-m_1 p_1 - m_2 p_2 + W = 0 \qquad (5.7)$$

Also $\Sigma X = 0$, i.e.

$$-l_1 p_1 - l_2 p_2 = 0 \qquad (5.8)$$

Eqns. (5.7) and (5.8) then give p_1 and p_2; hence δL_1, δL_2; and the displacements δx and δy are calculated from eqns. (5.5) and (5.6).

5.5 General Use of Geometrical Relations

The three-dimensional framework of Fig. 5.3 can be dealt with in the same way.

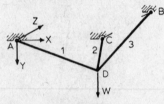

Fig. 5.3

Let the three direction cosines for each member be $l_1 m_1 n_1$, $l_2 m_2 n_2$, and $l_3 m_3 n_3$. Then the three compatibility equations are

$$\delta L_i = l_i \, \delta x + m_i \, \delta y + n_i \, \delta z \qquad (i = 1, 2, 3)$$

The three equilibrium equations at joint D are

$$\sum l_i p_i = W_x$$
$$\sum m_i p_i = W_y$$
$$\sum n_i p_i = W_z$$

and the three force-deformation relations are

$$(\delta L)_i = \left(\frac{pL}{AE} \right)_i$$

It is now possible to proceed from joint to joint throughout a large structure, but the arithmetic is troublesome and this method is, in fact, of little practical value.

The importance of these examples lies in the demonstration that, in order to retain linearity,

    (*a*) the compatibility equations omitted second-order terms, and

    (*b*) the equilibrium equations neglected the deformations entirely.

These simplifications are made throughout orthodox methods and are generally acceptable, although there are problems where they are not permissible, e.g. suspension bridge cables. Where this "first-order" approach is not adequate, the structure behaves non-linearly, the principle of superposition is invalid, and the problems require special treatment.

Furthermore, the success of this approach depends upon being able to obtain values for the stress resultants in each member, i.e. the frames must be statically-determinate as discussed in Chapter 1.

5.6 Non-linear Behaviour

The two-bar structure shown in Fig. 5.4 (*a*) is an example of simple non-linear elasticity. Under zero load, bars AB and AC are collinear, i.e. *h* is zero.

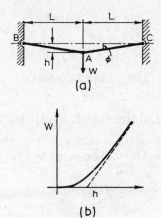

(a)

(b)

Fig. 5.4

When loaded as shown, the bar extensions and the deflexion of A are related geometrically by the compatibility equation—

$$\cos \phi = \frac{L}{L + e}$$

where *e* is the extension of each bar, i.e.

$$\frac{e}{L} = \frac{1 - \cos \phi}{\cos \phi} \qquad (5.9)$$

The usual elastic law gives

$$e = \frac{pL}{AE}$$

or, using eqn. (5.9),

$$p = \frac{AE}{\cos \phi}(1 - \cos \phi)$$

Vertical equilibrium of joint A then gives

$$W = 2p \sin \phi$$
$$= 2AE(1 - \cos \phi) \tan \phi$$

This gives the non-linear load-deflexion relationship sketched in Fig. 5.4 (*b*), but it can be tidied up a little. For small values of ϕ,

$$1 - \cos \phi = 2 \sin^2 \tfrac{1}{2}\phi \simeq \tfrac{1}{2}\phi^2$$
$$\tan \phi \simeq \phi$$

i.e. $\phi = \sqrt[3]{(W/AE)}$

and $h = L\phi$
$$= L\sqrt[3]{(W/AE)}$$

so that the flexibility (h/W) is not a constant.

5.7 Williot Displacement Diagram

The following graphical treatment was developed in 1877 by the Frenchman Williot and further extended by Otto Mohr in 1887.

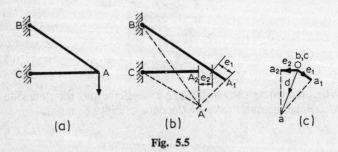

(a)   (b)   (c)

Fig. 5.5

Consider the simple frame in Fig. 5.5 (*a*), where AB extends e_1 while AC shortens e_2. Then, if the members are disconnected at A as in Fig. 5.5 (*b*), the ends of BA and CA will separate to A_1 and A_2. The ends of the altered lengths can be reconnected by swinging BA_1 about B, and CA_2 about C until they intersect at A', which then defines the final position of the frame.

Since the extensions are of the order of 5×10^{-4} times the member length, it is not possible to draw this diagram to a suitable scale and

Williot separated it from the structure by drawing a separate diagram and assuming, for small movements, that the arcs A_1A' and A_2A' are equivalent to the tangents, thus giving rise to Fig. 5.5 (c).

The Williot diagram (Fig. 5.5 (c)) is therefore constructed as follows. Point B is fixed in space at the origin O, and, relative to B, point A moves down and to the right, parallel to BA, for a distance e_1 due to the extension of AB, i.e. line Oa_1. Point A then swings perpendicular to BA along line a_1a to a.

Point C is also fixed at the origin, so that, relative to C, point A moves to the left parallel to AC a distance e_2 due to the shortening of AC, line Oa_2. Point A then swings perpendicular to AC along line a_2a to a.

The resultant movement of point A in space is the vector Oa in magnitude and direction. This is a neat and simple construction but great care must be taken to ensure that vectors of relative movements due to changes in length are drawn in the correct directions.

The essential pre-requisite is knowledge of the movement of two starting points (or they must be fixed in position) before a third point connected to both of them can be located and the process continued from joint to joint around the frame.

The construction also requires that the changes e in length L of the members be known; such changes, for example, as may be due to

(a) Loads giving rise to internal stress resultants p, where

$$e = \pm \frac{pL}{AE}$$

the $+$ indicating extension due to a load of $+p$ (tensile),
the $-$ indicating shortening due to a load of $-p$ (compressive).

(b) Temperature changes, where

$$e = \pm cLt$$

where c = coefficient of linear expansion,
t = temperature change.

(c) Errors or adjustments that have been made

$$e = \pm \lambda$$

where λ is the error in length or erection adjustment, etc.
The signs in each case are taken as in (a), i.e. $+$ for extensions and $-$ for contractions.

5.8 Mohr Correction Diagram

Consider now the braced girder in Fig. 5.6 (a). There are no two fixed points both connected to a third and providing a fixed reference from which to start drawing the Williot diagram.

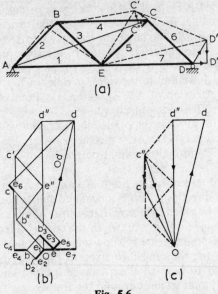

Fig. 5.6

Assuming that, say, AE remains horizontal the Williot diagram will show the relative change in shape of the structure. Then, numbering members as shown, diagram (*b*) can be drawn as follows.

Starting at an origin O, the point A is fixed, and then

relative to A, E moves e_1 parallel to AE, to e,
relative to A, B moves e_2 to b_2,
relative to E, B moves e_3 to b_3,
relative to A and E, b_2 and b_3 swing to b,
relative to B, C moves e_4 to c_4, and so on.

The final displacement of, for example, D is Od up and to the right and the distorted shape obtained has been superimposed (in dashed lines) on the original frame in Fig. 5.6 (*a*). This is, of course, wrong since D is constrained to move only horizontally. However the relative movements are correct and the diagram can be corrected by rotating the whole distorted frame as a rigid body about A to bring D back down to its roller bearing.

In this diagram DD′ is the displacement Od given by the Williot diagram, and D′D″ is the correction due to rotation about A, assumed normal to the original frame position AD rather than the more correct vector normal to the displaced position AD′. These vectors are added to give the true resultant displacement DD″ parallel to the roller path.

This vector addition can be done more conveniently on the Williot diagram as re-drawn in Fig. 5.6 (*c*). Vector Od″ is the rotation correction which, when *added* to Od, must give a resultant movement in the horizontal direction only (dd″); the correct sense is from d″ to d.

This can be done in Fig. 5.6 (*a*), for every point such as C on the frame but it is seen that the correction C′C″ (Fig. 5.6 (*a*)), taken perpendicular to AC (the original position), is proportional to the radius about A, i.e.

$$C'C'' = D'D''(AC)/(AD)$$

or
$$Oc'' = Od''(AC)/(AD)$$

and so on for each point.

If, therefore, this vector addition is carried out on the Williot diagram, it will result in a set of vectors all perpendicular to the radii about A on the original frame and all proportional to the respective radii; for example,

$$Oc'' = \frac{Od''}{AD} \times AC$$

$$Ob'' = \frac{Od''}{AD} \times AB, \quad \text{etc.}$$

The ends of these vectors will simply reproduce, on Od″ as a base, a scaled-down drawing of the original frame rotated about O by 90° in the opposite sense to the real rotation correction. This diagram was developed by Otto Mohr and is called the *Mohr correction diagram* The combined diagram is therefore named the *Williot-Mohr diagram*. The real displacements of points are then given by the vector measured *from the Mohr to the Williot diagrams*.

Going back to the Williot diagram Fig. 5.6 (*b*), we would now obtain the correction diagram by drawing dd″ parallel to the roller path since we know that D can have no vertical movement, i.e. d″ is now immediately located as also on the line through O perpendicular to AD. It is then a simple matter to construct on Od″ the scaled version of the frame.

Example 5.1
Figure 5.7 shows a relatively simple problem in that points A and B are pinned to a rigid foundation and their positions do not change and no Mohr correction diagram is required. This deflexion diagram has been drawn in terms of (pL/A) in mixed kN, m and mm² units, so that a final conversion of units and inclusion of E is required.

Example 5.2
The second worked example in Fig. 5.8 gives the changes in length on the structure diagram; this time a correction diagram is required.

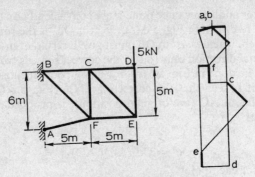

Member	pL/A
BC	+3·0
BF	+4·0
CF	−3·0
CE	+10·0
DE	−2·5
CD	0
AF	−3·0
EF	−2·0

Fig. 5.7

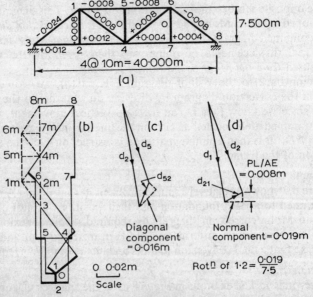

Fig. 5.8

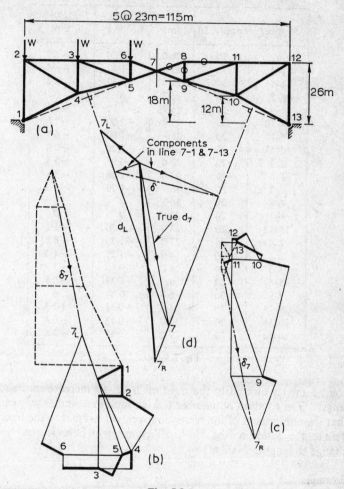

Fig. 5.9

In addition to the simple joint movements the change in length (m) of
the diagonal 5–2 and the rotation of the member 1–2 are determined.

The Williot diagram (Fig. 5.8 (*b*)) has been drawn with the origin
at joint 1 and member 1–2 as a reference direction.

In diagram (*c*) the displacement vector of joint 5 is d_5 down and to
the right, and when subtracted from d_2 gives the resultant vector d_{52}
which is the relative movement of the joints. This vector has two
components, one along the line 5–2 and one normal to this direction.
The component along 5–2 is the change in length of the diagonal.

Member	A mm² × 10	L (m)	P_0 kN × 10	P_0L/AE mm
1–2	30	26·0	−1·60	−6·7
1–4	30	25·9	−1·77	−6·5
4–5	30	23·8	−0·62	−2·2
5–7	30	16·7	0	0
7–9	30	16·7	−1·70	−3·3
9–10	30	23·8	−2·00	−7·1
10–13	30	25·9	−1·80	−6·7
12–13	30	26·0	+0·20	+0·9
2–3	20	23·0	−0·97	−5·5
3–6	20	23·0	−1·58	−8·8
6–7	20	16·7	−1·70	−5·0
7–8	20	16·7	0	0
8–11	20	23·0	0	0
11–12	20	23·0	+0·70	+1·1
2–4	20	27·0	+1·30	+8·5
3–4	20	14·0	−1·25	−4·2
3–5	20	24·0	+0·65	+3·9
5–6	20	8·0	−0·38	−0·8
8–9	20	8·0	0	0
9–11	20	24·0	+0·38	+2·3
10–11	20	14·0	−0·12	−0·4
10–12	20	27·0	−0·42	−2·8

Fig. 5.9 (*cont.*)

Joint 5 moves up and to the right of 2, i.e. an increase in diagonal length. The rotation of member 1–2 is found in a similar way except that the component of the relative movement normal to the line 1–2 defines the rotation (the "in-line" component here is simply the change in length, 0·008 m).

Example 5.3

The three-hinged arch example in Fig. 5.9 is rather more complicated. In order to start it is necessary to assume reference directions for each half of the arch. The separate deflexion diagrams (*b*) and (*c*) each assume that the end posts 1–2 and 12–13 remain vertical, so that these diagrams give different values for the movement of joint 7 and the Mohr correction diagram must adjust this incompatibility. The two displacement vectors are added in diagram (*d*) and, noting that this implies a separation at the hinge, the two halves of the arch can be re-joined by rigid body rotations about 1 and 13. This gives the correction vectors d_L and d_R perpendicular to 1–7 and 13–7 respectively. The final, true movement of joint 7 is d₇ (diagram (*d*)).

Knowing d_7 in magnitude and direction, separate Mohr correction diagrams can be drawn for each half; to do this, set off from point 7 of the original Williot diagrams vectors equal and opposite to d_7 to locate point 7 on the correction diagrams, which can then be completed to scale and in a position rotated through 90° to the structure opposite to the correction rotations. The accompanying table shows the method of working.

5.9 Principle of Virtual Work

(a) *Mass point.* Consider a mass point in Fig. 5.10 (a) under the action of a set of forces which have a resultant R. If the point undergoes a displacement d while the forces remain constant,

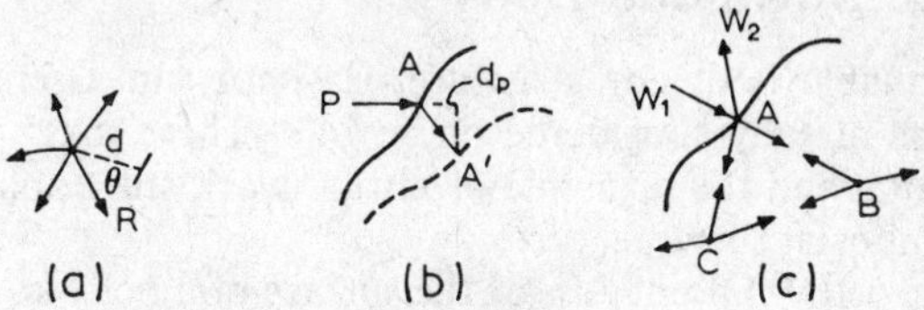

(a)   (b)   (c)

Fig. 5.10

then

$$\text{Work done} = R \times d \cos \theta$$

If the forces form an equilibrium system the resultant R is zero and the work done is also zero. This is another way of stating the conditions of equilibrium: "If a mass point suffers any arbitrary virtual displacement and the work done is zero then the point is in equilibrium."

(b) *Virtual displacement.* The displacement is called a virtual displacement because it need never actually occur; it is sufficient to visualize such a displacement and to see that if it did occur then the total work—virtual work—would be zero.

(c) *Equilibrium.* As a generalized statement of equilibrium the principle is inconvenient, since every possible virtual displacement would have to be investigated, but it will be seen that a general procedure can be based on the converse—

"When a system of forces which is in equilibrium undergoes any set of arbitrary, virtual, displacements the total work done is zero."

Note in this statement the essential requirement of equilibrium of the force system and note also that with real structures the statements only have meaning if these arbitrary virtual displacements are geometrically possible.

It is also essential to realize that the forces remain constant during the virtual displacements, i.e. the forces and displacements are not necessarily cause and effect, and may never occur together. It is sufficient that they each can occur.

(*d*) *Corresponding displacements.* Work is only done by a force moving along its line of action; so that forces are associated with their corresponding displacements and these may well be quite different from the total displacement of the point of action of a force. Thus in Fig. 5.10 (*b*) the point A on the boundary of a structure may move to A′, but the displacement corresponding to the force P is the component d_p.

5.10 Deformable Body

A deformable body, or structure, is simply an assembly of mass points, as already considered on p. 146. These mass points are in equilibrium and the principle of virtual work still applies, but there is one important new feature.

In Fig. 5.10 (*c*) point A is at the surface of a body so that some of the forces acting on it are external loads and some are from internal actions with neighbouring mass points. It then becomes convenient to separate the two sets of forces and the work done—

$$\text{Total work done, } W = W_e + W_i$$

where W_e = work done by external forces on mass point,
$\quad\quad W_i$ = work done by internal forces on mass point,
and for a whole structure, summing up for all the mass points,

$$\sum W = \sum W_e + \sum W_i = 0$$

The external virtual work is easily calculated—

$$W_e = \sum W \times d$$

where d is each of the corresponding displacements.

Internal virtual work done may not be so simple, but where the structure is composed of finite elements and joints between them we can consider the joints as mass points and calculate the virtual work done on them by the member forces.

Consider the tie bar AB in Fig. 5.11 (*a*), which forms one element of

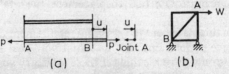

(a)   (b)

Fig. 5.11

the structure Fig. 5.11 (*b*). If the tensile stress resultant is p_1 and the bar extends an amount u_1, the work done by the force is $p_1 u_1$. Now considering the joint A, the force p_1 on the joint is equal and opposite to p_1 on the member, but the displacement of the joint is the same as that of the end of the member, i.e.

(Work done on joint) $=$ $-$(work done on member)

and for all members and all joints,

$$\text{Total virtual work} = \sum W_e + \sum W_i$$
$$= \sum W_e - \sum (\text{work done on members})$$
$$= \sum W_e - \sum pu$$

and so, since the total work is zero

$$\sum W_e = \sum pu$$

where p = member force (stress resultant), and
u = member deformation corresponding to p.

This provides a method for the calculation of the deformations of structures. It must be emphasized at this point that the virtual work is calculated for a system of *constant* loads moving through a set of *virtual* displacements so that the factor of $\frac{1}{2}$ does not appear as in the case of strain energy calculations (p. 126).

5.11 Deflexion of Braced Frames

The pin-jointed structure shown in Fig. 5.12 (*a*) is in equilibrium under the action of the following forces—
External loads: W_1, W_2, W_3, etc.
Internal member stress resultants: p_1, p_2, etc.
Reactions: R_1, R_2, R_3
The deformations are then
Joint displacements in various directions: d_1, d_2, etc.
Changes in lengths of members: u_{10}, u_{20}, etc.
It is convenient to call this equilibrium system of forces and displacements system I.

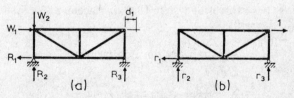

Fig. 5.12

Now consider the same structure with a single unit load in a direction corresponding to one of the displacements of system I, and call this system II (Fig. 12 (b)). The forces of this system are

External loads: single unit load

Internal stress resultants: p_{11}, p_{21}, p_{21}, etc.

Reactions: r_{11}, r_{21}, r_{31}.

The displacements of system I are a compatible set; using them as a set of virtual displacements for the forces of system II we get

$$\sum (\text{virtual work}) = 0$$

but

$$\sum W_e = 1 \times d_1$$

(since the reactions do no work), and

$$\sum W_i = -(p_{11}u_{10} + p_{21}u_{20} + p_{31}u_{30} + \cdots)$$
$$= -\sum p_1 u_0$$

(omiting the member identification subscript). Therefore by the principle of virtual work,

$$1 \times d_1 = \sum p_1 u_0$$

or

$$d_1 = \sum p_1 u_0$$

This is a very simple and convenient means of calculating the single displacement d_1, i.e. the displacement of system I corresponding to the unit load of system II. The detailed procedure is as follows—

(a) Analyse the forces p_0 in the structure of system I and calculate the changes in length u_0 of all the members due to any cause, e.g.

Due to stress resultants p_0, $\quad u_0 = p_0 L / EA$

Due to temperature change $\quad u_0 = cLt$

Due to wrong dimensions $\quad u_0 = e$

Due to adjustments $\quad u_0 = \lambda$

(Calling a tensile force positive, the elongations are positive and shortenings negative.)

(b) Apply a single unit load to the structure corresponding to the desired displacement d, and calculate the member stress resultants p_1.

(c) Use the principle of virtual work—

$$\sum W = \text{forces of system II} \times \text{displacements of system I}$$
$$= 0$$
$$W_e + W_i = 0$$
$$W_e = -W_i = \text{work done on members}$$

i.e. $\quad 1 \times \Delta = \sum p_1 u_0$

$$= \sum \frac{p_1 p_0 L}{EA} + \sum p_1 cLt + \sum p_1 e + \sum p_1 \lambda$$

Example 5.4

The complete calculations for the central deflexion of an N-girder
(Fig. 5.13) are recorded in the table. Columns 1 and 2 contain
structural data; column 3 gives the member forces p_0 due to
loading; corresponding changes in length u_0 are given in column 5.

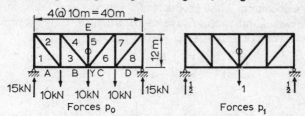

$$d_1 = \Sigma p_1 u = \Sigma p_1 p_0 L/AE + \Sigma p_1 \lambda$$

1	2	3	4	5	6	7	8
Member	L/A(mm)	p_0 (kN)	p_1	$p_0 L/A$	$p_1 p_0 L/A$	clt	$p_1 clt$
E1	12	−15	−0·50	−180	+90		
E2	6	−12·5	−0·416	−75	+31·2	+2·16	−0·899
E4	6	−16·67	−0·832	−100	+83·2	+2·16	−1·798
B3	6	12·5	+0·416	+75	+31·2		
12	12	19·9	+0·65	+234	+152·1		
23	12	−5·0	−0·50	−60	+30·0		
34	12	6·5	+0·65	+78	+50·7		
					$\Sigma = +468·4$		$\Sigma = −2·697$

$$d_1 = 468·4 \times 2/E \downarrow$$

Temp. effect: $d_1 = -2·697 \times 2 = -5·39$ mm ↑

Adjustment of top boom:

$$d_1 = 468·4 \times 2/E + [-0·416 - 0·832]2\lambda = 0$$

Therefore $\lambda = +750/E$ (elongation)

Fig. 5.13

Applying a unit load corresponding to the required displacement,
the internal forces p_1 are tabulated in column 4. Then, applying the
principle of virtual work, the products of columns 4 and 5 appear at
column 6 summing to +468·4 (kN/mm).

Due to the symmetry of the structure and the loading, it has been
necessary to record only one half of the structure (member 4–5 being
a zero member) and the deflexion is then

$$1 \times d_1 = \Sigma p_1 u_0$$

$$= +468·4 \times 2/E \left(\frac{\text{kN/mm}}{E}\right)$$

152

Deformation of Structures

(i.e. down, corresponding to the unit virtual load); the value of Young's Modulus $(200 \times 10^3\,\text{N/mm}^2)$ is inserted here since, for convenience of the arithmetic, it was omitted from column 5.

If the top boom only (members E2, E4, E5 and E7) is subject to a temperature rise of 20°C, each of these members will expand by the amounts in column 7, the products of columns 4 and 7 now forming the deflexion d_1 due to this rise in temperature, i.e. using the principle of virtual work (C.L.E $= 10\cdot8 \times 10^{-6}$ per °C),

$$1 \times d_1 = \sum p_1 u_0$$
$$= 2(-2\cdot697)$$
$$= -5\cdot394\,\text{mm} \quad \text{(up)}$$

In practice it is desirable to pre-camber the girder so that under load there is no sag, i.e. the centre-point and the supports are at the same level. This can be done by fitting the top boom members too long and so giving an initial upward camber equal to d_1.

Alternatively, let the members all be equally adjusted by an amount λ, then

$$d_1 = \sum p_1 u_0$$
$$= \sum p_1 \frac{p_0 L}{AE} + \sum p_1 \lambda$$
$$= 0$$

or $\qquad 0 = 936\cdot8/E + \lambda(2)(-0\cdot416 - 0\cdot832)$

therefore $\qquad \lambda = +\,750/E \quad \text{(lengthening)}$

Example 5.5
A second example (Fig. 5.14) is included to demonstrate a multiple-deflexion calculation. The deflexions d_A and d_C each require the application of a separate unit load. In each case $(p_1)_{BD}$ and $(p_2)_{BD}$ is zero, so that the deformations of BD and EF do not contribute to the deflexions and are not recorded. The values of p_1 and p_2 are used to derive the member forces due to the loads, i.e.

$$p_0 = 5p_1 + 20p_2\,\text{kN}$$

One common source of error in this type of calculation is confusion of the units; here the arithmetic has been kept simple by tabulating lengths in metres and omitting E so that the units conversions must be made in the final calculation.

Example 5.6
In the structure shown in Fig. 5.15 the change in length of the diagonal BE is required and the unit action corresponding to this

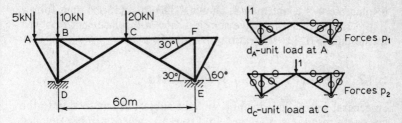

Member	L (m)	p_1	p_2	$p_0 = 5p_1 + 20p_2$	p_1p_0L	p_2p_0L
AC	40	$1/\sqrt{3}$		$5/\sqrt{3} + 0\ = 5/\sqrt{3}$	66·7	
AD	20	$-2/\sqrt{3}$		$-10/\sqrt{3} + 0\ = -10/\sqrt{3}$	133·3	
DC	$20\sqrt{3}$	$-1/3$	-1	$-5/3 - 20 = -65/3$	250·0	751·0
CE	$20\sqrt{3}$	$+1/3$	-1	$+5/3 - 20 = -55/3$	−212·0	635·0
					$\Sigma = +238$	$+1{,}386$

$$d_A = \frac{238 \times 10^6}{10^3 \times 200 \times 10^3} = 1\cdot09 \text{ mm} \downarrow$$

All areas = 10^3 mm²
$E = 200 \times 10^3$ N/mm²

$$d_C = \frac{1{,}386 \times 10^6}{10^3 \times 200 \times 10^3} = 6\cdot93 \text{ mm} \downarrow$$

Fig. 5.14

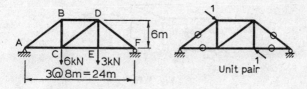

Area of upper and lower chords = 400 mm²
Area of web members = 300 mm²
$E = 200 \times 10^3$ N/mm²

Member	L (m)	A	p_0 (kN)	p_1	p_1p_0L/A
BD	8	4	−6·67	−0·8	10·67
CE	8	4	5·33	−0·8	−8·53
BC	6	3	5	−0·6	−6·00
DE	6	3	3	−0·6	−3·60
CD	10	3	1·67	+1	+5·57
					$\Sigma = -1\cdot89$

d corresponding to unit pair = $\Sigma p_0p_1L/AE = -\dfrac{1\cdot89 \times 10^6}{200 \times 10^5} = -0\cdot094$ mm

i.e. opposed to unit pair = separation.

Fig. 5.15

displacement is a unit pair as shown. The calculations then follow Fig. 5.15; the final negative sign indicates a displacement opposed to the unit pair, i.e. separation of B and E.

5.12 General Stress Resultants

In general a structural member may be subject at any section to the six stress resultants tabulated in Table 5.1, where a more detailed notation is used to distinguish the various actions rather than the general designation *p* (*see* Fig. 5.16).

Table 5.1

Type	System I (real loads, p_0)	System II (unit action, p_i)
Axial force	n_{x0}	n_x
Shear along y-axis	n_{y0}	n_y
Shear along z-axis	n_{z0}	n_z
Moment about x-axis (torque)	m_{x0}	m_x
Moment about y-axis	m_{y0}	m_y
Moment about z-axis	m_{z0}	m_z

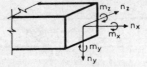

Fig. 5.16

Table 5.2

System II	Corresponding displacement	Type
n_x	$u_{x0} = n_{x0}(L/EA)$	Axial change in length L
n_y	$u_{y0} = \mu(n_{y0}\,\delta x/GA)$*	Shear deformation in length δx
n_z	$u_{z0} = \mu(n_{z0}\,\delta x/GA)$*	Shear deformation in length δx
m_x	$\theta_{x0} = m_{x0}\,\delta x/GJ$†	Torsion twist in length δx
m_y	$\theta_{y0} = m_{y0}\,\delta x/EI_y$	Change of slope in xz-plane over length δx
m_z	$\theta_{z0} = m_{z0}\,\delta x/EI_z$	Change of slope in xz-plane over length δx

* μ = shear correction factor (from p. 162). † J = effective torsional constant.

The displacements of system I corresponding to the actions of system II calculated by the methods of Chapter 3 are then as given in Table 5.2.

The full expression for internal virtual work is seldom necessary since it is usually the case that the virtual work contributed by one or two actions predominates, but the complete expression is developed below.

First consider a short length δx of one member subject to m_z only, then

$$\text{Virtual work done on element} = \text{stress resultant system II} \times \text{corresponding displacement system I}$$

or

$$-(\delta W_i) = m_z \theta_{z0}$$

$$= m_z \frac{m_{z0} \, \delta x}{EI}$$

therefore for whole member

$$-W_i = \int_0^L \frac{m_z m_{z0}}{EI} \, dx$$

and for all members

$$-W_i = \sum \int \frac{m_z m_{z0}}{EI} \, dx$$

For all six actions the expression becomes

$$-W_i = \sum \frac{n_x n_{x0} L}{EA} + \sum \int \frac{\mu n_y n_{y0}}{GA} \, dx + \sum \int \frac{\mu n_z n_{z0}}{GA} \, dx$$

$$+ \sum \int \frac{m_x m_{x0}}{GJ} \, dx + \sum \int \frac{m_y m_{y0}}{EI_y} \, dx + \sum \int \frac{m_z m_{z0}}{EI_z} \, dx$$

5.13 Evaluation of Integrals

The virtual work integrals can be evaluated by integration provided the functions m and m_0, etc., can be dealt with conveniently. Generally, the unit load functions n and m are simple and it is more convenient to treat the products as volume integrals and evaluate numerically; for a uniform number,

$$\int_0^L \frac{mm_0}{EI} \, ds = \frac{1}{EI} \int_0^L mm_0 \, ds$$

$$= \frac{1}{EI} \left[\text{Volume of solid prism of rectangular cross-section } m \times m_0 \right]$$

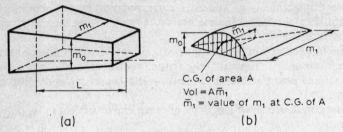

C.G. of area A
Vol $= A\bar{m}_1$
$\bar{m}_1$ = value of m_1 at C.G. of A

(a)   (b)

Fig. 5.17

This is shown in Fig. 5.17 (*a*) for two linear functions; and the volumes of a range of such solids is tabulated in Table 5.3. Also, provided that only one dimension (usually m_0) is non-linear (Fig. 5.17 (*b*)),

Volume = area of non-linear face
× thickness at C.G. of non-linear face

5.14 Non-linear Behaviour

The calculation of displacements by the above method based on the principle of virtual work is not restricted to linear elastic structures, i.e.

$$\sum W_e = \sum nu_0 + \sum m\phi_0$$

is perfectly general so long as we can obtain values for the member deformations u_0 and ϕ_0.

5.15 Reciprocal Theorem

The reciprocal theorem is one of the most powerful of the principles available in the analysis of structures and will be considered again in Chapter 11 on the use of models. Here it is demonstrated for a simple case only.

Consider a structure (Fig. 5.18 (*a*) and (*b*)) in equilibrium under

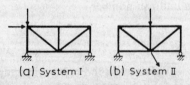

(a) System I   (b) System II

Fig. 5.18

Table 5.3

	rectangle m_i	triangle m_i	trapezoid (e, d)
rectangle m_o	$L\,(m_i m_o)$	$\dfrac{L}{2}(m_i m_o)$	$\dfrac{L}{2}(m_o)(e+d)$
triangle m_o (rising)	$\dfrac{L}{2}(m_i m_o)$	$\dfrac{L}{3}(m_i m_o)$	$\dfrac{L}{6}(m_o)(d+2e)$
triangle m_o	$\dfrac{L}{2}(m_i m_o)$	$\dfrac{L}{6}(m_i m_o)$	$\dfrac{L}{6}(m_o)(2d+e)$
trapezoid (a, b)	$\dfrac{L}{2}(m_i)(a+b)$	$\dfrac{L}{6}(m_i)(2a+b)$	$\dfrac{L}{6}[d(a+2b)+e(2a+b)]$
Parabolic	$\dfrac{L}{6}(m_i)(a+4b+c)$	$\dfrac{L}{6}(m_i)(a+2b)$	$\dfrac{L}{6}[e(a+2b)+d(2b+c)]$

two separate load systems, giving

> External forces P^{I} and P^{II}; internal forces n^{I} and n^{II}
> External displacement d^{I} and d^{II}
> Internal length changes u^{I} and u^{II}

Then, applying the principle of virtual work to the virtual work done by the forces of system I acting through the virtual displacements from system II, i.e.

$$P^{I}d^{II} + (W_i)_1 = 0$$

or
$$P^{I}d^{II} = -(W_i)_1 = \sum n^{I}u^{II}$$

$$= \sum n^{I}\left(\frac{n^{II}L}{AE}\right)$$

and repeating in the reverse way,

$$P^{II}d^{I} + (W_i)_2 = 0$$

$$P^{II}d^{I} = \sum n^{II}u^{I}$$

$$= \sum n^{II}\frac{n^{I}L}{AE} = \sum n^{I}\left(\frac{n^{II}L}{AE}\right)$$

i.e.
$$P^{I}d^{II} = P^{II}d^{I}$$

Then, for any two independent force systems on a linearly-elastic structure the work done by the external forces of system I acting through the corresponding displacements from system II is the same as the work done by the external forces of system II acting through the corresponding displacements from system I.

This general principle is very important, but at this stage it is sufficient to specialize this to the simple case of P^{I} and P^{II} being single actions of equal magnitude though not necessarily the same type, i.e. P^{I} may be a 10 kN load and P^{II} a 10 kN-m moment. Then

$$P^{I}d^{II} = P^{II}d^{I}$$

or since
$$P^{I} = P^{II}$$

$$d^{II} = d^{I}$$

That is to say, the displacement at I due to P at II is the same as displacement at II due to P at I and, if P is a force, d^{II} is a linear displacement; if P^{II} is a moment d^{I} is a rotation (Fig. 5.19 (*a*) and (*b*)).

This is more conveniently expressed in terms of flexibility coefficients by numbering the positions being considered. In Fig.

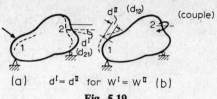

(a) $d^I = d^{II}$ for $W^I = W^{II}$ (b)

Fig. 5.19

5.19 (*a*), the rotational displacement at 2 due to the load P at 1 is

$$d_{21} = f_{21}P$$

and in Fig. 5.19 (*b*) in the same way the linear displacement at 1 is

$$d_{12} = f_{12}P$$

Then, by the special case of the Reciprocal Theorem,

$$d_{21} = d_{12}$$

i.e.

$$f_{21} = f_{12}$$

The constants f_{21} and f_{12} are simply flexibility coefficients as discussed on p. 134.

5.16 Beam Deflexion Examples

As a simple example of the numerical evaluation of the virtual work integral

$$d_1 = \int \frac{m_1 m_0}{EI}\, ds$$

consider the simply-supported beam in Fig. 5.20. The loaded structure and the m_0 diagram are as in (*a*) and (*b*); in order to calculate the load point deflexion apply a virtual unit load (*c*) and obtain the m_1 diagram (*d*).

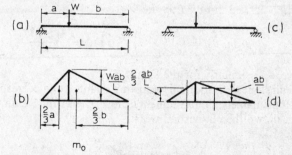

Fig. 5.20

Length a: $\displaystyle\int_0^a m_1 m_0 \frac{ds}{EI} = \Big($ area of m_0 diag. $\times$ value of m_1

at C.G. of m_0 area $\times \dfrac{1}{EI}\Big)$

$$= \left(\tfrac{1}{2}a \frac{Wab}{L}\right) \times \frac{2}{3}\left(\frac{ab}{L}\right)\frac{1}{EI}$$

$$= \frac{1}{3}\frac{Wa^3b^2}{EIL^2}$$

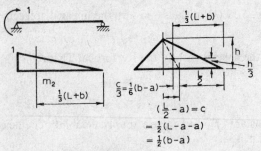

Fig. 5.21

Length b: $\displaystyle\int_0^b m_1 m_0 \frac{ds}{EI} = \left(\tfrac{1}{2}b \frac{Wab}{L}\right) \times \frac{2}{3}\left(\frac{ab}{L}\right)\frac{1}{EI}$

$$= \frac{1}{3}\frac{W}{EI}\frac{a^2b^3}{L^2}$$

Summing gives $\qquad d_1 = \dfrac{1}{3}\dfrac{W}{EI}\dfrac{a^2b^2}{L}$

Alternatively the formulae from Table 5.3 could be used, or the integration carried out directly, i.e.

$$\int_0^a m_1 m_0\, ds = \int_0^a \left(\frac{Wb}{L}s\right)\left(\frac{b}{L}s\right) ds = \frac{Wa^3b^2}{3L^2}$$

as above, and similarly for the length b.

The end slope θ_A is found by applying a unit action corresponding to the slope, namely a unit moment as in Fig. 5.21 to give an m_2 diagram as shown then by virtual work,

$$1 \times \theta_A = \int m_2 m_0 \frac{ds}{EI}$$

In this case some saving of arithmetic is achieved by noticing that m_2 is a single linear function and writing

$$\int_0^L m_2 m_0 \, ds = \begin{pmatrix} \text{area of } m_0 \text{ diagram} \times \\ \text{value of } m_2 \text{ at C.G. of } m_0 \text{ area} \end{pmatrix}$$

$$= \left(\tfrac{1}{2} L \, \frac{Wab}{L} \right) \times \frac{1}{L} \frac{1}{3} (L + b)$$

$$= \frac{Wab}{6} \left(1 + \frac{b}{L} \right)$$

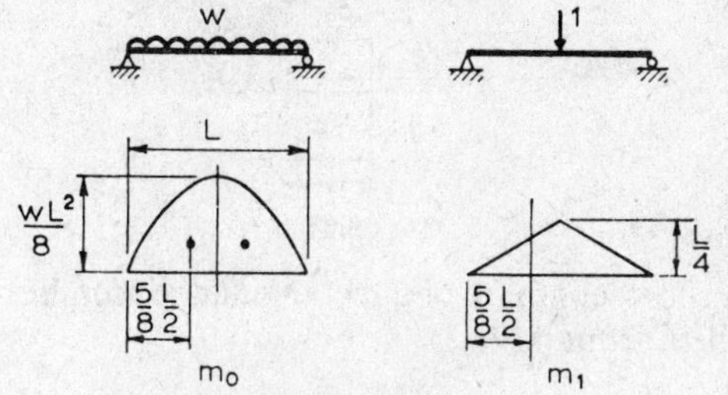

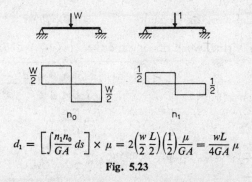

$$EId = 2 \left[\frac{2}{3} \frac{wL^2}{8} \frac{L}{2} \right] \left[\frac{5}{8} \frac{L}{4} \right] = \frac{5}{384} wL^4$$

Fig. 5.22

The central deflexion of a uniformly-loaded beam is readily cal-culated as in Fig. 5.22. However, it is apparent that in order to calculate deflexions at other points, the unsymmetrical m_1 diagram will require the calculation and locations of the centroids of m_0 areas forming segments of a parabola.

Shear deflexions as calculated in Fig. 5.23 will require correction for the shape of the beam cross-section (p. 129), because the internal

$$d_1 = \left[\int \frac{n_1 n_0}{GA} \, ds \right] \times \mu = 2 \left(\frac{w}{2} \frac{L}{2} \right) \left(\frac{1}{2} \right) \frac{\mu}{GA} = \frac{wL}{4GA} \mu$$

Fig. 5.23

virtual work due to the unit load shears acting through the real displacements is not calculable unless the shear strains due to n_0 and
the shear stresses due to n_1 can be calculated. This follows from the
fact that the shear stresses and hence the shear strains are not linearly
distributed across a section.

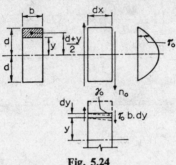

Fig. 5.24

Consider a short length δx of a rectangular-section beam shown in
Fig. 5.24. Then, from p. 76,

$$\tau_0 = \frac{n_0 Q}{I \times b} = \frac{3}{2} \frac{n_0}{A} \frac{(d^2 - y^2)}{d^2}$$

where $A = 2bd$. Therefore

$$\text{Shear strain} = \gamma_0 = \frac{\tau_0}{G} = \frac{3}{2} \frac{n_0}{GA} \frac{(d^2 - y^2)}{d^2}$$

Shear stress at height y above the neutral axis due to the unit load is

$$\tau_1 = \frac{n_1 Q}{Ib} = \frac{3}{2} n_1 \frac{(d^2 - y^2)}{2bd^3}$$

Therefore the shear force on element δy deep is $\tau_1 b \, \delta y$ and so

Virtual work on element $\qquad = \gamma \, \delta x \tau_1 b \, \delta y$

Total virtual work on length $\delta x = 2\int_0^d \frac{}{}(\gamma \, \delta x \tau_1 b) \, dy$

$$= \frac{n_1 n_0}{GA} \frac{9}{4} \int_0^d \frac{(d^2 - y^2)^2}{d^5} \delta x \, dy$$

$$= \frac{n_1 n_0}{GA} \times \frac{6}{5} \delta x$$

Virtual work over whole span $\quad = \frac{6}{5} \int_0^L \frac{n_1 n_0}{GA} dx$

and, for the central load, $\qquad n_0 = \tfrac{1}{2}W \quad\text{and}\quad n_1 = \tfrac{1}{2}$

Virtual work $\qquad\qquad = \dfrac{6}{5}\cdot\dfrac{WL}{4GA}$

i.e.   Correction factor $\mu \qquad = \tfrac{6}{5}$

Correction factors for other shapes may be obtained in the same way; for an I-section it will usually be sufficient to take μ as unity and to consider n_0 and n_1 to be uniformly distributed over the web area only.

5.17 Combined Bending and Axial Stresses

In the structure shown in Fig. 5.25 the shortening of the strut BD will contribute to the vertical deflexion at C. In Fig. 5.25 the vertical deflexion at C is obtained from a unit virtual load as shown and, neglecting the effect of the axial force in AB, the full calculations are given.

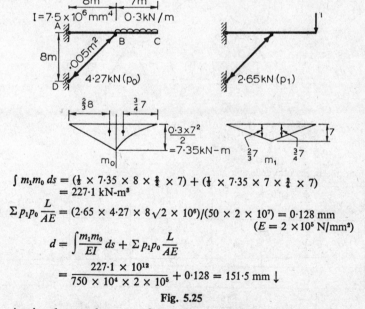

$\int m_1 m_0\, ds = (\tfrac{1}{2}\times 7\cdot 35\times 8\times \tfrac{2}{3}\times 7)+(\tfrac{1}{3}\times 7\cdot 35\times 7\times \tfrac{3}{4}\times 7)$
$\qquad\qquad = 227\cdot 1 \text{ kN-m}^3$

$\Sigma\, p_1 p_0\,\dfrac{L}{AE} = (2\cdot 65\times 4\cdot 27\times 8\sqrt{2}\times 10^6)/(50\times 2\times 10^7) = 0\cdot 128 \text{ mm}$
$\qquad\qquad\qquad\qquad\qquad\qquad (E = 2\times 10^5 \text{ N/mm}^2)$

$d = \int\dfrac{m_1 m_0}{EI}\,ds + \Sigma\, p_1 p_0\,\dfrac{L}{AE}$

$\quad = \dfrac{227\cdot 1\times 10^{12}}{750\times 10^4\times 2\times 10^5} + 0\cdot 128 = 151\cdot 5 \text{ mm}\downarrow$

Fig. 5.25

Again, the usual source of error lies in confusing units: using the virtual work equation as a formula

$$d = \int\dfrac{m_1 m_0}{EI}\,ds$$

disguises the real equation which is

$$1 \text{ (kN)} \times d = \int \frac{m_1 m_0}{EI}\, ds$$

or

$$d = \int \frac{m_1 m_0\, ds/EI}{1 \text{ (kN)}} \quad \text{(i.e. } m_1 = \text{kN-m per kN)}$$

and this final division by a unit force action must not be forgotten when checking the units of the answer.

A second example is shown in Fig. 5.26 and this time the effects of

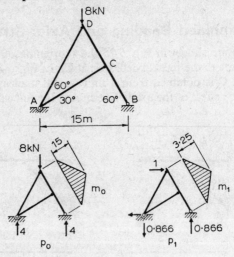

$I_{BCD} = 15 \times 10^6 \text{ mm}^4; \quad A_{BCD} = 4 \times 10^3 \text{ mm}^2$
$A_{AD} = A_{AO} = 2 \times 10^3 \text{ mm}^2$
$E = 2 \times 10^5 \text{ N/mm}^2$

Find the horizontal deflexion at D.

Member	L (m)	A	p_0 (kN)	p_1	$p_1 p_0 L/A$
AD	15	2	-6.93	$+0.5$	-26.0
AC	13	2	$+4.00$	$+0.87$	$+22.5$
BD	15	4	-3.46	-0.75	$+9.75$
					$\Sigma = 6.25$

$$\int m_1 m_0\, ds = 2 \left[\frac{15 \times 7.5}{2} \times \frac{2}{3} \times 3.25 \right] = 244 \text{ kN-m}^3$$

Therefore

$$d = \frac{244 \times 10^{12}}{1{,}500 \times 10^4 \times 2 \times 10^5} + \frac{6.25 \times 10^{12}}{10^3 \times 2 \times 10^5} = 81.3 + 0.0312 = 81.33 \text{ mm}$$

Fig. 5.26

both moments and axial forces in the continuous member BCD are included. Note here the separation of the direct force calculations, which are tabulated, from the bending deformations. It is usually more convenient to do these calculations separately and the chances of errors are reduced.

5.18 Virtual Displacement Systems

The method developed in 5.11 using the technique of a unit load to produce a force system should be more precisely called a Virtual Force System. A parallel technique can then be called a Virtual Displacement System.

For example Fig. 5.27 (a) shows a structure in equilibrium under a given load system. If this structure is subjected to the 'virtual' displacement system of Fig. 5.27 (b) the total virtual work is, of course, zero.

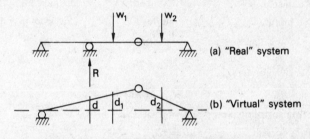

Fig. 5.27

i.e.   Ext. Vir. Work $= W_1 d_1 + W_2 d_2 + Rd$.

           Int. Vir. Work $= 0$ (Rigid body displacement)

Then $W_1 d_1 + W_2 d_2 + Dd = 0$

Then $$R = \frac{-ZW_1 d_1}{d}$$

The important feature is the development of virtual force systems and virtual displacement systems so that the procedures are not restricted to developing these systems on or in the 'original' structure itself.

For example the real displacements of the structure shown in Fig. 5.28 (a) are, say,

$d = $ Virt. deflection in the left span

$\theta = $ Change in angle of the left lower chord

$u = $ Internal deformations

Then, choosing the different structure and virtual force system in Fig. 28 (*b*) and applying the principle of virtual work.

$$1 \times d = \Sigma . p . u \quad \text{hence } d = \Sigma pu \text{ (as before)}.$$

Alternatively choosing a simple beam as the 'virtual' system in Fig. 5.28 (*c*) the virtual force corresponding to the change of slope, θ, is the bending moment, m, in the beam.

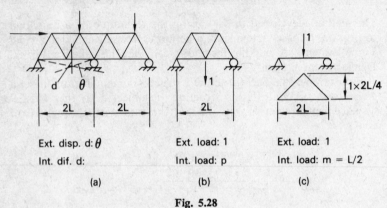

Ext. disp. d: θ	Ext. load: 1	Ext. load: 1
Int. dif. d:	Int. load: p	Int. load: m = L/2
(a)	(b)	(c)

Fig. 5.28

i.e. $1 \times d = m . \theta$

from this simple case: $m = 1 \times L/2$

$$\theta = 2d/L$$

a result easily checked by geometry.

This simple demonstration is not developed further it serving only to indicate the generality of virtual work procedures.

5.19 Examples for Practice

1. The pin-jointed truss loaded as shown in Fig. 5.29 is hinged to a rigid support at the left-hand end and has a roller bearing supported on an inclined surface at the other. The table gives the values of the forces in the members, in terms

Member	A1	B3	D2	D1	D4	C4	12	34	23
P/W	-3	-2	$+1$	$+\sqrt{2}$	0	$-2\sqrt{2}$	-1	$+1$	$-\sqrt{2}$

of *W*. Determine the maximum permissible value of *W* if the movement on the inclined surface is limited to 50 mm. Area of all members is 2,700 mm². ($E = 2 \times 10^5$ N/mm².)
(*Ans.* *W* = 300 kN.)

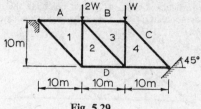

Fig. 5.29

2. The hexagonal frame shown in Fig. 5.30 is supported on a roller bearing at *A* and pinned at *B*. The loads shown produce stresses of 120 N/mm² in tension and 75 N/mm² in compression. Determine the total movement of the point *C*. ($E = 2 \times 10^5$ N/min².)
Ans. (8 mm upwards; 14 mm to left.)

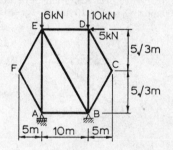

Fig. 5.30

3. The frame shown in Fig. 5.31 is made from a material having working stresses in tension and compression of 75 and 45 N/mm² respectively. Calculate the cross-sectional areas of the members. Calculate, also, the horizontal movement of *E* under these conditions of stress. ($E = 2 \times 10^5$ N/mm².)
(*Ans.* 49 mm.)

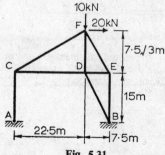

Fig. 5.31

4. The structure shown in Fig. 5.32 consists of members of equal cross-section. It is simply supported at B and D and carries vertical loads *V* at A and E.

Determine the vertical deflexion of G and calculate the magnitude of forces *P*, applied by diagonals GD and GB to halve this deflexion.
(*Ans.* $74LV/18AE$; $370V/3,240$.)

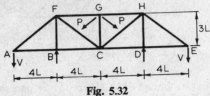

Fig. 5.32

5. A beam ABC is supported at A and by the strut BD at B as shown in Fig. 5.33. The connexions at A, B, and D may be taken as pin joints. The load carried is 15 kN/m distributed over AB. Find the vertical deflexion at C. For ABC, area = 2,500 mm², $I = 2 \times 10^6$ mm⁴; for BD, area = 1,500 mm². ($E = 2 \times 10^5$ N/mm².)
(*Ans.* 17·5 mm.)

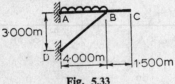

Fig. 5.33

6. Determine the adjustment required in the length of members EF and FG of the frame shown in Fig. 5.34, such that A, C, D, B all lie at the same level. The

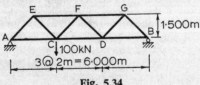

Fig. 5.34

structure is loaded as shown and all members except the lower boom are subject to a temperature rise of 30°C. All areas are 1,000 mm², *E* is 2×10^5 N/mm², and the coefficient of expansion is = 12×10^{-6} per deg C.
(*Ans.* −5·0 mm; −7·8 mm.)

6

Statically-indeterminate Structures

6.1 Introduction

In most of the structures so far dealt with the internal stress resultants can be found from the external loads by the equations of equilibrium, i.e. the structures are statically-determinate. Alternatively, they have been simple members such that, if indeterminate, the equations of compatibility could be set up and solved readily.

In practice, most structures are highly indeterminate; for example, the multi-storey frame shown in Fig. 6.1 (a) has 18 indeterminate internal forces. This can be seen by cutting the structure back to the

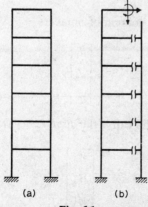

(a)   (b)

Fig. 6.1

statically-determinate set of simple cantilevers shown in Fig. 6.1 (*b*). In order to do this, six cuts have to be made and at each cut 3 internal stress resultants have been released, making a total of 18 statically indeterminate actions, i.e. there are 18 "releases."

If the number of such releases is small, say up to three or four, the setting up of the equations of compatibility and their solution can best be done by hand. For larger numbers than this it is essential to develop and use Matrix Notation in order to keep control of the very large amounts of data.

6.2 Matrix Notation

It is not thought desirable to add standard mathematical material to a text on the Theory of Structures, but it is necessary for a reader to be able to follow and use the text without breaking off to read up the mathematics in detail. Therefore a simple demonstration of the concise nature of a matrix statement is given below and the bare essentials which should allow the text to be understood will be found in the Appendix.

Consider any structure under the action of any set of loads,

$$W_1 \text{ at point 1}$$
$$W_2 \text{ at point 2} \cdots \text{etc.}$$

Then, using flexibility coefficients as on p. 134, we could write

$$\left. \begin{aligned} d_1 &= f_{11}W_1 + f_{12}W_2 + f_{13}W_3 + \cdots \text{etc.} \\ d_2 &= f_{21}W_1 + f_{22}W_2 + f_{23}W_3 + \cdots \text{etc.} \\ d_3 &= f_{31} \cdots \text{etc.} \end{aligned} \right\} \qquad (6.1)$$

Now define a vertical column of numbers

$$\begin{bmatrix} d_1 \\ d_2 \\ d_3 \end{bmatrix}$$

as a column matrix and call it the *displacement matrix*, and another column of numbers

$$\begin{bmatrix} W_1 \\ W_2 \\ W_3 \end{bmatrix}$$

as the load matrix. It is now possible to condense the statement of the set of deflexion equations as follows—

$$\begin{bmatrix} d_1 \\ d_2 \\ d_3 \end{bmatrix} = \begin{bmatrix} f_{11} & f_{12} & f_{13} \\ f_{21} & f_{22} & f_{23} \\ f_{31} & f_{32} & f_{33} \end{bmatrix} \begin{bmatrix} W_1 \\ W_2 \\ W_3 \end{bmatrix} \qquad (6.2)$$

Still to be defined is the array of three rows and three columns of the flexibility coefficients and this, in the same way, can be called the *flexibility matrix*.

This presentation of the original equations (6.1) is meaningless on its own unless we have a set of rules to translate it back to the original form. Here we require a rule for matrix multiplication and this will be found in the Appendix (p. 438); it must be noted that the order of multiplication is important, i.e. it is not commutative.

We can bring about a further condensation if we denote each matrix by a single letter and omit the brackets, namely

$$d = fW$$

and we understand d, f, and W each to represent a matrix. Thus
  (i) a matrix is simply an array or tabulation of numbers (elements), i.e. should not be confused with a determinant, and
  (ii) there does not require to be any relationship between the elements.

Some facility with simple matrix operations is necessary and the multiplication rule should be used to verify that the matrix equation (6.2) does expand to the original set of equations (6.1).

6.3 Member Actions

Before considering the setting up of matrix statements in any greater detail it is necessary to restate the actions and deformations to be considered. There is no generally agreed set of symbols; considering the typical member shown in Fig. 6.2, the nomenclature used is shown in Table 6.1.

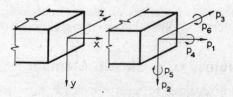

Fig. 6.2

Table 6.1

P		P	u
Stress resultant	Definition	End action	Displacement
n_x	Axial force along x-axis	N_x	u_x
n_y	Shear force along y-axis	Q_y	u_y
n_z	Shear force along z-axis	Q_z	u_z
m_x	Moment about x-axis	M_x	θ_x
m_y	Moment about y-axis	M_y	θ_y
m_z	Moment about z-axis	M_z	θ_z

A distinction has been made between Stress Resultants and End Actions because the former are defined on the positive face of a section (para 2.7, p. 24), but the latter take their sign from a single set of general axes. They are force actions proper and have been designated with capital letters.

The stress resultant set p is then represented by a column matrix or vector—

$$p = \{n_x n_y n_z m_x m_y m_z\}$$

where the curly bracket is used to save space since this vector should be written as a vertical column. Similarly the end action set is

$$P = \{N_x Q_y Q_z M_x M_y M_z\}$$

where Q has been used for the shear forces as a fairly common convention.

In general discussion this nomenclature does prove cumbersome and it may be convenient just to use number subscripts (it is, of course, not so easy to see at a glance the relation between the number and the type of action); the vectors p and P are then

$$p = \{p_1 p_2 p_3 p_4 p_5 p_6\}$$
and
$$P = \{P_1 P_2 P_3 P_4 P_5 P_6\}$$

and the displacements

$$u = \{u_1 u_2 u_3 u_4 u_5 u_6\}$$

The displacements u are sets of displacements corresponding to associated actions and follow the same sign convention as the actions.

6.4 Flexibility of a Simple Member

If discussion is restricted to a plane member, displacements in the x,y-plane only, and using number subscripts, then (Fig. 6.2 and

Chap. 5),

$$p_3 = p_4 = p_5 = 0$$

i.e.

$$\begin{bmatrix} u_1 \\ u_2 \\ u_6 \end{bmatrix} = \begin{bmatrix} f_{11} & f_{12} & f_{16} \\ f_{21} & f_{22} & f_{26} \\ f_{61} & f_{62} & f_{66} \end{bmatrix} \begin{bmatrix} p_1 \\ p_2 \\ p_6 \end{bmatrix}$$

or, simply,

$$u = fp$$

From Fig. 6.3 (*a*), and for $p_1 = 1$ and $p_2 = p_6 = 0$,

$$u_1 = f_{11} \quad \text{(by definition of flexibility)} = L/AE$$

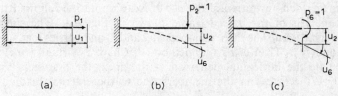

(a)   (b)   (c)

Fig. 6.3

Also, since an axial force produces only an axial change in length, there are no displacements corresponding to p_2 or p_6, i.e. $u_2 = u_6 = 0$. Therefore

$$f_{21} = f_{61} = 0$$

Similarly, in Fig. 6.3 (*b*), for $p_2 = 1$ and $p_1 = p_6 = 0$ and neglecting shear deformations,

$$u_1 = f_{12} = 0$$
$$u_2 = f_{22} = L^3/3EI \quad \text{(p. 116)}$$
$$u_6 = f_{62} = L^2/2EI \quad \text{(p. 116)}$$

In this case the unit action corresponding to p_2 does produce a displacement corresponding to p_6 (a change in slope), i.e.

$$f_{62} \neq 0$$

but

$$f_{12} = 0 \quad \text{(no change in length)}$$

and from Fig. 6.3 (*c*)

$$u_1 = f_{16} = 0$$
$$u_2 = f_{26} = L^2/2EI \quad (=f_{62} \text{ by reciprocal theorem, p. 155})$$
$$u_6 = f_{66} = L/EI \quad \text{(p. 116)}$$

The complete flexibility matrix is then

$$\begin{bmatrix} L/AE & 0 & 0 \\ 0 & L^3/3EI & L^2/2EI \\ 0 & L^2/2EI & L/EI \end{bmatrix}$$

the diagonal L/AE, $L^3/3EI$ and L/EI is called the *leading diagonal*; following from the reciprocal theorem (p. 155) the flexibility matrix is symmetrical about this diagonal.

Note that all elements on a leading diagonal are positive.

6.5 Simple Braced Frame

A further demonstration of the use of matrix notation follows from the analysis of the braced frame shown in Fig. 6.4. The simple triangular frame has three members 1, 2 and 3, and carries loads W_1, W_2 and W_3. Then using the methods of Chap. 5 (virtual work) to calculate the horizontal deflexion at the apex, the first step is to analyze the forces in the members due to the external loading—

$$\text{Member forces} = p_{10}, p_{20}, p_{30}$$

Next the frame is analysed subject to a single unit horizontal load, i.e. a load corresponding to the required deflexion—

$$\text{Member forces} = p_{11}, p_{21}, p_{31}$$

This procedure is repeated with unit loads corresponding to the vertical deflexion at the apex and to the horizontal deflexion at the roller support. The usual way of recording the results is in the form of a table as shown below the figure.

The calculation of d_1 (horizontally at the apex) is written out in full and is followed by the same calculation written out in matrix form. Note that, in order for the matrices to be conformal for multiplication, the matrices must be written as a row and as a column matrix or vector. Repeating this for d_2 and d_3 and collecting the three equations shows the condensation achieved by matrix notation, particularly the final statement (i)—

$$d = B_0'p$$

where each matrix (or vector) is represented by a single symbol. The reason for calling this matrix the transpose of a matrix B_0 will be seen below. In addition, comparison of the full matrix statement with the table shows that the matrix notation simply follows the same pattern and is, effectively, a standardized, alternative form of tabulation.

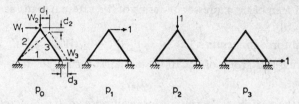

p_0 p_1 p_2 p_3

By virtual work $d_i = \Sigma\, p_i p_0 L/AE$

Member	p_0	p_1	p_2	p_3
1	p_{10}	p_{11}	p_{12}	p_{13}
2	p_{20}	p_{21}	p_{22}	p_{23}
3	p_{30}	p_{31}	p_{32}	p_{33}

$L/AE = \text{constant} = 1$

$$d_1 = p_{11}p_{10} + p_{21}p_{20} + p_{31}p_{30} \quad \text{or} \quad d_1 = [p_{11}p_{21}p_{31}]\begin{bmatrix} p_{10} \\ p_{20} \\ p_{30} \end{bmatrix}$$

$$d_2 = p_{12}p_{10} + p_{22}p_{20} + p_{32}p_{30} \quad \text{or} \quad d_2 = [p_{12}p_{22}p_{32}]\begin{bmatrix} p_{10} \\ p_{20} \\ p_{30} \end{bmatrix}$$

$$d_3 = p_{13}p_{10} + p_{23}p_{20} + p_{33}p_{30} \quad \text{or} \quad d_3 = [p_{13}p_{23}p_{33}]\begin{bmatrix} p_{10} \\ p_{20} \\ p_{30} \end{bmatrix}$$

$$\begin{bmatrix} d_1 \\ d_2 \\ d_3 \end{bmatrix} = \begin{bmatrix} p_{11} & p_{21} & p_{31} \\ p_{12} & p_{22} & p_{32} \\ p_{13} & p_{23} & p_{33} \end{bmatrix}\begin{bmatrix} p_{10} \\ p_{20} \\ p_{30} \end{bmatrix}$$

$$d = B_0' \quad p \qquad \text{(i)}$$

$$p_1 = p_{11}W_1 + p_{12}W_2 + p_{13}W_3$$
$$p_2 = p_{21}W_1 + p_{22}W_2 + p_{23}W_3 \quad \text{or} \quad \begin{bmatrix} p_1 \\ p_2 \\ p_3 \end{bmatrix} = \begin{bmatrix} p_{11} & p_{12} & p_{13} \\ p_{21} & p_{22} & p_{23} \\ p_{31} & p_{32} & p_{33} \end{bmatrix}\begin{bmatrix} W_1 \\ W_2 \\ W_3 \end{bmatrix}$$
$$p_3 = p_{31}W_1 + p_{32}W_2 + p_{33}W_3$$

$$p = B_0 \quad W \qquad \text{(ii)}$$

i.e. $$d = B_0'B_0W \qquad \text{(iii)}$$

Fig. 6.4

Continuing the study of this structure, the stress resultant or force in each member is expressed in terms of the external loads as shown in eqn. (ii), i.e.

Total force in member $1 = p_1 = p_{11}W_1 + p_{12}W_2 + p_{13}W_3$, etc.

Rewriting this group of equations in matrix form leads to

$$p = B_0 W$$

where the matrix B_0 is simply the transpose of the matrix B_0' of eqn. (i), i.e. the rows and columns are interchanged.

This is a very important result which will be further discussed; here it allows the replacement of the p-matrix in (i) to give a final statement (iii) of the frame deflexions in terms of the external loads and member properties only.

Writing

$$F = B_0' B_0$$

gives $\qquad\qquad d = FW$

i.e. $\qquad$ Displacements = flexibility × loads

or $\qquad\qquad F$ = structure flexibility matrix

6.6 Contragredient Transformation: the Principle of Contragredience

Consider the two structural systems in Fig. 6.5, where system I represents an equilibrium system of external forces W and internal stress resultants p, and system II is a set of compatible displacements such that the external d corresponds to W and the internal u to p. Then, by the principle of virtual work, if the forces of system I act through the displacements of system II (p. 146),

$$\sum Wd = \sum pu \qquad\qquad (6.3)$$

In the more general case there will be multiple actions and W, p, and u can all be represented by column matrices—

$$W = \{W_1\ W_2\ W_3 \cdots W_n\} \text{ etc.}$$

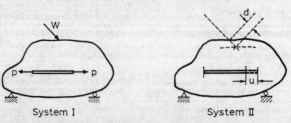

System I   System II

Fig. 6.5

and written in curly brackets to save vertical space. However, as shown by the matrix multiplication rule on p. 438, two-column matrices are not conformal for multiplication and it is necessary to transpose W and p to get

$$[W_1 \; W_2 \; W_3 \cdots] \begin{bmatrix} d_1 \\ d_2 \\ d_3 \\ \cdot \\ \cdot \\ \cdot \end{bmatrix} = [p_1 \, p_2 \cdots] \begin{bmatrix} u_1 \\ u_2 \\ u_3 \\ \cdot \\ \cdot \\ \cdot \end{bmatrix}$$

or
$$W'd = p'u$$

which is the matrix form of eqn. (6.3).

Also, by statics, we can express the p's in terms of the loads W, e.g.,

$$\begin{aligned} p_1 &= p_{11} && \text{for} && W_1 = 1 \\ &= p_{12} && \text{for} && W_2 = 1 \\ && \text{etc.} \end{aligned}$$

so that
$$\begin{aligned} p_1 &= p_{11}W_1 + p_{12}W_2 + p_{13}W_3 + \cdots \\ p_2 &= p_{21}W_1 + p_{22}W_2 + \cdots \\ p_3 &= p_{31}W_1 + \cdots \\ & \qquad\qquad \text{etc.} \end{aligned}$$

or, in matrix form,

$$\begin{bmatrix} p_1 \\ p_2 \\ \cdot \\ \cdot \\ \cdot \\ p_m \end{bmatrix} = \begin{bmatrix} p_{11} & p_{12} & \cdots & p_{1n} \\ & & & \\ & & & \\ & & & \\ & & & \\ p_{m1} & p_{m2} & \cdots & p_{mn} \end{bmatrix} \begin{bmatrix} W_1 \\ W_2 \\ \cdot \\ \cdot \\ \cdot \\ W_n \end{bmatrix}$$

i.e.
$$p = BW$$

where B is a $(m \times n)$ matrix obtained by statics, which transforms the external loads W into the internal stress resultants p, i.e. B is a *transformation matrix*.

Substituting this in the virtual work equation,

$$W'd = p'u$$
gives
$$W'd = (BW)'u$$

and from the rules of matrix algebra the transpose of a product is the reverse product of the transposes, i.e.

$$(BW)' = W'B'$$

so that we get

$$W'd = W'B'u$$

or

$$d = B'u$$

Thus if

$$p = BW$$

then

$$d = B'u$$

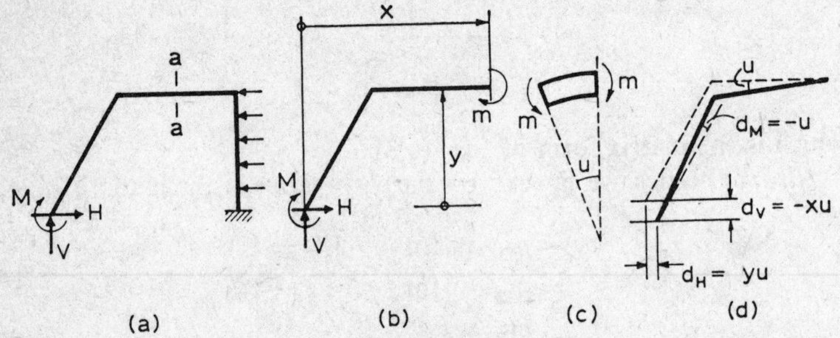

Fig. 6.6

where the displacements u are arbitrary, compatible, and correspond to p, and the displacements d are arbitrary, compatible, and correspond to W. This is called a *contragredient transformation* and we have this general result—

> If two statically-equivalent force systems (p and W) are connected by the transformation B, then any two systems of corresponding displacements (u and d) are connected in the reverse order by the transformation B'.

It is important to realize that there has been no restriction to elasticity or linearity, only to the equilibrium of the p, W system and, of course, d and u need not be related in any way to W and p.

At the section aa of the structure in Fig. 6.6 (a) the bending moment m is shown in (b) to be found from

$$m + Vx - Hy + M = 0$$

or

$$m = [-x, +y, -1] \begin{bmatrix} V \\ H \\ M \end{bmatrix}$$

i.e.

$$m = BW$$

where

$$B = [-x, +y, -1]$$

A displacement u corresponding to the moment m is shown in Fig. 6.6 (c) and the resulting distortion of the frame in (d). Thus

$$d_M = -u$$
$$d_H = +yu$$
$$d_V = -xu$$

or, in matrix form,

$$\begin{bmatrix} d_V \\ d_H \\ d_M \end{bmatrix} = \begin{bmatrix} -x \\ y \\ -1 \end{bmatrix} [u]$$

i.e.
$$d = B'u$$

6.7 Braced-frame Deflexion

The braced frame in Fig. 6.7 is readily analysed to give the internal actions p in terms of the external loads W—

$$p_i = p_{i1}W_1 + p_{i2}W_2 + p_{i3}W_3$$

(in fact p_{ij} = value of p_i for $W_j = 1$). In matrix notation,

$$p = B_0 W$$

Then, by contragredience,

$$d = B_0' u$$

where u is the change in length of each member.

If we require the displacements due to the actual loads,

$$u_i = p_i L_i / A_i E_i = f_i p_i$$

Where $f_i = L_i / A_i E_i$ = extensional flexibility, i.e.

$$u_1 = f_1 p_1 = f_1(p_{11}W_1 + p_{12}W_2 + p_{13}W_3)$$
$$u_2 = f_2 p_2 = f_2(p_{21}W_1 + p_{22}W_2 + p_{23}W_3)$$
$$\text{etc.}$$

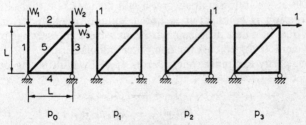

Fig. 6.7

or, using matrices,

$$
\begin{bmatrix} u_1 \\ u_2 \\ u_3 \\ u_4 \\ u_5 \end{bmatrix} = \begin{bmatrix} f_1 & 0 & 0 & 0 & 0 \\ 0 & f_2 & 0 & 0 & 0 \\ 0 & 0 & f_3 & 0 & 0 \\ 0 & 0 & 0 & f_4 & 0 \\ 0 & 0 & 0 & 0 & f_5 \end{bmatrix} \begin{bmatrix} p_{11} & p_{12} & p_{13} \\ p_{21} & p_{22} & p_{23} \\ p_{31} & p_{32} & p_{33} \\ p_{41} & p_{42} & p_{43} \\ p_{51} & p_{52} & p_{53} \end{bmatrix} \begin{bmatrix} W_1 \\ W_2 \\ W_3 \end{bmatrix}
$$

i.e.
$$u = F_0 B_0 W$$

where F_0 is a diagonal matrix of member flexibilities and is called the *unassembled flexibility matrix*. Then, since

$$d = B_0' u$$

we get
$$d = B_0' F_0 B_0 W \qquad (6.4)$$

and finally
$$d = FW$$

where F is the assembled flexibility matrix $= (B_0' F_0 B_0)$ for a given load system, i.e. displacements = flexibilities × loads.

Evidently, since u (change in length of a member) is perfectly general, it can be due to any cause, such as temperature, as well as to an elastic change in length. Thus

$$
\begin{aligned}
u_1 &= u_{10} + u_{1e} \\
&= p_1 L_1 / A_1 E_1 + u_{1e}
\end{aligned}
$$

(where u_{10} = extension due to loads)
Then, in general,

$$d = B_0' F_0 B_0 W + B_0' u_e$$

The effect of the matrix multiplications $B_0' f B_0$ is thus to assemble the member flexibilities f in the correct way and to transform a member extension into its effect on the joint displacements. Working backwards it is apparent that all that has been done is to re-develop the method of Chap. 5 (p. 148); the equations are identical, and the example in the following paragraph will demonstrate that although this notation is compact it is not a suitable method for manual calculation of small problems. It is, however, an essential part of the efficient arrangements for computer calculations.

Now, partly expanding matrix eqn. (6.4) by multiplying out, we shall get

$$
\begin{aligned}
d_1 = & \; p_{11} f_1 p_{11} W_1 + p_{11} f_1 p_{12} W_2 + \cdots \\
& + p_{21} f_2 p_{21} W_1 + p_{21} f_2 p_{22} W_2 + \cdots
\end{aligned}
$$
etc.

Then, since $p_1 = p_{11}W_1 + p_{12}W_2 + \cdots$

$$d_1 = p_{11}f_1p_{10}$$
$$+ p_{21}f_2p_{20}$$
$$\text{etc.}$$

i.e. $d_1 = \sum p_0 p_1 L / AE$ as on p. 149, where p_0 is the force in any bar due to all loads W, and p_1 that due to a unit load corresponding to W_1.

Example 6.1

Derive the deflexions in Fig 6.7.

Using Fig. 6.8, evaluating the elements of matrix B_0 in $p = B_0 W$ gives

$$\begin{bmatrix} p_1 \\ p_2 \\ p_3 \\ p_4 \\ p_5 \end{bmatrix} = \begin{bmatrix} -1 & 0 & 0 \\ 0 & 0 & 0 \\ 0 & -1 & -1 \\ 0 & 0 & 0 \\ 0 & 0 & \sqrt{2} \end{bmatrix} \begin{bmatrix} W_1 \\ W_2 \\ W_3 \end{bmatrix}$$

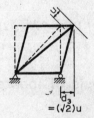

$$u = p_5 L_5 / AE = \sqrt{2}\, W_3 \sqrt{2}\, L / AE$$

Therefore $\qquad d_3 = \sqrt{2}\, u = 2\sqrt{2}\, W_3 L / AE$

Fig. 6.8

Since, by contragredience,

$$d = B_0' u$$

and also $\qquad u = fp$

i.e. $\qquad d = B_0' F_0 B_0 W$

we get

$$\begin{bmatrix} d_1 \\ d_2 \\ d_3 \end{bmatrix} = \begin{bmatrix} -1 & 0 & 0 & 0 & 0 \\ 0 & 0 & -1 & 0 & 0 \\ 0 & 0 & -1 & 0 & \sqrt{2} \end{bmatrix} \begin{bmatrix} f_1 & 0 & 0 & 0 & 0 \\ 0 & f_2 & 0 & 0 & 0 \\ 0 & 0 & f_3 & 0 & 0 \\ 0 & 0 & 0 & f_4 & 0 \\ 0 & 0 & 0 & 0 & f_5 \end{bmatrix} \begin{bmatrix} -1 & 0 & 0 \\ 0 & 0 & 0 \\ 0 & -1 & -1 \\ 0 & 0 & 0 \\ 0 & 0 & \sqrt{2} \end{bmatrix} \begin{bmatrix} w_1 \\ w_2 \\ w_3 \end{bmatrix}$$

For the sake of demonstration, assume all bars except 5 to be rigid, i.e.

$$f_1 = f_2 = f_3 = f_4 = 0$$
$$f_5 = (\sqrt{2})L/AE$$

Then, multiplying $B'f$ first,

$$\begin{bmatrix} d_1 \\ d_2 \\ d_3 \end{bmatrix} = \begin{bmatrix} 0 & 0 & 0 & 0 & 0 \\ 0 & 0 & 0 & 0 & 0 \\ 0 & 0 & 0 & 0 & (\sqrt{2})f_5 \end{bmatrix} \begin{bmatrix} -1 & 0 & 0 \\ 0 & 0 & 0 \\ 0 & -1 & -1 \\ 0 & 0 & 0 \\ 0 & 0 & \sqrt{2} \end{bmatrix} \begin{bmatrix} W_1 \\ W_2 \\ W_3 \end{bmatrix}$$

Next, multiplying $(B'F_0)B$ we get

$$\begin{bmatrix} d_1 \\ d_2 \\ d_3 \end{bmatrix} = \begin{bmatrix} 0 & 0 & 0 \\ 0 & 0 & 0 \\ 0 & 0 & 2f_5 \end{bmatrix} \begin{bmatrix} W_1 \\ W_2 \\ W_3 \end{bmatrix}$$

i.e.
$$d_1 = 0$$
$$d_2 = 0$$
$$d_3 = 2f_5 W_3 = 2W_3(\sqrt{2})L/AE$$

The geometry of Fig. 6.8 shows this to be correct.

6.8 General Analysis of Indeterminate Structures

The analysis of indeterminate beam structures was developed in Chapter 4, the fundamental requirements being—

(a) static equilibrium,
(b) compatibility of deformations and, the link between the two,
(c) force—deformation relations.

It is now possible to develop two general techniques following from the two ways of expressing the force-deformation relations.

6.9 Force-deformation Relations

(a) *Flexibility*

$$\text{Flexibility} = \text{deformation per unit force}$$
$$= f$$

Therefore in Fig. 6.9 deflexion of the spring under a force W is

$$d = fW$$

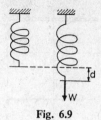

Fig. 6.9

(b) Stiffness

$$\text{Stiffness} = \text{force per unit deformation}$$
$$= k$$

In Fig. 6.9 the force required to produce a deflexion d is

$$W = kd$$

Example 6.2. Flexibility Method

Reconsider the two-span continuous beam (Fig. 6.10 (a)), already analysed on p. 118.

(i) Release the reaction (r_1) at B to reduce the structure to a statically-determinate beam (Fig. 6.10 (b)), then the bending moment at any section $= m_0$ and the deflexion corresponding to r_1 is

$$d_{10} = -\frac{w(2L)^4}{EI} \cdot \frac{5}{384} \quad \text{(p. 116)}$$

The subscript zero indicates the reduced structure and a positive deflexion is upwards corresponding to r_1.

(ii) Apply a unit action (Fig. 6.10 (c)) corresponding to r_1, i.e. a unit force at B upwards. The beam now deflects a distance f_{11}, i.e.

$$f_{11} = \frac{1 \times (2L)^3}{48EI}$$

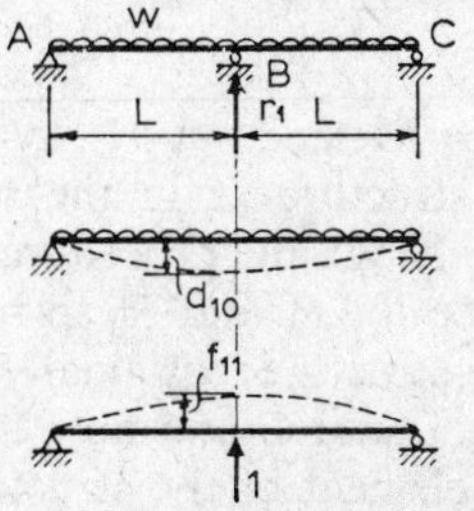

```
A  ╍╍╍╍╍╍ w ╍╍╍╍╍╍ C        (a)
        B
   L    r₁   L

   ╍╍╍╍╍╍╍╍╍╍╍╍              (b)
        d₁₀

        f₁₁                  (c)
          ↑
          1
```

Fig. 6.10

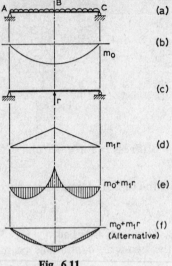

Fig. 6.11

Bending moments = moments due to $r = 1$ on the
                 reduced structure
             = m_1

(iii) The compatibility condition is that the final deflexion at B
is zero, i.e.

$$d_{10} + f_{11}r_1 = 0$$

$$r_1 = \frac{-d_{10}}{f_{11}} = \tfrac{5}{4}wL$$

The final bending moments are then (Fig. 6.11)

$$m = m_0 + m_1r_1$$

Example 6.3. Stiffness Method

Analysing the same structure as in the previous paragraph, and
shown again in Fig. 6.12 (a), the procedure is almost the reverse.

   (i) Add constraints R_1, R_2 and R_3 to remove all the degrees of
freedom of the structure: originally the beam ABC was free
to rotate at A, B and C and now these movements are prevented. Each element of the structure is now isolated (Fig.
6.12 (b)).

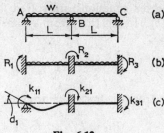

Fig. 6.12

(ii) Analyse the isolated elements by any convenient means (e.g. Section 4.5 (*d*) p. 119). In this example we can use a previous result for a fixed beam (as Fig. 6.12 (*b*)) and find—

$$M_{AB} = \text{moment on end } A \text{ of member AB}$$
$$= M_{AB}^F$$

where the superscript F indicates a fixed end moment, i.e.

$$R_{10} = \text{initial value of } R_1$$
$$= M_{AB}^F = -wL^2/12$$
$$R_{20} = M_{BA}^F + M_{BC}^F = 0 \quad (\text{in this case } M_{BA}^F = -M_{BC}^F)$$
$$R_{30} = M_{CB}^F = +wL^2/12$$

(iii) Give unit deformations corresponding to each constraint in turn, all other constraints remaining (Fig. 6.12 (*c*)), i.e. cause unit rotation corresponding to R_1, then

Forces required per unit displacement = stiffness coefficients
$$k_{11} = \text{force at 1 due to unit displacement at 1}$$
$$k_{21} = \text{force at 2 due to unit displacement at 1}$$

(iv) Finally for the equilibrium condition the constraints must be removed, i.e. the structure allowed to deform. If the final deformations at A, B and C are θ_1, θ_2 and θ_3, the associated moment at A will be

$$R_1 = k_{11}\theta_1 + k_{12}\theta_2 + k_{13}\theta_3$$

and similarly for R_2 and R_3, so that finally

$$R_{10} + k_{11}\theta_1 + k_{12}\theta_2 + k_{13}\theta_3 = 0$$
$$R_{20} + k_{21}\theta_1 + k_{22}\theta_2 + k_{23}\theta_3 = 0$$
$$R_{30} + k_{31}\theta_1 + k_{32}\theta_2 + k_{33}\theta_3 = 0$$

These three equations are solved for the deformations θ_1, θ_2 and θ_3. In this case, from symmetry,

$$\theta_2 = 0$$

$$\theta_1 = -\theta_3$$

so that, from the first equilibrium equation,

$$R_{10} + k_{11}\theta_1 + k_{12}\theta_2 + k_{13}\theta_3 = 0$$

Since (p. 120) $k_{11} = 4EI/L$ and $k_{13} = 0$,

$$-\frac{wL^2}{12} + \frac{4EI}{L}\theta_1 = 0$$

Therefore $$\theta_1 = \frac{wL^3}{48EI}$$

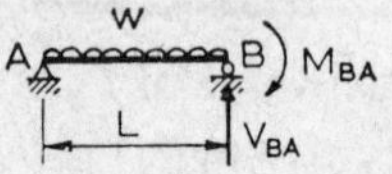

Fig. 6.13

(v) Final actions on the ends of each element are then

$$M_{AB} = M_{AB}^F + \text{effect of rotations}$$

$$= M_{AB}^F + k_{11}\theta_1 = 0$$

(in this case)

and $$M_{BA} = M_{BA}^F + k_{21}\theta_1 \qquad (k_{21} = 2EI/L, \text{ p. 120})$$

$$= +wL^2/12 + wL^2/24 = wL^2/8 = -M_{BC}$$

from symmetry. Considering then the equilibrium of AB (Fig. 6.13),

$$\Sigma M_A = 0$$

i.e. $$\tfrac{1}{2}wL^2 + M_{BA} - V_{BA}L = 0$$

$$V_{BA} = \tfrac{5}{8}wL = V_{BC}$$

from symmetry. Therefore, for both spans,

$$V_B = V_{BA} + V_{BC} = \tfrac{5}{4}wL$$

6.10 Comparison between the Flexibility and the Stiffness Methods

The two typical approaches to the analysis of an indeterminate structure are thus—

Flexibility Method (*Force Method*)	**Stiffness Method** (*Displacement Method*)
Indeterminate constraints are *released* and the resulting deformation discontinuities calculated. These redundant actions are then replaced to restore the continuity and the resulting compatibility equations solved for the redundant force actions.	Additional restraints are *added* to fix all degrees of freedom and the values of these restraints calculated. The restraints are then removed to allow deformations and restore equilibrium. The resulting equilibrium equations are solved for the displacements and subsequently the force actions are determined.

Evidently the methods are related and in some cases one will be more convenient than the other, though the two can be mixed. The frame shown in Fig. 6.14 has five bars, i.e. three statically-indeter-

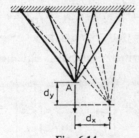

Fig. 6.14

minate bar forces (or, more concisely, "three redundants") but the joint A has only two unknown displacement components d_x and d_y so that it would seem advantageous to use the stiffness method.

Conversely the frame in Fig. 6.15 has twelve possible joint

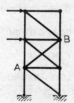

Fig. 6.15

displacements but only a single redundant (AB, say) and the flexibility method is indicated.

In general, the choice is not as simple as this and depends on experience. The student however, will have no difficulty since problems for hand calculation can be readily classified.

6.11 Matrix Formulation

The flexibility solution of Example 6.2 was given as

$$d_{10} + f_{11}r_1 = 0$$

and, extending this to include more than one release, say r_1, r_2 and r_3, the cross flexibilities

f_{12} = displacement corresponding to r_1 due to unit r_2
f_{13} = displacement corresponding to r_1 due to unit r_3
(or f_{ij} = displacement corresponding to r_i due to unit r_j)

would also be required and we shall get three equations of compatibility—

$$d_{10} + f_{11}r_1 + f_{12}r_2 + f_{13}r_3 = 0$$
$$d_{20} + f_{21}r_1 + f_{22}r_2 + f_{23}r_3 = 0$$
$$d_{30} + f_{31}r_1 + f_{32}r_2 + f_{33}r_3 = 0$$

These can clearly be written as a matrix equation—

$$\begin{bmatrix} d_{10} \\ d_{20} \\ d_{30} \end{bmatrix} + \begin{bmatrix} f_{11} & f_{12} & f_{13} \\ f_{21} & f_{22} & f_{23} \\ f_{31} & f_{32} & f_{33} \end{bmatrix} \begin{bmatrix} r_1 \\ r_2 \\ r_3 \end{bmatrix} = 0$$

or simply
$$d_0 + Fr = 0$$

The flexibility solution from Example (6.2) is then obtained by solving the equations or by premultiplying by the inverse matrix F^{-1}, i.e.

$$F^{-1}d_0 + F^{-1}Fr = 0$$

and, since F^{-1} is defined by the relation $F^{-1}F = 1$ (unit matrix, p. 437)

$$r = -F^{-1}d_0$$

In the same way the stiffness-method equilibrium equations on p. 183 can be written as matrices—

$$R_0 + Kd = 0$$

and the solution for the stiffness method becomes

$$d = -K^{-1}R_0$$

For small problems, inversion of the matrices is not efficient; it is better to solve the equations directly.

The two methods can then be compared as in Table 6.1.

Table 6.1

Flexibility	Stiffness
Unknown redundants r	Unknown displacements d
Flexibility matrix F	Stiffness matrix K
Displacements due to $r = Fr$	Forces due to $d = Kd$
Displacements due to loads $= d_0$	Forces due to loads $= R_0$
For compatibility $d_0 + Fr = 0$	Equilibrium $R_0 + Kd = 0$
Solve for r	Solve for d

6.12   Examples for Practice

1. Analyse the beam of Fig. 6.16 by the flexibility method and draw the bending moment diagram.

Fig. 6.16

2. Analyse the structure of Fig. 6.17 by the stiffness method using the rotation at joint B as the only degree of freedom. Draw the bending moment diagram and evaluate all reactions.

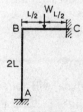

Fig. 6.17

3. Repeat Example 2 using the flexibility method.
   (*Ans. $EI\theta = WL^2/48$.*)

7
Flexibility Method

7.1 Indeterminate Structures

Simple criteria by which the statical determinacy of braced frames can be examined have been presented in Chapter 1, but in the more general case of structural members carrying end actions in addition to simple axial forces, e.g. bending and shear, the criteria for determinacy are not so easily formulated.

With experience a structure can be examined by trial and cut back step by step to give a statically-determinate, stable, reduced frame, and the number of separate actions which have been released can then be counted. It will be seen later that the exercise of some thought during this "releasing" process can make a difference to the subsequent arithmetical labour.

7.2 Release Systems

The choice of releases which will reduce a structure to a stable, statically-determinate form is arbitrary but it is important to realize that the general member may be subjected to six end actions and associated stress resultants (*see*, for example, Table 6.1, p. 170 and Fig. 7.1). When making releases any one or all of these actions may be released. For example, in Fig. 7.2 (*b*) the releases at the end B of the fixed beam AB are p_2 and p_6, whereas an alternative set in Fig. 7.2 (*c*) is p_6 at A and p_6 at B. The former release system cuts the structure back to a simple cantilever and the latter leaves it a simply-supported beam.

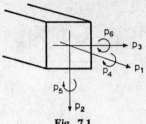

Fig. 7.1

In this discussion the individual releases have been defined by the term p with a subscript, but this notation will be restricted, in what follows, to refer in general to stress resultants using any convenient subscripts and released actions will be defined by the general term r, with subscripts as convenient.

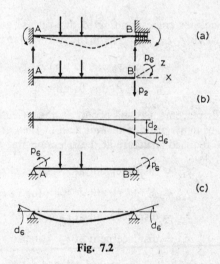

Fig. 7.2

7.3 Statement of the Flexibility Method

In Chapter 6 (p. 181) the steps in the analysis of a structure by the flexibility method were shown to be as follows.

Referring to the doubly-redundant braced frame of Fig. 7.3 (a)—

(1) Sufficient releases r_1 and r_2 are made to reduce the structure to a stable, statically-determinate form as in Fig. 7.3 (b). The member forces are now p_0 and displacements d_{10} and d_{20} will occur at the releases r_1 and r_2 respectively.

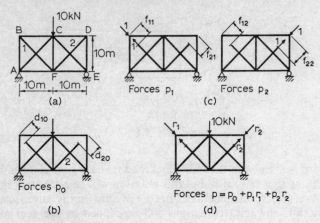

Fig. 7.3

(2) Unit actions corresponding to one release at a time are then applied (Fig. 7.3 (c)). The member forces are defined as

$$p_1 \text{ due to unit } r_1$$

$$p_2 \text{ due to unit } r_2$$

and displacements will occur at the releases. Since these displacements are due to unit actions they are the flexibilities of the reduced structure at the releases, that is

$$f_{11} = \text{displacement at } r_1 \quad (\text{due to } r_1 = 1)$$

$$f_{21} = \text{displacement at } r_2 \quad (\text{due to } r_1 = 1)$$

$$f_{12} = \text{displacement at } r_1 \quad (\text{due to } r_2 = 1)$$

$$f_{22} = \text{displacement at } r_2 \quad (\text{due to } r_2 = 1)$$

(3) For compatibility, i.e. zero displacement at the releases, the total final displacements due to loads and the *r* forces are then

$$d_{10} + f_{11}r_1 + f_{12}r_2 = 0$$

$$d_{20} + f_{21}r_1 + f_{22}r_2 = 0$$

These equations are then solved for r_1 and r_2.

(4) The coefficients in the above equations are found individually by the methods of Chap. 5, so that, keeping to the notation *p* for all member stress resultants instead of the special axial

force notation n we have

$$d_{10} = p_1 p_0 L/AE \quad ; \quad d_{20} = p_2 p_0 L/AE$$
$$f_{11} = p_1{}^2 L/AE \quad ; \quad f_{12} = p_1 p_2 L/AE$$
$$f_{21} = p_1 p_2 L/AE \quad ; \quad f_{22} = p_2{}^2 L/AE$$

(5) The final member forces are (Fig. 7.3 (d))

$$p = p_0 + p_1 r_1 + p_2 r_2$$

7.4 Flexibility Method: Matrix Notation

A statically-indeterminate structure such as that shown in Fig. 7.3 can conveniently be reduced to a stable statically-determinate form by releasing the axial forces in two members as shown there. The releases r_1 and r_2 are then represented by the column matrix

$$r = \begin{bmatrix} r_1 \\ r_2 \end{bmatrix}$$

Considering the reduced frame of Fig. 7.3 (b), this is seen to be a solution of the special case for which r is zero, i.e. it is a *Particular Solution* and displacements will occur at the releases.

The internal forces due to the loads are readily calculated as

$$p_0 = \begin{bmatrix} p_{10} \\ p_{20} \\ p_{30} \\ \text{etc.} \end{bmatrix}$$

where the number subscripts are the numbers of the bars; as on p. 173, this can be written

$$p_0 = B_0 W$$

where the matrix B_0 transforms the external loads W into the member forces and is obtained by statics as in Fig. 6.4 (p. 173); thus the ijth term, say p_{ij}, is

$$p_{ij} = \text{force in member } i \text{ due to } W_j = 1$$

The *Complementary Solution*, which together with the Particular Solution above forms the solution to the complete problem, is obtained by replacing the releases as in Fig. 7.4. Considering this figure, the external forces r and the internal member forces p_r are an equilibrium system and

$$p_r = Br$$

where the matrix B is similar to B_0 except that the ijth element is now

$$p_{ij} = \text{force in member } i \text{ due to } r_j = 1$$

and the final member forces are

$$p = B_0 W + Br$$

$$= [B_0 B] \begin{bmatrix} W \\ r \end{bmatrix}$$

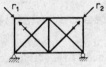

Fig. 7.4

by contragredience
$$\begin{bmatrix} d \\ d_r \end{bmatrix} = \begin{bmatrix} B_0' \\ B' \end{bmatrix} [u]$$

where $[u]$ = element deformations corresponding to the stress resultants $[p]$ but not necessarily due to them, e.g. may be due to temperature changes (see later).

$[d_r]$ = displacements at releases corresponding to the releases $[r]$ and due to the element deformations $[u]$.

$[d]$ = structural displacements corresponding to the loads $[W]$ and also due to the element deformations $[u]$, if the element deformations due to the actions $[p]$ are given by

$$[u] = [f][p]$$

where $[f]$ is the element flexibility (p. 178)

we get
$$\begin{bmatrix} d \\ d_r \end{bmatrix} = \begin{bmatrix} B_0' \\ B \end{bmatrix} [f][p]$$

and since
$$[p] = [B_0 B] \begin{bmatrix} W \\ r \end{bmatrix}$$

it follows that
$$\begin{bmatrix} d \\ d_r \end{bmatrix} = \begin{bmatrix} B_0' \\ B' \end{bmatrix} [f][B_0 B] \begin{bmatrix} W \\ r \end{bmatrix}$$

multiplying out and omitting brackets.

$$\begin{bmatrix} d \\ d_r \end{bmatrix} = \begin{bmatrix} B_0' f B_0 & B_0' f B \\ B' f B_0 & B' f B \end{bmatrix} \begin{bmatrix} W \\ r \end{bmatrix} \tag{7.1}$$

$$= \begin{bmatrix} f_{11} & f_{12} \\ f_{21} & f_{22} \end{bmatrix} \begin{bmatrix} W \\ r \end{bmatrix} \tag{7.2}$$

where f_{11} = direct flexibility of the structure corresponding to $[W]$ and due to actions $[W]$.

f_{22} = direct flexibility of structure corresponding to the releases $[r]$ and due to the actions $[r]$.

$f_{12} = f_{21}$ = cross flexibilities of structure corresponding to $[W]$, or $[r]$, and due to actions $[r]$, or $[W]$.

Then for compatibility the displacements corresponding to the releases are zero.

i.e. $d_r = 0$

$$f_{21}W + f_{22}r = 0$$
$$r = -f_{22}^{-1}f_{21}W$$

and the final stress resultants are known.

$$p = [B_0 B] \begin{bmatrix} W \\ r \end{bmatrix}$$

also the displacement can be calculated from 7.2

$$\begin{aligned} d &= f_{11}W + f_{12}r \\ &= f_{11}W - f_{12}f_{22}^{-1}f_{21}W \\ &= [f_{11} - f_{12}f_{22}^{-1}f_{21}]W \\ &= [F][W] \end{aligned}$$

where F is finally the flexibility matrix for the structure.

7.5 Sign Convention

The significant sign convention follows directly from the use of the principal of virtual work or directly from contragredience, i.e. considering contragredience the displacement set must correspond to the force set.

Member deformations correspond to member stress resultants, or

Internal virtual work = $u \times p$

so that the same sign convention must apply to both (p. 169). For example, if a tensile stress resultant is called positive, then an elongation must be a positive deformation. Similarly, considering the external loads and deflexions, whatever sign convention is adopted for the loads will also apply to the deflexions.

Example 7.1. Braced Frame (use of matrix notation)
The arithmetical procedure to analyse the forces in the typical, redundant, pin-jointed braced frame of Fig. 7.3 is then as follows and in Table 7.1 (p. 197)—

(1) Reduce the structure to determinate form by releasing forces r_1 and r_2 in members BF and FD respectively (Fig. 7.3 (b)).

(2) Analyse the reduced frame to get the p_0 actions (Table 7.1, column 3).

(3) Apply a unit pair corresponding to r_1 on the reduced frame and analyse (Fig. 7.3 (c)) to get the forces p_1 which form part of the B matrix (Table 7.1, column 4).

(4) Repeat for a unit pair corresponding to r_2, giving p_2 (Table 7.1, column 5).

(5) Form the unassembled stiffness matrix—

$$F_0 = \begin{bmatrix} f_1 & & & \\ & f_2 & & \\ & & f_3 & \\ & & & \text{etc.} \end{bmatrix}$$

Since in this example $L/AE = $ constant, we have

$$f_1 = f_2 = \text{etc.}$$

$$F_0 = \begin{bmatrix} 1 & & & \\ & 1 & & \\ & & 1 & \\ & & & \text{etc.} \end{bmatrix} L/AE$$

(6) Evaluate $B'F_0p_0$, where the matrix B is simply columns 4 and 5 of Table 7.1; since here F_0 contains only units on the diagonal (it is a unit matrix)

$$B'F_0 = B'(L/AE)$$

and

$$B'F_0p_0 = B'p_0(L/AE)$$

i.e.

$$\begin{bmatrix} -1/\sqrt{2} & -1/\sqrt{2} & 0 & -1/\sqrt{2} & 1 & 1 & -1/\sqrt{2} & 0 & 0 & 0 & 0 \\ 0 & -1/\sqrt{2} & -1/\sqrt{2} & 0 & 0 & 0 & 0 & -1/\sqrt{2} & 1 & 1 & -1/\sqrt{2} \end{bmatrix} \begin{bmatrix} 0 \\ 0 \\ 0 \\ 0 \\ 0 \\ -5\sqrt{2} \\ 5 \\ 0 \\ -5\sqrt{2} \\ 0 \\ 5 \end{bmatrix} L/AE$$

$$= \begin{bmatrix} -10 \cdot 60 \\ -10 \cdot 60 \end{bmatrix} L/AE$$

(7) Evaluate

$$B'F_0B = B'B(L/AE)$$

$$= \begin{bmatrix} -1/\sqrt{2} & -1/\sqrt{2} & 0 & -1/\sqrt{2} & 1 & 1 & -1/\sqrt{2} & 0 & 0 & 0 & 0 \\ 0 & -1/\sqrt{2} & -1/\sqrt{2} & 0 & 0 & 0 & 0 & -1/\sqrt{2} & 1 & 1 & -1/\sqrt{2} \end{bmatrix} \begin{bmatrix} -1/\sqrt{2} & 0 \\ -1/\sqrt{2} & -1/\sqrt{2} \\ 0 & -1/\sqrt{2} \\ -1/\sqrt{2} & 0 \\ 1 & 0 \\ 1 & 0 \\ -1/\sqrt{2} & 0 \\ 0 & -1/\sqrt{2} \\ 0 & 1 \\ 0 & 1 \\ 0 & -1/\sqrt{2} \end{bmatrix} L/AE$$

$$= \begin{bmatrix} 4 \cdot 0 & 0 \cdot 5 \\ 0 \cdot 5 & 4 \cdot 0 \end{bmatrix} L/AE$$

(8) Then

$$B'F_0p_0 + B'F_0Br = 0$$

becomes

$$\begin{bmatrix} -10 \cdot 60 \\ -10 \cdot 60 \end{bmatrix} + \begin{bmatrix} 4 & 0 \cdot 5 \\ 0 \cdot 5 & 4 \end{bmatrix} \begin{bmatrix} r_1 \\ r_2 \end{bmatrix} = 0$$

i.e.

$$d_0 + F \quad r = 0$$

(9) Invert F to get (p. 441)

$$F^{-1} = \begin{bmatrix} 0 \cdot 254 & -0 \cdot 032 \\ -0 \cdot 032 & 0 \cdot 254 \end{bmatrix}$$

then

$$r = -F^{-1}d_0$$

or

$$\begin{bmatrix} r_1 \\ r_2 \end{bmatrix} = \begin{bmatrix} 0 \cdot 254 & -0 \cdot 032 \\ -0 \cdot 032 & 0 \cdot 254 \end{bmatrix} \begin{bmatrix} +10 \cdot 60 \\ +10 \cdot 60 \end{bmatrix}$$

$$= \begin{bmatrix} +2 \cdot 356 \\ +2 \cdot 356 \end{bmatrix}$$

(10) For every member,

$$p = p_0 + r_1p_1 + r_2p_2$$

and all member forces can now be calculated.

Example 7.2. Braced Frames (orthodox hand calculation)
The same structure is more simply dealt with by setting up the

compatibility equations in full—

$$d_{10} + f_{11}r_1 + f_{12}r_2 = 0$$
$$d_{20} + f_{21}r_1 + f_{22}r_2 = 0$$

The deflexion calculation procedures of Chap. 5 can then be used directly, and the equations become

$$\sum p_1 p_0 L/AE + r_1 \sum p_1^2 L/AE + r_2 \sum p_1 p_2 L/AE = 0$$
$$\sum p_2 p_0 L/AE + r_1 \sum p_1 p_2 L/AE + r_2 \sum p_2^2 L/AE = 0$$

and since L/AE is here constant, the equations become

$$\sum p_1 p_0 + r_1 \sum p_1^2 + r_2 \sum p_1 p_2 = 0$$
$$\sum p_2 p_0 + r_1 \sum p_1 p_2 + r_2 \sum p_2^2 = 0$$

The results are then collected in Table 7.1. Summing the relevant

Table 7.1

1	2	3	4	5	6	7	8	9	10
Member	L/AE	p_0	p_1	p_2	$p_1 p_0$	$p_2 p_0$	p_1^2	p_2^2	$p_1 p_2$
BC	L/AE	0	$-1/\sqrt{2}$	0	0	0	0.5	0	0
CF	L/AE	0	$-1/\sqrt{2}$	$-1/\sqrt{2}$	0	0	0.5	0.5	0.5
DE	L/AE	0	0	$-1/\sqrt{2}$	0	0	0	0.5	0
AB	L/AE	0	$-1/\sqrt{2}$	0	0	0	0.5	0	0
BF	L/AE	0	1	0	0	0	1	0	0
AC	L/AE	$-5\sqrt{2}$	1	0	$-5\sqrt{2}$	0	1	0	0
AF	L/AE	5	$-1/\sqrt{2}$	0	$-5/\sqrt{2}$	0	0.5	0	0
CD	L/AE	0	0	$-1/\sqrt{2}$	0	0	0	0.5	0
CE	L/AE	$-5\sqrt{2}$	0	1	0	$-5\sqrt{2}$	0	1	0
DF	L/AE	0	0	1	0	0	0	1	0
EF	L/AE	5	0	$-1/\sqrt{2}$	0	$-5/\sqrt{2}$	0	0.5	0
									0
				Σ	-10.60	-10.60	4	4	0.5

columns, the equations are

$$-10.60 + 4r_1 + 0.5r_2 = 0$$
$$-10.60 + 0.5r_1 + 4r_2 = 0$$
$$r_1 = r_2 = +2.356$$

It is in this form that the calculations are best carried out for small problems.

7.6 Lack of Fit

The expression *lack of fit* is used to express any differences in the sizes of a member other than those due to elastic deformations. In a determinate structure, for example the braced frame of Fig. 7.5 (*a*), it is possible to fit the structure together without force even if the members are of incorrect lengths, whereas for the indeterminate frame of Fig. 7.5 (*b*) the addition of a second diagonal member adds a constraint to the geometry and it cannot be fitted without force unless it has exactly the correct length. Such lack of fit may be due to several causes, but will cause self-straining of the structure independent of external loading. The effects of lack of fit due to temperature changes, etc., in the deflexions of determinate structures

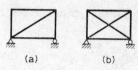

(a) (b)

Fig. 7.5

has already been discussed in Chap. 5 ; these effects can readily be included in the deflexion calculations as follows. From p. 192 the structure deformations are

$$\begin{bmatrix} d \\ d_r \end{bmatrix} = \begin{bmatrix} B'_0 \\ B' \end{bmatrix} [u]$$

and $[u]$ now becomes element deformations due to lack of fit as well as to the stress resultants, e.g.

$$\begin{bmatrix} d \\ d_r \end{bmatrix} = \begin{bmatrix} B'_0 \\ B' \end{bmatrix} [u + \lambda]$$

where $[\lambda]$ is described in Table 7.2.

Table 7.2

Elements of B	Lack of fit λ
Axial forces	Changes in length, e.g. temp. chg. t giving $\lambda = clt$
Bending moments	Changes in slope, e.g. temp. gradt t/h giving $\delta\lambda = (2ct/h)\,\delta s$ (h = beam depth)
Torques	Changes in angle of twist
Shears	Shear deformations

Proceeding as on p. 192 eqn. 7.2 now becomes

$$\begin{bmatrix} d \\ d_r \end{bmatrix} = \begin{bmatrix} f_{11} & f_{12} \\ f_{21} & f_{22} \end{bmatrix} \begin{bmatrix} W \\ r \end{bmatrix} + \begin{bmatrix} B'_0 B \\ B' \end{bmatrix} [\lambda]$$

and eqn. 7.3 becomes

$$f_{21}W + f_{22}r + B'\lambda = 0$$

the new compatibility condition from which the releases $[r]$ can be evaluated.

In non-matrix form for a doubly-redundant braced frame the lack of fit modifies the compatibility equations of p. 196 to become

$$\sum \frac{p_1 p_0 L}{AE} + r_1 \sum \frac{p_1^2 L}{AE} + r_2 \sum \frac{p_1 p_2 L}{AE} + \sum p_1 \lambda = 0$$

$$\sum \frac{p_2 p_0 L}{AE} + r_1 \sum \frac{p_1 p_2 L}{AE} + r_2 \sum \frac{p_2^2 L}{AE} + \sum p_2 \lambda = 0 \qquad (7.3)$$

Equations for other types of structure follow by analogy.

Example 7.3

The treatment of lack of fit is shown by the example in Fig. 7.6. A roof truss is shown pinned at A to a rigid abutment but pin-connected to the top of a flexible column at D. If the column cap spreads

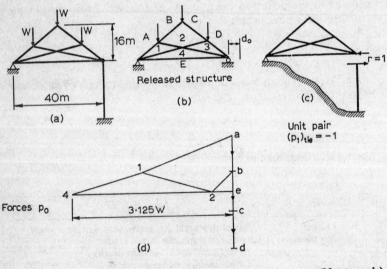

Area of all ties = 2×10^3 mm²   (Not to scale)
$E = 2 \times 10^5$ N/mm²

Fig. 7.6

outwards 1·5 mm/kN it is required to determine the permissible loading such that the movement at the column cap does not exceed 2·5 mm. The dimensions and the cross-sectional areas of all members are given in Fig. 7.6.

SOLUTION

Consider the horizontal reaction between the column and the truss to be the redundant action, allow the release as in Fig. 7.6 (*b*), and analyse the reduced frame for the p_0 forces.

Apply the unit pair at the release as in Fig. 7.6 (*c*) and obtain the p_1 forces.

The spread of the column cap can be considered to be a change in length of the foundation or as a change of length of an imaginary element between the roof truss and column cap. Considering the former the stress resultants in this member must be included in the analysis.

$$[p_0] = \begin{bmatrix} ? \\ ? \\ ? \\ ? \\ ? \\ ?? \\ ? \\ +3{\cdot}124 \\ -3{\cdot}125 \end{bmatrix} [W] \quad \text{i.e. } p = B_0 W$$

$$p_1 = \begin{bmatrix} 0 \\ 0 \\ 0 \\ 0 \\ 0 \\ 0 \\ -1 \\ +1 \end{bmatrix} [r] \quad \text{i.e. } p = Br$$

Then from eqn 7.1

$$f_{21}W + f_{22}r + B'\lambda = 0$$
$$f_{21}B'fB_0 = -3{\cdot}125L/AE$$
$$f_{22} = B'fB = L/AE$$

also $\quad \lambda = +1{\cdot}5r$ (an extension of the foundation)

so that $\quad B'\lambda = +1{\cdot}5$

i.e. $\quad -3{\cdot}125L/AE \, W + r \, L/AE = 1{\cdot}5r = 0$

$$r = 0{\cdot}196W.$$

For a total yield of 2·5 mm

$$r = 2{\cdot}5/1{\cdot}5 = 1{\cdot}67 \text{ kN}$$
$$W = 1{\cdot}67/0{\cdot}196 = 8{\cdot}53 \text{ kN}.$$

The unit pair stresses only one member, the bottom tie AB, so that p_1 is zero for all other members and this is the only p_0 force required and the arithmetic is simple. In this example a convenient way in which to analyse the reduced structure is by a force diagram (Fig. 7.6 (d)). Note also the description of the three loads by a single parameter, a scalar, W. This is an example of a 'generalized load'.

Example 7.4

A more orthodox example is shown in Fig. 7.7, where the structure has one redundant member. The figure shows the release chosen and the reduced frame so that the member axial forces from (b) and (c) are the elements of the B_0 and B matrices respectively and the matrix

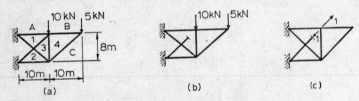

Release 1–3 (r_1)   Unit pair $r_1 = 1$
Stress resultants p_0 Stress resultants p_1
Displacement d_{10} Displacement f_{11}
Compatibility $d_{10} + f_{11}r_1 = 0$, $d_{10} = \Sigma\, p_1 p_0 L/AE$, $f_{11} = \Sigma\, p_1^2 L/AE$.

Member	L	p_0	p_1	$p_1 p_0 L$	$p_1^2 L$	$r_1 p_1$	$p = p_0 + r_1 p_1$
A-1	10	+6·25	+0·78	48·80	6·10	+9·60	+15·85
3-4	8	−10·00	+0·62	−50·00	3·13	+7·69	−2·31
2-C	10	−25·00	+0·78	−195·0	6·10	+9·60	−15·40
1-2	12·8	24·00	−1·00	−307·0	12·80	−12·30	+11·70
1-3 (r_1)	12·8	—	−1·00	—	12·80	−12·30	−12·30
AE (all members) = constant Σ				−503·2	40·93		

$$-503 \cdot 2/AE + r_1(40 \cdot 93/AE) = 0$$

so $r_1 = +\,12 \cdot 3$ kN (1–3 in compression)

Final member forces $= p = p_0 + r_1 p_1$
To find the adjustment at 1–3 for $r_1 = 8$ kN,

$$d_{10} = \Sigma\, p_1 p_0 L/AE + \Sigma\, p_1 \lambda$$

i.e. $\dfrac{-503 \cdot 2}{EA} + (-1\lambda) + (40 \cdot 93/EA)(8) = 0$ *(watch the units)*

so $\lambda = -175 \cdot 7/EA$ m

Since tensile forces are positive, λ is a shortening.

Fig. 7.7

operations of eqn 7.1 are readily carried out. In the case of such a small structure the matrix operations are clumsy and the orthodox, tabular, form of hand calculation is shown in Fig. 7.7.

Considering 1–3 as the redundant, the release is made by cutting 1–3, and the p_0 forces are calculated and tabulated. The unit pair is applied as shown, i.e. assuming 1–3 is in compression, and the p_1 forces are tabulated. The calculations of $(p_1 p_0 L)$ and $(p_1^2 L)$ are carried out using the length in metre units and the units corrected (not necessary in this case) in the final evaluation to give $r_1 = 12\cdot3$ kN, positive, therefore, in the assumed direction.

If the compressive force in 1–3 is restricted to 8 kN the initial length can be adjusted, i.e. rewriting the compatibility equation to include the effect of lack of fit (λ) of 1–3 only gives

$$\lambda = -175\cdot7/EA \text{ m}$$

The negative sign means that the member 1–3 is in compression and so it undergoes a shortening.

7.7 Bending Deformations

The foregoing is completely general in application and the assembly of a structure flexibility matrix is dealt with in exactly the same way. The difference being that the element flexibility matrix from p. 172 and omitting the change in length is now a 2 × 2, e.g.

$$\begin{bmatrix} u_2 \\ u_6 \end{bmatrix} = \begin{bmatrix} L^2/3 & L/2 \\ L/2 & 1 \end{bmatrix} \frac{L}{EI} \begin{bmatrix} p_2 \\ p_6 \end{bmatrix} \qquad (7.4)$$

where p_2 is the end shear and p_6 the end bending moment.

These are still related to loads as before by

$$[p] = [B_0][w]$$

the only differences being the larger number of terms in $[p]$ and $[B_0]$.

7.8 Beam Element Flexibility

On page 172 the flexibility was defined in terms of the end deformations of a cantilver and alternative formulation is shown in Fig. 7.8.

In this case, using say the procedure as in 5.16 (page 158)

$$f_{11} = L^2/3EI = f_{22}$$
$$f_{21} = L^2/6EI = f_{12}$$

so that the flexibility matrix becomes

$$\begin{bmatrix} u_1 \\ u_2 \end{bmatrix} = \begin{bmatrix} 2 & 1 \\ 1 & 2 \end{bmatrix} L^2/6EI \begin{bmatrix} p_1 \\ p_2 \end{bmatrix}$$

It should now be clearly recognized that there is no such thing as a unique structure or element flexibility matrix, its form depends on the deformations—degrees of freedom—defined. The requirements follow from the method of calculating the flexibility coefficients, i.e. apply an action corresponding to one of the degrees of freedom only, actions at all other freedoms to be zero.

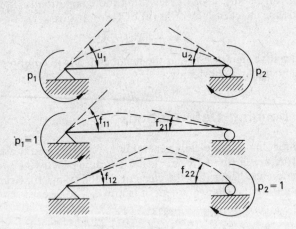

Fig. 7.8

Thus the freedoms defined must and can only correspond to independent force actions and in the case of a beam element as in Fig. 7.8 there are only four stress resultants possible, viz. moment and shear at each end. Since there are two equations of equilibrium there then remain two independent forces so that the flexibility matrix must be 2×2 matrix but the form may be as either on page 172 or as above.

The choice is one of convenience only—either may be used.

Example 7.5

The following example demonstrates the application of the procedures of eqn 7.2 on page 192 to bending elements. Consider the fixed ended beam in Fig 7.9 (*a*) to consist of two parts between the nodes or joints numbered 1 to 3.

Define the element flexibilities as in Fig. 7.9 (*b*), i.e.

element 1–2
$$\begin{bmatrix} u_1 \\ u_2 \end{bmatrix} = \begin{bmatrix} 2 & 1 \\ 1 & 2 \end{bmatrix} \frac{(L/2)^2}{6EI} \begin{bmatrix} p_1 \\ p_2 \end{bmatrix}$$

element 2–3
$$\begin{bmatrix} u_3 \\ u_4 \end{bmatrix} = \begin{bmatrix} 2 & 1 \\ 1 & 2 \end{bmatrix} \frac{(L/2)^2}{6EI} \begin{bmatrix} p_3 \\ p_4 \end{bmatrix}$$

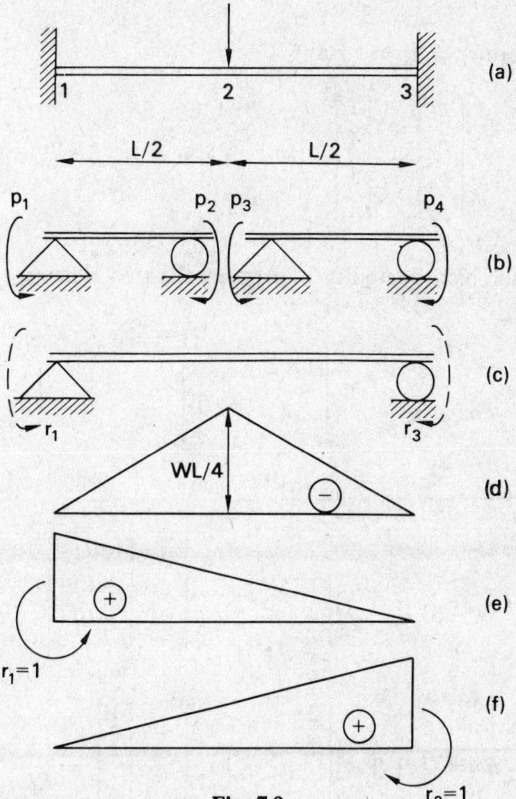

Fig. 7.9

Define the releases at the end moment r_1 and r_3, which, due to symmetry in this case are equal, so that the reduced structure and the bending moment diagram are as in Fig. 7.9 (*c*) and (*d*).
Clearly internal actions [*p*] are then

$$\begin{bmatrix} p_1 \\ p_2 \\ p_3 \\ p_4 \end{bmatrix} = \begin{bmatrix} 0 \\ -L/4 \\ -L/4 \\ 0 \end{bmatrix} W \qquad \text{i.e. } [p] = [B_0][W]$$

Due to the releases the bending moment diagrams as in Fig. 7.9 (e) and (f) and the stress resultants are

$$\begin{bmatrix} p_1 \\ p_2 \\ p_3 \\ p_4 \end{bmatrix} = \begin{bmatrix} 1 & 0 \\ \frac{1}{2} & \frac{1}{2} \\ \frac{1}{2} & \frac{1}{2} \\ 0 & 1 \end{bmatrix} \begin{bmatrix} r_1 \\ r_3 \end{bmatrix} \quad \text{i.e. } p = Br$$

or since here $r_1 = r_3 = r$ (say)

$$\begin{bmatrix} p_1 \\ p_2 \\ p_3 \\ p_4 \end{bmatrix} = \begin{bmatrix} 1 \\ 1 \\ 1 \\ 1 \end{bmatrix} [r] \quad \text{i.e. } p = Br$$

The unassembled flexibility matrix for the two element structure is

$$f = \begin{bmatrix} 2 & 1 & & \\ 1 & 2 & & \\ & & 2 & 1 \\ & & 1 & 2 \end{bmatrix} \frac{(L/2)^2}{6EI}$$

and from eqn (7.2) on page 192,

$$f_{21} = B'fB_0 = [1\ 1\ 1\ 1] \begin{bmatrix} 2 & 1 & & \\ 1 & 2 & & \\ & & 2 & 1 \\ & & 1 & 2 \end{bmatrix} \frac{L^2}{24EI} \begin{bmatrix} 0 \\ -L/4 \\ -L/4 \\ 0 \end{bmatrix} = \frac{L^3}{8EI} = f_{12}$$

$$f_{22} = B'fB = [1\ 1\ 1\ 1] \begin{bmatrix} 2 & 1 & & \\ 1 & 2 & & \\ & & 2 & 1 \\ & & 1 & 2 \end{bmatrix} \frac{L^2}{24EI} \begin{bmatrix} 1 \\ 1 \\ 1 \\ 1 \end{bmatrix} = \frac{L^2}{EI}$$

i.e. $\qquad \left(\dfrac{L^3}{8EI}\right) W + \dfrac{L^2}{EI}(r) = 0 \qquad$ (eqn 7.3 page 192)

or $\qquad\qquad\qquad\qquad r = \dfrac{WL}{8}$

which is the well known standard result. The reader should now recalculate r using the alternative flexibility matrix form.

7.9 Loads not at Joints

On page 175 where the principle of contragredience is developed the basic virtual work expression is

$$[W]'[d] = [p]'[u]$$

i.e. $[p]'[u]$ must evaluate the total internal virtual work done as the structure loads displace through the displacements $[d]$.

In the case of elements with loads carried internally, e.g. Fig. 7.9 the stress resultants do not vary uniformly from end to end of an element so that defining p's and u's at the ends only is insufficient to define the internal virtual work. The foregoing therefore applies only to structures loaded at joints or defined node points at which element flexibilities are defined. Where loads are not at nodes it is necessary to superpose solutions.

Example 7.6 (Bending flexibility)

The two-panel vierendeel girder shown in Fig. 7.10 is an example of the large group of structures whose bending flexibilities predominate, that is to say the effects of axial and shear strains on the force actions is negligible.

Members BA are uniformly loaded, i.e. loading is not solely at joints, it is necessary to proceed as follows, Fig. 7.10 (b). Completely fix joints A and B, i.e. remove all degrees of freedom at these joints and members BA become fixed ended beams and require end moments and shears to be supplied by the supports. These actions can be calculated by the methods of Chapter 4, e.g. Table 4.2 on page 120.

i.e. M_B^F = Fixing moment at joint B

$$= \frac{wL^2}{12} = \frac{1 \times 10^2}{12} = 8 \cdot 33 \text{ kN m}.$$

M_A^F = Fixing moment at joint A

$\quad$ = 0, since M^F balance at each side of the joint or, in general

M^F = Σ Fixed end moments for all members meeting at the joint.

also: $$Q_B = \frac{wL}{2} = 5 \text{ kN}$$

$$Q_A = 2 \times \frac{wL}{2} = 10 \text{ kN}$$

Actions M_B^F, M_A^F and Q_A are zero for the real structure, i.e., there are no restraints corresponding to these forces. Thus the equal and opposite actions M_B and W_A in Fig. 7.10 (c) must be added.

The structure and loading of Fig. 7.10 (c) is now analysed (loaded at the joints) and the two solutions superposed.

The structure forms two closed rings, so that cutting each ring (say in the end verticals) forms a branched structure of statically-determinate cantilevers. Each cut releases three actions, a moment, a thrust and a shear, so that six releases have been made and the structure has six redundants. Due to the symmetry of the structure and the loading all actions are symmetrical and the left-hand releases will be the same as the right-hand set. Also, due to symmetry, there will be no rotation at the centre joints, which can thus be considered fixed and only one half of the structure need be analysed, thereby reducing the redundants to three. Choose the moment, thrust and shear at the end of BC as the redundants (r_1, r_2 and r_3). Three unit actions are then applied separately and the four loading cases to be analysed are as shown.

Define element flexibilities and internal stress resultants as shown, then set up the force transformations

$$[p] = [B_0 \mid B] \begin{bmatrix} W \\ \hline r \end{bmatrix}$$

This is shown in Fig. 7.10 where the elements of the B_0 and B matrices are simply the bending moment values at the element ends taken from the bending moment diagrams shown.

The remaining calculations set out in Fig. 7.10 follow the procedure already established and should be verified by the reader. One point requiring firm understanding is the necessity to have a clear definition of the element stress resultants used to define the element flexibility matrices. This is open to some choice and the reader should re-analyse this structure using alternative flexibility matrices.

Example 7.7 (Direct and Bending stresses)

The two-span trussed beam of Fig. 7.11 is a simple example of a structure where both axial forces and moments must be considered. The beam ABC has a large cross-sectional area, so that axial changes in length are negligible, but while the ties AD, etc. have no bending, they have a small cross-sectional area and their extensions cannot be neglected.

Inspection of the structure for the degree of redundancy shows that if both the props are cut the two-span continuous beam ABC remains and this can be reduced to a determinate structure by one

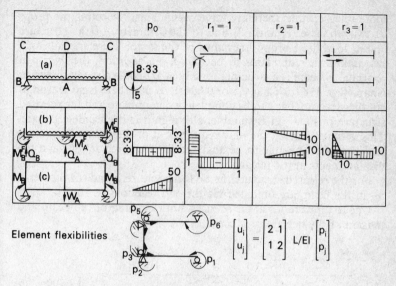

Element flexibilities

$$\begin{bmatrix} u_i \\ u_j \end{bmatrix} = \begin{bmatrix} 2 & 1 \\ 1 & 2 \end{bmatrix} L/EI \begin{bmatrix} p_i \\ p_j \end{bmatrix}$$

$$\begin{bmatrix} p_1 \\ p_2 \\ p_3 \\ p_4 \\ p_5 \\ p_6 \end{bmatrix} = \begin{bmatrix} 50 & 8 \cdot 33 & -1 & 10 & 10 \\ 0 & 8 \cdot 33 & -1 & 0 & 10 \\ 0 & 0 & -1 & 0 & 10 \\ 0 & 0 & -1 & 0 & 0 \\ 0 & 0 & 0 & 0 & 0 \\ 0 & 0 & 0 & 10 & 0 \end{bmatrix} \begin{bmatrix} Q_A \\ M_B \\ r_1 \\ r_2 \\ r_3 \end{bmatrix}$$

f_{21} (eqn 7.2) $= B'fB_0 =$

$$\begin{bmatrix} -1 & -1 & -1 & -1 & 0 & 0 \\ 10 & 0 & 0 & 0 & 0 & 10 \\ 10 & 10 & 10 & 0 & 0 & 0 \end{bmatrix} \begin{bmatrix} 2 & 1 & & & & \\ 1 & 2 & & & & \\ & & 2 & 1 & & \\ & & 1 & 2 & & \\ & & & & 2 & 1 \\ & & & & 1 & 2 \end{bmatrix} \begin{bmatrix} 50 & 8 \cdot 33 \\ 0 & 8 \cdot 33 \\ 0 & 0 \\ 0 & 0 \\ 0 & 0 \\ 0 & 0 \end{bmatrix} L/EI$$

$$= \begin{bmatrix} -150 & -150 \\ 1000 & 250 \\ 1500 & 500 \end{bmatrix}$$

f_{22} (eqn 7.2) $= B'fB = \begin{bmatrix} 12 & -30 & -90 \\ -30 & 200 & 300 \\ -90 & 300 & 800 \end{bmatrix}$

Hence from (eqn 7.3)

$$\begin{bmatrix} r_1 \\ r_2 \\ r_3 \end{bmatrix} = \begin{bmatrix} 240 & 3 \cdot 4 \\ & 55 \cdot 6 \\ & -317 \cdot 7 \end{bmatrix}$$

Fig. 7.10

more release: thus there are three redundants. Selecting the props as two of these (r_1 and r_2), due to the symmetry of the structure and its load these prop forces must be the same: there are thus only two unknown redundants to be found. In addition, the two-span continuous beam is a structure readily analysed as already shown in Chap. 4, p. 118 (or Chaps. 8 or 9 later), so that there is no reason to cut the structure back to be completely determinate and the released structure of Fig. 7.11 is suitable, although it is itself indeterminate. This is quite general, the only requirement of the reduced structure is that it must be able to be analysed for the force actions and the displacements at the releases.

For the released structure the tie forces are zero and the moments m_0 in the beam are obtained via the end reactions as shown.

Due to unit actions at the releases, both bending of the beam and tension of the ties are produced.

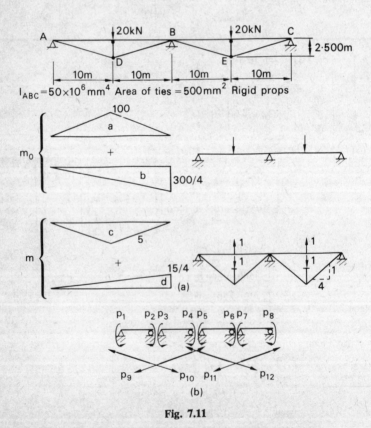

Fig. 7.11

In addition, since the prop forces are equal $(r_1 = r_2)$, it is convenient to retain symmetry by applying two unit actions simultaneously.

This means that a new redundant called a *generalized force* is defined. This consists of the pair of releases r_1 and r_2 together and the corresponding stress resultant system is as shown.

Defining element stress resultants as in Fig. 7.11 (*b*) the force transformation

$$[p] = [B_0 B] \begin{bmatrix} W \\ \overline{r} \end{bmatrix}$$

is then

$$
\begin{bmatrix} p_1 \\ p_2 \\ p_3 \\ p_4 \\ p_5 \\ p_6 \\ p_7 \\ p_8 \\ p_9 \\ p_{10} \\ p_{11} \\ p_{12} \end{bmatrix}
=
\begin{bmatrix}
0 & 0 \\
-62{\cdot}5 & 3{\cdot}125 \\
-62{\cdot}5 & 3{\cdot}125 \\
75 & -3{\cdot}75 \\
75 & -3{\cdot}75 \\
-62{\cdot}5 & 3{\cdot}125 \\
-62{\cdot}5 & 3{\cdot}125 \\
0 & 0 \\
0 & 2{\cdot}06 \\
0 & 2{\cdot}06 \\
0 & 2{\cdot}06 \\
0 & 2{\cdot}06
\end{bmatrix}
\begin{bmatrix} W \\ r \end{bmatrix}
$$

and the unassembled element flexibility matrix is

$$
[f] =
\begin{bmatrix}
2 & 1 \\
1 & 2 \\
& & 2 & 1 \\
& & 1 & 2 \\
& & & & 2 & 1 \\
& & & & 1 & 2 \\
& & & & & & \cdot618 \\
& & & & & & & \cdot618 \\
& & & & & & & & \cdot618 \\
& & & & & & & & & \cdot618
\end{bmatrix}
\frac{L^2}{6EF}
$$

Hence from eqn 7.2 and 7.3

$$r = 17{\cdot}8 \text{ kN}$$

The reader should verify this by carrying out the matrix operations and repeating using an alternative flexibility matrix for the bending elements.

Symmetry in this case, as in example 7.6, would have allowed only half of the structure to be analysed, i.e. zero rotation at B. The full structure has been retained for the example to include the ideas of an indeterminate 'released' structure, a 'generalized' release system and loading system. Note also the omission of the axial stress resultants from the flexibilities of the beams and the props, changes in length of the former being neglected and the latter being rigid have no change in length.

7.10 Mixed Release Systems

It has already been pointed out that there are alternative possible release systems and that an intelligent choice helps to simplify the arithmetical work, while a bad choice produces ill-conditioned equations and leads to inaccurate solutions.

If the selected release system is

$$r = \{r_1 \, r_2 \cdots\}$$

and an alternative system is

$$q = \{q_1 \, q_2 \cdots\}$$

then by statics we can get q from r; let $q = Hr$ (as p. 240).

One check on the correctness of the alternative is that the reverse must be possible, i.e. it must be possible to obtain r from q by statics,

$$r = H^{-1}q$$

Therefore $\qquad\qquad |H| \neq 0$

The compatibility equations already used are

$$d_0 + Fr = 0$$
$$B'F_0p_0 + B'F_0Br = 0 \qquad (7.5)$$

Then if the q-system is used they would have a different form, e.g.

$$p = Aq \text{ instead of } p = Br$$

or $\qquad\qquad p = AHr$

Thus $\qquad\qquad B = AH$

and $\qquad\qquad B' = (AH)' = H'A' \quad \text{(reversal rule)}$

Using this in eqn. (7.5),

$$H'A'F_0p_0 + H'A'F_0AHr = 0$$
$$A'F_0p_0 + A'F_0Aq = 0$$

The flexibility matrix $F = A'F_0A$ is now that for the q-system, but the load term—stress resultants p_0—in the reduced structure are for the r-system of releases. Thus

The stress resultants due to loads on the reduced structure may be calculated for any combination of release systems and none of these need necessarily be used for the flexibility matrix corresponding to the selected redundants.

The final actions are then

$$p = p_0 + Aq$$

Fig. 7.12 shows an example of how this can simplify the analysis. The cut at the top of the left-hand column allows this column to carry all the horizontal load as a simple cantilever and gives a simple bending moment diagram. The different release system of the right-hand sketch then allows the vertical loading to be carried on a simply-supported beam, which again gives a simple bending moment diagram.

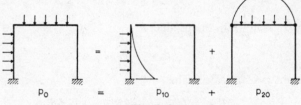

$$p_0 \qquad = \qquad p_{10} \qquad + \qquad p_{20}$$

Fig. 7.12

7.11 Scaling

It will sometimes be found that the numerical values of the B-matrix vary widely, from very small to very large. This *ill-conditioning* leads to inconvenient arithmetic and loss of accuracy so that it may be advantageous to scale the values.

(*a*) Writing in full the transformation,

$$p = Br$$

gives

$$\begin{bmatrix} p_1 \\ p_2 \\ p_3 \end{bmatrix} = \begin{bmatrix} p_{11} & p_{12} & p_{13} & \cdots \\ p_{21} & p_{22} & \cdot & \cdots \\ \cdot & \cdot & \cdot & \cdots \end{bmatrix} \begin{bmatrix} r_1 \\ r_2 \\ r_3 \end{bmatrix}$$

and if scaled values of the releases $\bar{r}_j = S_j r_j$ are used, each of the elements p_{ij} is multiplied by the scale factor S_j, and a matrix of the scaled elements $= \bar{p}_{ij} = S_j p_{ij}$ results, i.e.

$$\bar{B} = B[S]^D$$

i.e.

$$p = BS\bar{r}$$

where $[S]^D$ is a diagonal matrix of the scale factors. Also

$$\bar{B}' = (BS)' = S'B' \quad \text{by the reversal rule}$$
$$= SB'$$

(since S is a diagonal matrix, $S = S'$).

(b) Using scaled values in the matrix product

$$d = B'F_0 Br$$

gives a scaled value

$$\bar{d} = S'B'F_0 BSr$$
$$= (S'FS)r$$
$$= \bar{F}r$$

A scaled flexibility matrix is then

$$\bar{F} = S'FS = SFS$$

(c) Also, using $\bar{B}$ in the product

$$d_0 = B'F_0 p_0$$

again gives a scaled value

$$\bar{d}_0 = S'B'F_0 p_0 = S(B'F_0 p_0) = Sd_0$$

(d) Using these same scale factors in the compatibility equation

$$d_0 + Fr = 0$$

gives

$$Sd_0 + (SF)r = 0$$

and by manipulation as follows—

$$Sd_0 + (SFS)S^{-1}r = 0$$

since $SS^{-1} = 1$ (unit matrix). Therefore

$$d_0 + \bar{F}\bar{r} = 0$$

where there are now scaled values of the releases

$$\bar{r} = S^{-1}r$$

or the true values are $\quad r = S\bar{r}$

(e) Final internal actions are

$$p = p_0 + Br$$

and, using scaled values,

$$\bar{p} = p_0 + \bar{B}\bar{r}$$
$$= p_0 + BSS^{-1}r = p_0 + Br = p$$

i.e. use of scaled values $\bar{B}$ and $\bar{r}$ gives the final actions correctly

7.12 Deflexions of Indeterminate Structures

This problem has already been dealt with on page 194, eqn 7.2. It is further discussed to demonstrate the 'mixing' of release systems and the calculations are carried out by the 'hand' methods of 5.12 and 5.13.

Consider the symmetrical rigid frame of Fig. 7.13 (a), where it is required to calculate the rotational displacement of joint C. Assume that the frame has been analysed for the internal forces by forming first of all the reduced frame Fig. 7.13 (b) to give actions p_0 (or m_0) and that the release system chosen for the set of redundants is that in

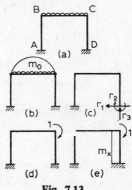

Fig. 7.13

Fig. 7.13 (c) giving $p_r = Br$. Note here that the reduced frame used for the p_0 actions (Fig. 7.13 (b)) is not actually statically-determinate for any other type of loading.

Following the virtual work procedure to obtain the rotation at C, apply a unit moment to the joint and choose any convenient reduced structure, say Fig. 17.13 (d) and (e). Thus, for a set of displacements apply a set of unit forces $X = \{x_1 \; x_2 \; x_3\}$ and the internal actions p_x in any reduced structure are formed by statics to follow the transformation $p = CX$.

Now, since the ps are stress resultants and the Xs external forces forming two statically equivalent systems, by contragredience the displacements corresponding to X, i.e. the structure displacements d, and those corresponding to p, i.e. the internal deformations u, are connected by

$$d = C'u$$

If
$$\begin{aligned} u &= \text{total member deformations} \\ &= f(\text{total member forces}) \\ &= f(p_0 + Br) \end{aligned}$$

then
$$d = C'fp_0 + C'fBr \qquad (7.6)$$

the important feature being that any arbitrary reduced frame can be used for the set of unit actions giving transformation C; it may be different from that used for the p_0 actions and from that used to define the redundants r and matrix B.

Equation (7.6) expands from the matrix form in the same way as the compatibility equations to give

$$\begin{aligned} d &= d_{x0} + d_x \\ &= \text{deflexions at } x \text{ due to the } p_0 \text{ actions} \\ &\quad + \text{deflexions at } x \text{ due to the } r \text{ releases} \end{aligned}$$

and

$$d_a = \int \frac{m_a m_0}{EF}\, ds + r_1 \int \frac{m_a m_1}{EF}\, ds + r_2 \int \frac{m_a m_2}{EF}\, ds$$

$$d_b = \int \frac{m_b m_0}{EI}\, ds + r_1 \int \frac{m_b m_1}{EF}\, ds + r_2 \int \frac{m_0 m_2}{EF}\, ds$$

etc.

where m_a, m_b, etc. are the elements of the matrix C; they are the values of force actions (in this case moment) throughout the structure due to the set of forces X used to determine the displacements.

Example 7.8. Deflexion of Indeterminate Structures

The two-pinned portal frame shown in Fig. 7.14 carries a single concentrated load on the beam and it is desired to find the sway, i.e. the horizontal movement at B, for unyielding supports and for the case of one support which spreads horizontally.

In the case of this simple structure there is little real advantage to be gained by mixing release systems. The simplest released structure for the p_0 forces is that obtained by releasing the horizontal reaction and this is also the simplest system of redundants. However, in order to demonstrate the mixed-system idea a different released structure is used for the unit action corresponding to the required displacement, the X system, i.e. horizontal unit load at B. This release system, a hinge at the mid-point of the beam BC, gives a skew-symmetrical moment system

$$m_x = C \cdot X$$

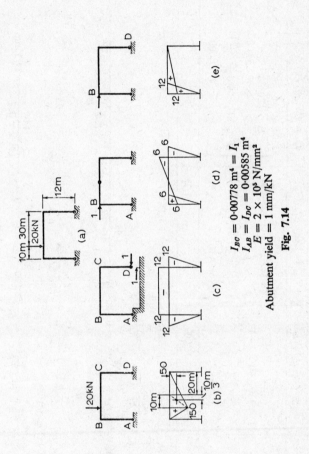

$I_{BC} = 0.00778 \text{ m}^4 = I_1$
$I_{AB} = I_{DC} = 0.00585 \text{ m}^4$
$E = 2 \times 10^5 \text{ N/mm}^2$
Abutment yield = 1 mm/kN

Fig. 7.14

$\dfrac{1}{EI}\int mm\,\dfrac{I_1}{I}\,ds$	Member	I_1/I	$d_{10} = \int (m_1 m_0 I/EI)\,ds$	$f_{11} = \int (m_1^2\,EI)\,ds$	$d_{B0} = \int (m_B m_0\,EI)\,ds$	d_{B1} (Fig. (e))
	AB	4/3	$1\times150\times40\times(-12)$ $=-36,000$	$(-12)\dfrac{12}{2}\cdot\dfrac{2}{3}(-12)[4/3]$ $=768$		$\dfrac{12\times12}{2}\cdot\dfrac{2}{3}\cdot(-12)\dfrac{4}{3}$ $=-768$
	BC	1		$(-12)40(-12)[1] = 5,760$	$\left.\begin{array}{l}\dfrac{150\times40}{2}\left(\dfrac{6}{20}\cdot\dfrac{10}{3}\right)\\=3,000\\ \text{(Fig. (e))}\ 3,000\left(\dfrac{12}{40}\cdot\dfrac{0}{3}\right)\end{array}\right\}=21,000$	$\dfrac{12\times40}{2}(-12)$ $=-2,880$
	CD	4/3		(as for AB) $=768$		
		Σ	$-36,000$	$7,296$		$-3,648$
	(kN-m^3)		$-36,000$	$7,296$		

No yield: $d_{10} + f_{11}r_1 = 0$, i.e. $-36,000 + 7,296r_1 = 0$ so $r_1 = +4.95$ kN

Effect of yield:
$$\left[\sum\int\frac{m_1 m_0}{EI}\,ds + \sum(p\lambda)\right] + f_{11}r_1 = 0$$

i.e. $\dfrac{-36,000}{EI_1} + (1)(10^{-3}\times r_1) + \dfrac{7,296}{EI_1}r_1 = 0$ so $r_1 = \dfrac{4.95}{1+\dfrac{10^{-3}EI_1}{7,296}} = \dfrac{4.95}{1+\dfrac{10^{-3}\times2\times10^8\times0.00778}{7,296}} = \dfrac{4.95}{1.213} = 4.08$ kN

Deflexion at B (no yield):

$d = C'F_0 p_0 + C'F_0 Br = \int\dfrac{m_b m_0}{EI}\,ds + r_1\int\dfrac{m_b m_1}{EI}\,ds = \dfrac{3,000\times10^3}{2\times10^8\times0.00778} = 1.93$ mm

Deflexion at B (yield at D) (Fig. (e)):

$d = \int\dfrac{m_e m_0}{EI}\,ds + r_1\int\dfrac{m_b m_1}{EI}\,ds = \dfrac{1}{EI}(21,000 - 3,648(4.08)) = \dfrac{21 - 14.9}{2\times10^2\times0.00778} = 3.93$ mm

Fig. 7.14 (contd.)

where the elements of the matrix C are the values of bending moment at each point in the structure.

The initial analysis of the value of H, the horizontal reaction, for the yield and no-yield cases follows previous examples; as for the compatibility equation, the deflexion equation expanded from the matrix form becomes

$$d = \int m_x \frac{m_0}{EI} \, ds + r \int \frac{m_x m_1}{EI} \, ds$$

and the integrals of the products can be evaluated in the usual way using Table 5.3, p. 156, or as in Fig. 7.14. The integral $\int m_x m_1 \, ds$ is zero here due to the antisymmetry of m_x and the symmetry of m_1. This means, of course, that the value of the support reaction has no effect on the absolute displacement of B.

The effect on this of yield of the support is interesting; it is not reflected simply in the effect of the changed value of m_1 since this does not appear in the deflexion calculation above and yet if only one abutment moves it is intuitively obvious that point B must also move. This means that the choice of a released structure for the X-system of unit loads is not unrestricted; if displacements are to be imposed the X-system must possess the necessary degree of freedom and then it is necessary to go back to the released structure of Fig. 7.14 (*c*), shown again in (*e*), where support D is free to move horizontally.

7.13 Examples for Practice

($E = 2 \times 10^5$ N/mm², C.L.E. $= 12 \times 10^{-6}$ per °C)

1. The given frame is simply-supported and loaded with a single 240 kN load as shown in Fig. 7.15. Compare the stresses in the two diagonals. All members are 2×10^3 mm² in area. Assume pin joints.

   What would be the effect on these stresses of a rise in temperature of 20 degC in member EF relative to the other members?

   (*Ans.* $p_{EC} = 56.5$ kN; with temperature, $p_{EC} = 42.4$ kN.)

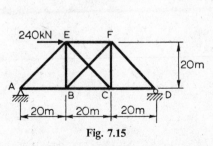

Fig. 7.15

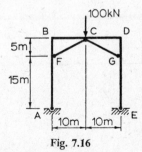

Fig. 7.16

2. A knee-braced portal ABCDE (Fig. 7.16) supports a vertical load of 100 kN placed at C on the continuous horizontal member BCD.

220 *Flexibility Method*

Considering bending effects only, determine the horizontal thrust at A, the load in the member CF and the bending moment at C.
(*Ans.* $H = 0.8$ kN.)

3. A two-hinged portal frame of span L and height h carries a U.D.L. of w per unit length on the beam.

   If one support yields horizontally by an amount equal to k times the horizontal thrust, show that

$$H\left(k + \frac{\frac{2}{3}h^3 + h^2L}{EI}\right) = \frac{hwL^3}{12EI}$$

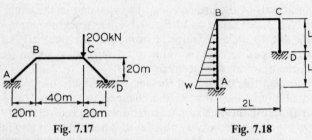

Fig. 7.17   Fig. 7.18

4. Calculate the horizontal reaction at D for the structures shown in Figs. 7.17 and 7.18.
   (*Ans.* $H = 10$ kN; $H = (104/345)wL$.)

8

Stiffness Method

8.1 Introduction

The essential parts of this chapter for a first reading are the development and applications of the slope-deflexion equations, since these are used as the basis of the moment-distribution procedure in Chapter 9. Treatment is therefore restricted to the large group of single-plane

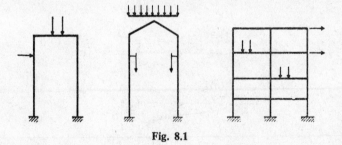

Fig. 8.1

structures where the deformations depend mainly on the flexural stiffness of the elements and the effects of axial forces and shear forces are negligible. Typical structures of this type are shown in Fig. 8.1. A useful generic description is *rigid-frame* structures, the features being rigid joints to transmit bending moments and members capable of carrying loads other than at the joints.

8.2 Sign Convention

A structure such as Fig. 8.2 (*a*) deforms in the manner shown by the dotted lines, so that a typical member AB becomes deformed as in Fig. 8.2 (*b*), and the end actions are a moment, and shear $p = \{MQ\}$.

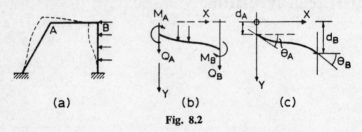

Fig. 8.2

Then, with the right-handed set of axes at the left-hand end, clockwise moments and downward shear forces on the member ends are positive and the corresponding positive deformations are also clockwise, as in Fig. 8.2 (*c*).

8.3 Slope-Deflexion Equations

The first step in the procedure for analysing a beam such as AB in Fig. 8.3 (*a*) by the stiffness method is to remove all degrees of freedom

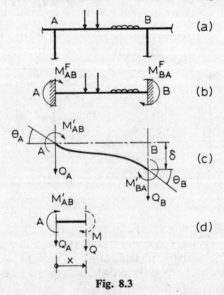

Fig. 8.3

by adding constraints which rigidly fix both ends, i.e. by applying the fixed-end moments M^F_{AB} and M^F_{BA} in Fig. 8.3 (*b*). It is convenient to introduce a specific notation for the end actions, e.g.

M^F_{AB} = fixed-end moment on end A of the member AB (positive when clockwise on the member)

These fixed end moments, which are a particular solution to the problem, are readily calculated for simple cases by the methods of

Table 8.1

Loading	M^F_{AB}	M^F_{BA}
	$\dfrac{-WL^2}{12}$	$\dfrac{+WL^2}{12}$
	$\dfrac{-WL}{8}$	$\dfrac{+WL}{8}$
	$\dfrac{-Wab^2}{L^2}$	$\dfrac{+Wa^2b}{L^2}$
	$(M^F_{AB} - \tfrac{1}{2}M^F_{BA})$	

Chap. 4, although for more complex structures and loadings it may be more suitable to apply the general methods of this or the previous chapter. Four standard results are given in Table 8.1 for convenience.

The second step is to remove the constraints and allow the structure to deform, as in Fig. 8.3 (*c*). This introduces additional end actions M'_{AB} and M_{BA} so that the final actions are

$$M = M' + M^F$$

The slope-deflexion equations relate the additional moments M' to the deformations, i.e. define the end stiffness of a member.

In Fig. 8.3 (*d*), taking the origin of coordinates at the left-hand end,

$$(\Sigma M)_A = 0$$

$$M'_{AB} - Q_A x + M = 0$$

i.e. $$M = Q_A x - M'_{AB} \tag{8.1}$$

The differential equation of equilibrium with these coordinates and sign convention (p. 109, Chap. 4) is

$$EI \frac{d^2y}{dx^2} = M$$

Therefore, from eqn. (8.1),

$$EI \frac{d^2y}{dx^2} = Q_A x - M'_{AB}$$

Integrating once,

$$EI \frac{dy}{dx} = \tfrac{1}{2} Q_A x^2 - M'_{AB} x + C_1 \tag{8.2}$$

At $x = 0$, $\dfrac{dy}{dx} = \theta_A$, and therefore

$$C_1 = EI\theta_A$$

At $x = L$, $\dfrac{dy}{dx} = \theta_B$, and therefore eqn. (8.2) becomes

$$EI\theta_B = \tfrac{1}{2} Q_A L^2 - M'_{AB} L + EI\theta_A$$

$$Q_A = \frac{2M'_{AB}}{L} + \frac{2EI\theta_B}{L^2} - \frac{2EI\theta_A}{L^2} \tag{8.3}$$

Therefore from eqns. (8.2) and (8.3),

$$EI \frac{dy}{dx} = M'_{AB}\left(\frac{x^2}{L} - x\right) + EI\theta_A\left(1 - \frac{x^2}{L^2}\right) + EI\theta_B \frac{x^2}{L^2}$$

Integrating again,

$$EIy = M'_{AB}\left(\frac{x^3}{3L} - \frac{x^2}{2}\right) + EI\theta_A\left(x - \frac{x^3}{3L^2}\right) + EI\theta_B \frac{x^3}{3L^2} + C_2$$

At $x = 0$, $y = 0$, i.e.

$$C_2 = 0$$

At $x = L$, $y = \delta$, i.e.

$$EI\delta = -\frac{M'_{AB}L^2}{6} + \frac{2}{3} LEI\theta_A + \frac{1}{3} LEI\theta_B$$

or

$$M'_{AB} = \frac{2EI}{L}\left[2\theta_A + \theta_B - \frac{3\delta}{L}\right] \qquad (8.4)$$

similarly

$$M'_{BA} = \frac{2EI}{L}\left[2\theta_B + \theta_A - \frac{3\delta}{L}\right] \qquad (8.5)$$

For equilibrium, ΣM about $B = 0$

$$M'_{AB} - Q_A L + M'_{BA} = 0$$

Substituting (8.4) and (8.5) for M'_{AB} and M'_{BA} and rearranging,

$$Q_A = \frac{6EI}{L^2}\left(\theta_A + \theta_B - \frac{2\delta}{L}\right) = -Q_B \qquad (8.6)$$

Finally, including the fixed end moments

$$M_{AB} = \frac{2EI}{L}\left(2\theta_A + \theta_B - \frac{3\delta}{L}\right) + M_{AB}^F \qquad (8.7)$$

$$M_{BA} = \frac{2EI}{L}\left(2\theta_B + \theta_A - \frac{3\delta}{L}\right) + M_{BA}^F \qquad (8.8)$$

8.4 Special, or Degenerate Cases

It is of interest and useful for calculation purposes to develop from eqns. (8.7) and (8.8) three special cases as follows. In each case there is no lateral load, so that $M_{AB}^F = M_{BA}^F = 0$.

(a) Fig. 8.4 (a), $\theta_B = 0 = \delta/L$

i.e. from (8.7),
$$M_{AB} = \frac{4EI\theta_A}{L} \qquad (8.9)$$

from (8.8),
$$M_{BA} = \frac{2EI\theta_A}{L} = \tfrac{1}{2}M_{AB} \qquad (8.10)$$

(b) Fig. 8.4 (b) $\theta_A = 0 = \theta_B$

i.e. from (8.7),
$$M_{AB} = -\frac{6EI\delta}{L^2} \qquad (8.11)$$

from (8.8),
$$M_{BA} = -\frac{6EI\delta}{L^2} \qquad (8.12)$$

This is called a case of *pure sway* or rotation of the member axis by an angle δ/L (positive clockwise).

(c) Fig. 8.4 (c) $M_{BA} = 0$ (pin end at right)

i.e. from (8.8) $M_{BA} = \dfrac{2EI}{L}\left(2\theta_B + \theta_A - \dfrac{3\delta}{L}\right) = 0$

Therefore $\dfrac{4EI\theta_B}{L} = -\dfrac{2EI\theta_A}{L} + \dfrac{6EI\delta}{L^2}$

Then from (8.7), using the above,

$$M_{AB} = \dfrac{3EI}{L}\left(\theta_A - \dfrac{\delta}{L}\right) \qquad (8.13)$$

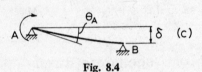

Fig. 8.4

Similarly, for a pin end at the left,

$$M_{BA} = \dfrac{3EI}{L}\left(\theta_B - \dfrac{\delta}{L}\right) \qquad (8.14)$$

Example 8.1

As an example of the stiffness method of analysis using the slope-deflexion equations, consider the two-span continuous beam ABC in Fig. 8.5 (a). From symmetry it is seen that there will be no rotation of the beam over the central support B, so that the problem reduces to the propped cantilever of Fig. 8.5 (b) with only one degree of freedom θ_A.

This freedom is removed by adding a restraint at A to give the orthodox fixed beam in Fig. 8.5 (c) and the fixed-end moments can be calculated or taken from Table 8.1. Therefore

$$R_{10} = M^F_{AB} = -wL^2/12$$

Release of the constraint requires a force k_{AA} to produce a unit rotation at A as shown; that is, the stiffness of beam AB at end A is

$$k_{AA} = 4EI/L$$

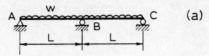

for $\theta_A = 1$, $\theta_B = 0 = \delta$ from eqn. (8.9). Then, for equilibrium,

$$R_{10} + k_{AA}\theta_A = 0$$

$$\frac{-wL^2}{12} + \frac{4EI\theta_A}{L} = 0$$

$$\frac{EI\theta_A}{L} = \frac{wL^2}{48}$$

From the slope-deflexion equation (8.7), for BC

$$M_{BC} = \frac{2EI}{L}\left(2\theta_B + \theta_A - \frac{3\delta}{L}\right) + M_{BC}^F$$

$$= \frac{2EI}{L}(\theta_A) + \frac{wL^2}{12}$$

$$= \frac{2wL^2}{48} + \frac{wL^2}{12} = \frac{wL^2}{8}$$

In Fig. 8.5 (f), and taking moments about end B,

$$Q_A L - \frac{wL^2}{2} + \frac{wL^2}{8} = 0$$

Therefore $\qquad Q_A = \tfrac{3}{8}wL = Q_C$ (from symmetry)

Therefore total reaction at B is

$$2wL - Q_A - Q_B = \tfrac{5}{4}wL$$

A more formal presentation is given below; it sets out the three fundamental statements of—

1. Force-deformation relations (stiffness),
2. Compatibility,
3. Equilibrium,

in the order in which orthodox calculations are carried out for small problems.

1. *Stiffness*

Member AB.

$$M_{AB} = \frac{2EI}{L}\left(2\theta_{AB} + \theta_{BC} - \frac{3\delta}{L}\right) + M_{AB}^F$$

but $\delta = 0$ (since A, B and C remain at the same level) and $M_{AB}^F = -wL^2/12$ (Table 8.1, p. 222). Therefore

$$M_{AB} = \frac{2EI}{L}(2\theta_{AB} + \theta_{BC}) - \frac{wL^2}{12}$$

and the stiffness M_{BA} is

$$M_{BA} = \frac{2EI}{L}(2\theta_{BA} + \theta_{AB}) + \frac{wL^2}{12}$$

Member BC. In the same way,

$$M_{BC} = \frac{2EI}{L}(2\theta_{BC} + \theta_{CB}) - \frac{wL^2}{12}$$

$$M_{CB} = \frac{2EI}{L}(2\theta_{CB} + \theta_{BC}) + \frac{wL^2}{12}$$

2. *Compatibility*

$\theta_{BC} = \theta_{BA} = \theta_B$ and, from symmetry, $\theta_{BA} = \theta_{BC} = 0$.

3. *Equilibrium*

At joint A there is no external couple, therefore

$$M_{AB} = 0$$

$$0 = \frac{2EI}{L}(2\theta_A) - \frac{wL^2}{12}$$

$$EI\theta_A = wL^3/48$$

Since, in this case, $\theta_B = 0$ and $\delta = 0$, θ_A is the only unknown deformation, back substitution in the stiffness equations gives

$$M_{BA} = \frac{2EI\theta_A}{L} + \frac{wL^2}{12} = \frac{wL^2}{8}$$

For equilibrium at joint B, $M_{BA} + M_{BC} = 0$. Therefore

$$M_{BC} = -\frac{wL^2}{8}$$

The remaining internal actions are determined from equilibrium considerations as before. This is, of course, a particularly simple example and use could be made of the special cases of p. 224 and the special cases of M^F from Table 8.1, p. 222. Thus, knowing that $M_{AB} = M_{CB} = 0$,

$$M_{BA} = M_{BA}^F - \tfrac{1}{2}M_{AB}^F$$

$$M_{BC} = M_{BC}^F - \tfrac{1}{2}M_{CB}^F$$

and, since $\theta_B = 0$, we get directly

$$M_{BA} = +\frac{wL^2}{8} = -M_{BC}$$

Example 8.2

A slightly different form of structure is shown in Fig. 8.6 (*a*). Using the slope–deflexion equations (8.7) directly (p. 224), considering member AB,

$$M_{AB} = \frac{2EI}{L}\left(2\theta_A + \theta_B - \left(\frac{3\delta}{L}\right)_{AB}\right) + M_{AB}^F$$

But $\theta_A = 0$, since built-in end,

$\delta_{AB} = 0$, neglecting change in length of BC,

$M_{AB}^F = 0$, since no lateral load on AB.

Therefore $$M_{AB} = \frac{2EI\theta_B}{L} \qquad (8.15)$$

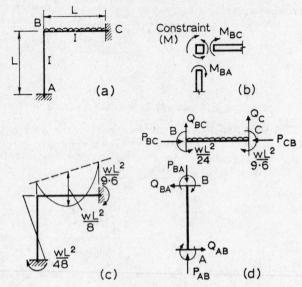

Fig. 8.6

Similarly
$$M_{BA} = \frac{4EI\theta_B}{L} \qquad (8.16)$$

Considering member BC,

$$M_{BC} = \frac{2EI}{L}\left(2\theta_B + \theta_C - \left(\frac{3\delta}{L}\right)_{BC}\right) + M_{BC}^F$$

$\theta_C = 0$, since built-in end,

$\delta_{BC} = 0$, neglecting change in length of AB,

$M_{BC}^F = -\dfrac{wL^2}{12}$ from Table 8.1.

Therefore
$$M_{BC} = \frac{4EI\theta_B}{L} - \frac{wL^2}{12} \qquad (8.17)$$

Similarly
$$M_{CB} = \frac{2EI\theta_B}{L} + \frac{wL^2}{12} \qquad (8.18)$$

Now to calculate the constraint required at B (Fig. 8.6 (b)). $M = M_{BA} + M_{BC} = 0$, since there is no constraint. Alternatively, for equilibrium of the joint B we get the same statement, i.e. using eqns. (8.16) and (8.17),

$$\frac{4EI\theta_B}{L} + \frac{4EI\theta_B}{L} - \frac{wL^2}{12} = 0$$

or

$$EI\theta_B = \frac{wL^3}{96}$$

(Arithmetically it is convenient always to associate $EI\theta$ in order to keep the figures a reasonable value, for θ itself is very small.) Then from eqns. (8.15) to (8.18),

$$M_{AB} = \frac{wL^2}{48} \qquad M_{BA} = \frac{wL^2}{24}$$

$$M_{BC} = -\frac{wL^2}{24} \qquad M_{CB} = \frac{5wL^2}{48} = \frac{wL^2}{9\cdot 6}$$

The final bending moment diagram drawn on the tension side of the members is shown in Fig. 8.6 (c) and the final calculations of the remaining member end actions can be carried out by statics. Isolating each member in turn as in Fig. 8.6 (d),

$$(\textstyle\sum M_B)_{BC}: \quad \frac{wL^2}{9\cdot 6} - Q_C L - \frac{wL^2}{24} + \frac{wL^2}{2} = 0$$

therefore $Q_C = 27/48 \ wL$

$$(\textstyle\sum y)_{BC}: \quad Q_{BC} - wL + Q_C = 0$$

therefore $Q_{BC} = 21/48 \ wL$

$$(\textstyle\sum M_A)_{AB}: \quad \frac{wL^2}{24} - Q_{BA} L + \frac{wL^2}{48} = 0$$

therefore $Q_{BA} = \dfrac{wL}{16}$

Then

$$(\textstyle\sum x = 0): \quad Q_{AB} = -Q_{BA} = -P_{CB} = P_{BC}$$
$$(\textstyle\sum y = 0): \quad P_{AB} = -Q_{CB} = -P_{BA} = Q_{BC}$$

8.5 The Sway Equation

The two previous examples set up the orthodox procedure—
  (i) Write down the end stiffnesses of each member using the slope-deflection equations.
  (ii) Use the compatibility condition imposed by the rigid joints, i.e.

$$\theta_{BA} = \theta_{BC} = \theta_{Bn}$$

(iii) Use the joint equilibrium equations

$$\sum M_{joint} = 0$$

This provides the same number of equations as the unknown joint rotations, but if the member *sway* terms (i.e. δ/L) are non-zero, additional equations are required for these additional unknowns.

The uniform-section fixed-base portal frame shown in Fig. 8.7 (a)

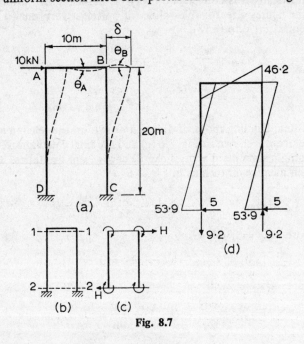

Fig. 8.7

has three degrees of freedom, namely

$$\theta_A, \quad \theta_B \quad \text{and} \quad \delta$$

The beam AB is stiffer than the columns in that, if

$$(I/L)_{DA} = (I/L)_{BC} = k$$

$$(I/L)_{AB} = 2k$$

Member stiffnesses

$$M_{AB} = \frac{2EI}{L}\left(2\theta_A + \theta_B - \frac{3\delta}{L}\right); \quad \text{and} \quad \frac{\delta}{L} = 0$$

$$= 4Ek(2\theta_A + \theta_B)$$

and $\qquad M_{BA} = 4Ek(2\theta_B + \theta_A)$

$$M_{AD} = \frac{2EI}{L}\left(2\theta_A + \theta_D - \frac{3\delta}{L}\right); \qquad \text{and} \qquad \theta_D = 0$$

$$= 2Ek\left(2\theta_A - \frac{3\delta}{L}\right)$$

$$M_{BC} = \frac{2EI}{L}\left(2\theta_B + \theta_C - \frac{3\delta}{L}\right); \qquad \text{and} \qquad \theta_C = 0$$

$$= 2Ek\left(2\theta_B - \frac{3\delta}{L}\right)$$

$$\text{etc.}$$

Joint equilibrium

$$M_{AD} + M_{AB} = 0$$

$$M_{BC} + M_{BA} = 0$$

But from symmetry it is seen that $\theta_A = \theta_B$, so that only one equation is required, i.e.

$$4Ek(2\theta_A + \theta_B) + 2Ek\left(2\theta_A - \frac{3\delta}{L}\right) = 0$$

Sway equation

The final equation is obtained by considering the equilibrium of the free body formed by sections 1–1 and 2–2 just below the column tops and just above the column bases, as shown in Figs. 8.7 (*b*) and (*c*). Taking moments about the base of the frame,

$$M_{DA} + M_{AD} + M_{CB} + M_{BC} + Hh = 0$$

(i.e. the sum of the column end moments + shear × height). Using the slope-deflexion equations, including those for M_{DA} and M_{AB} which are not given above, the sway equation becomes

$$3Ek\theta_A + 3Ek\theta_B - 4Ek\frac{3\delta}{L} = \tfrac{1}{2}Hh$$

The two equations are then

$$6Ek\theta_A + 2Ek\theta_B - 3Ek\frac{\delta}{L} = 0$$

$$3Ek\theta_A + 3Ek\theta_B - 4\left(3Ek\frac{\delta}{L}\right) = \tfrac{1}{2}Hh$$

Solving these gives

$$Ek\theta_A = 0{\cdot}0385 \times \tfrac{1}{2}Hh$$

$$3Ek\delta/L = 0{\cdot}3080 \times \tfrac{1}{2}Hh$$

Back substitution in the original slope–deflexion equations gives

$$H = 10 \text{ kN},$$

$$M_{BA} = 46 \cdot 2 \text{ kN-m} = -M_{BL}$$

$$M_{AB} = 46 \cdot 2 \text{ kN-m} = -M_{AD}$$

$$M_{DA} = M_{CB} = -53 \cdot 9 \text{ kN-m}$$

The final moments are shown in Fig. 8.7 (*d*).

8.6 General Use of Slope-Deflexion Equations

Two problems for practice solution will be found on p. 254, but in view of the fact that larger structures of this type would normally be analysed using the methods of Chap. 9 or using a computer, these problems are restricted to the illustration of principles.

8.7 Secondary Stresses

Earlier chapters have postulated idealized, frictionless, pin-jointed frames, and it is obvious that in practice such a frame does not exist. In fact orthodox methods of connecting members result in joints which are essentially rigid. However, the members of such structures are usually so flexible in bending that bending stiffnesses do not influence the behaviour of the structure, which behaves, within practical limits, as if the members were actually pin-connected and subject to axial forces only.

The presence of the stiff joints will cause bending actions and additional bending stresses in the members; since they do not affect overall structural behaviour appreciably they are called *secondary stresses*. Then, assuming that the primary axial forces and these secondary bending actions are mutually independent, the slope-deflexion equations provide a method of calculating the secondary moments. This is only of practical interest for the study of the principles involved and for special simple cases, because no advantage is gained by separating these effects when using general computer programmes for the analysis of structures.

If the simple braced frame of Fig. 8.8 (*a*) is loaded as shown and analysed as a pin-jointed structure for the primary forces, the Williot diagram can be drawn as sketched in Fig. 8.8 (*b*). An exaggerated diagram of the distorted structure is shown dotted in Fig. 8.8 (*a*) and the rotations of the members measured from the Williot diagram are as shown in the table p. 235. Consider the frame

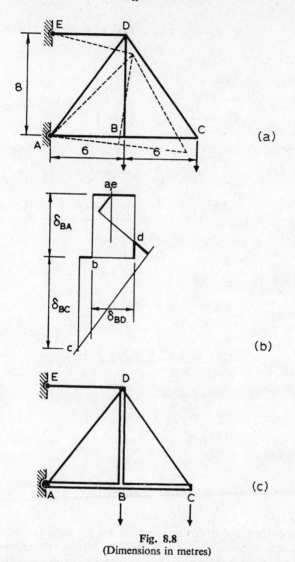

Fig. 8.8
(Dimensions in metres)

to be fabricated in such a way that joint B is rigid (Fig. 8.8 (c)). Then, assuming independence of the primary and secondary actions, the deformations shown by the Williot diagram are unaltered and, for equilibrium of joint B,

$$M_{BA} + M_{BD} + M_{BC} = 0 \qquad (8.19)$$

where $M_{BA} = \dfrac{3EI_1}{L_1}\left(\theta_B - \left(\dfrac{\delta}{L}\right)_{AB}\right)$ (from eqn. (8.4) p. 224)

$$M_{BD} = \dfrac{3EI_2}{L_2}\left(\theta_B - \left(\dfrac{\delta}{L}\right)_{BD}\right)$$

$$M_{BC} = \dfrac{3EI_1}{L_1}\left(\theta_B - \left(\dfrac{\delta}{L}\right)_{BC}\right)$$

and the member rotations are obtained from the Williot diagram.

Member	BA	BD	BC	Units
$10^{-1}\,E\delta/L^2$	$+2\cdot4$	$+0\cdot506$	$+1\cdot33$	N/mm³

If $I_1 = I_2 = 8 \times 10^6$ mm⁴, eqn. (8.19) becomes

$$\dfrac{3EI\theta_B}{10^3}\left(\dfrac{1}{6} + \dfrac{1}{6} + \dfrac{1}{8}\right) - \dfrac{3I}{10}(2\cdot4 + 1\cdot33 + 0\cdot506) = 0$$

i.e. $EI\theta_B = 7\cdot4 \times 10^9$ (mm²N)

Back substituting,

$$M_{BA} = \dfrac{3 \times 7\cdot4 \times 10^9}{6 \times 10^3 \times 10^6} - \dfrac{3 \times 8 \times 10^6 \times 2\cdot4}{10 \times 10^6}$$

$$= -2\cdot06 \text{ kN-m}$$

and $M_{BD} = +1\cdot56$ kN-m

This process can evidently be extended to frames with more than one rigid joint.

8.8 Introduction to the General Stiffness Method

Direct use of the slope-deflection methods has been shown to lead to a method of analysis of the type of structure which is dependent for stability primarily on the flexural stiffness of the members and joints, but a general approach to the stiffness method of analysis provides a much more powerful technique. The rest of this chapter is devoted to the beginnings of this development. These procedures only show to advantage on difficult structures and using computer calculation, but the development here is restricted to cases which allow hand calculation.

Repeating (from p. 182) the basis of the stiffness method, but using a slightly more complicated structure to work on (Fig. 8.9 (*a*)), each member of the rigid frame structure in the figure is isolated by the additional constraints R_{01}, R_{02} and R_{03} (called the R_0 set), which are defined in relation to a set of axes XY and correspond to the independent degrees of freedom of the structure. Thus

R_{01} prevents rotation at joint B,

R_{02} prevents rotation at joint C,

R_{03} prevents lateral sway of the frame.

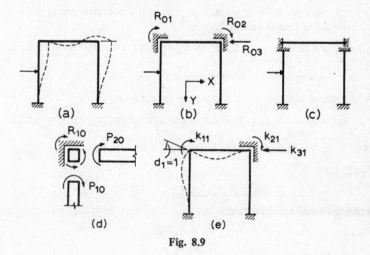

Fig. 8.9

These isolated members are not themselves statically-determinate, but they can be dealt with quite simply as before (e.g. p. 120 and Table 8.1) for the end actions. Several members will meet at a joint and in order to write down the equilibrium equations for a joint all forces on it must be described in the same coordinate system as for the slope-deflexion equations for Fig. 8.3. If the member end actions are defined relative to the axes XY in Fig. 8.9 (*b*), and are called the P_0 set of forces then, at any constraint, for equilibrium, (Fig. 8.9 (*d*))

$$R_0 = P_0$$

and this force system is a particular solution to the problem.

The real solution to the real problem (i.e. no constraints) is then found by adding to this system the complementary solution obtained by releasing the constraints. In terms of the slope-deflexion equations,

Particular solution (with constraints)

$$M_{AB} = M_{AB}^F$$

Complementary solution (release of constraints)

$$M_{AB} = (2EI/L)(2\theta_A + \theta_B - 3\delta/L)$$

and the complete solution is, as before, (8.7)

$$M_{AB} = (2EI/L)\left(2\theta_A + \theta_B - \frac{3\delta}{L}\right) + M_{AB}^F$$

Thus, in Fig. 8.9 (*e*), a unit displacement corresponding to R_{01} requires an external action k_{11}, i.e.

k_{11} = force corresponding to R_{01} to produce a unit displacement corresponding to R_{01}
   = *direct* stiffness at restraint 1

Actions will also be produced at restraints 2 and 3, i.e.

k_{21} = force at release 2 due to unit displacement corresponding to R_{01}
   = *cross* stiffness at restraint 2

Similarly,

k_{31} = *cross* stiffness at restraint 3

Then for a final displacement d_1,

$$R_1 = k_{11}d_1$$
$$R_2 = k_{21}d_1$$
$$R_3 = k_{31}d_1$$

and in a similar way for displacements at restraints 2 and 3,

$$R_1 = k_{11}d_1 + k_{12}d_2 + k_{13}d_3$$
$$R_2 = k_{21}d_1 + k_{22}d_2 + k_{23}d_3$$
$$R_3 = k_{31}d_1 + k_{32}d_2 + k_{33}d_3$$

These can be written in matrix form

$$\begin{bmatrix} R_1 \\ R_2 \\ R_3 \end{bmatrix} = \begin{bmatrix} k_{11} & k_{12} & k_{13} \\ k_{21} & k_{22} & k_{23} \\ k_{31} & k_{32} & k_{33} \end{bmatrix} \begin{bmatrix} d_1 \\ d_2 \\ d_3 \end{bmatrix}$$

or $$R = Kd$$

Finally, for zero constraints and using matrix notation,

$$R_0 + R = 0$$

or

$$R_0 + Kd = 0$$ $$\left.\begin{array}{c}\\\\\end{array}\right\} \quad (8.20)$$

where K is the matrix of stiffness coefficients at the restraints, or the *structure stiffness matrix*. The eqns. (8.20) are then solved for the displacements as with the slope–deflexion equations, or the matrix K can be inverted to give

$$d = -K^{-1}R_0$$

Knowing the joint displacements, i.e. the displacements of the member ends, the member end actions can then be calculated from the individual member stiffnesses, e.g. the slope–deflexion equations (8.7) and (8.8) (p. 224).

8.9 Member Stiffness

By using the slope–deflexion equations and the rules for matrix multiplication, the reader should verify that, in the "member" coordinate system of Fig. 8.10, the stiffness matrix of member AB

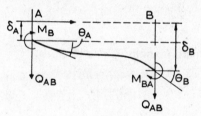

Fig. 8.10

in the matrix equation $P_{AB} = K_{AB}u_{AB}$ is

$$\begin{bmatrix} M_{AB} \\ Q_{AB} \\ M_{BA} \\ Q_{BA} \end{bmatrix} = \begin{bmatrix} 4EI/L & 6EI/L^2 & 2EI/L & -6EI/L^2 \\ 6EI/L^2 & 12EI/L^3 & 6EI/L^2 & -12EI/L^3 \\ 2EI/L & 6EI/L^2 & 4EI/L & -6EI/L^2 \\ -6E/L^2 & -12EI/L^3 & -6EI/L^2 & 12EI/L^3 \end{bmatrix} \begin{bmatrix} \theta_A \\ \delta_A \\ \theta_B \\ \delta_B \end{bmatrix}$$

Note here that the end action column vector

$$P = \{M_{AB}Q_{AB}M_{BA}Q_{BA}\}$$

does not include the axial forces N_{AB} and N_{BA} since axial changes in length were ignored in formulating the equations; and that the letter u is used for the matrix $\{\theta_A \delta_A \theta_B \delta_B\}$ at the deformations since these are *member* deformations and not necessarily joint displacements d. It is also important to realize that one set of axes only is referred to and that the matrix elements are described relative to this set of member axes.

8.10 Local Stiffness Transformations

It will be observed that the stiffness matrix K_{AB} can be partitioned as shown by the dotted lines into four sub-matrices and written

$$\begin{bmatrix} P_A \\ \hline P_B \end{bmatrix} = \begin{bmatrix} k_{AA} & \vdots & k_{AB} \\ \hline k_{BA} & \vdots & k_{BB} \end{bmatrix} \begin{bmatrix} u_A \\ \hline u_B \end{bmatrix}$$

where the 2×2 sub-matrices k_{ij} are the local stiffness matrices; for example,

$$k_{AA} = \begin{bmatrix} 4EI/L & 6EI/L^2 \\ 6EI/L^2 & 12EI/L^3 \end{bmatrix}$$

This describes the actions at end A

$$P_A = \begin{bmatrix} M_{AB} \\ Q_{AB} \end{bmatrix}$$

due to member deformations u_A at end A, i.e.

$$u_A = \begin{bmatrix} \theta_A \\ \delta_A \end{bmatrix}$$

so that k_{AA} is the direct stiffness at end A. Similarly, k_{AB} is the cross stiffness at A due to deformations at end B, k_{BB} is the direct stiffness at B, and k_{BA} is the cross stiffness at B. This notation is rather more concise.

In cases where all six end actions (Fig. 6.2) are to be considered, each sub-matrix k_{ij} will be a 6×6 and the full matrix K a 12×12. This additional complication does not affect the general matrix statements and is not required as long as discussion is restricted to plane frames where axial changes in length of the members is not important.

Continuing to restrict discussion to the case of plane frames and flexural deformations only, the four end actions

$$P = \{M_A Q_A M_B Q_B\}$$

(i.e. omitting the second subscript in the case of a single member) are not independent. For example, from Fig. 8.10,

$$\sum Y = 0, \quad \text{i.e.} \quad Q_A = -Q_B \tag{8.21}$$

$$\sum M \text{ about B}, \quad M_A + M_B - Q_A L = 0 \tag{8.22}$$

giving two equations of static equilibrium. This means that only two values are arbitrary, or that the deformed shape of a beam can be completely defined by a suitably chosen pair of end actions or end deformations. It follows that the local stiffness matrices are also not independent, e.g.

$$M_B = -M_A + Q_A L \qquad \text{and} \qquad Q_B = -Q_A$$

or, in matrix form,

$$\begin{bmatrix} M_B \\ Q_B \end{bmatrix} = \begin{bmatrix} -1 & L \\ 0 & -1 \end{bmatrix} \begin{bmatrix} M_A \\ Q_A \end{bmatrix}$$

or

$$P_B = H P_A$$

where

$$H = \begin{bmatrix} -1 & L \\ 0 & -1 \end{bmatrix}$$

Since, for $u_B = 0$,

$$\begin{bmatrix} M_A \\ Q_A \end{bmatrix} = k_{AA} u_A$$

or

$$P_A = k_{AA} u_A$$

it follows that

$$P_B = H k_{AA} u_A$$

But

$$P_B = k_{BA} u_A$$

(because $u_B = 0$), and therefore

$$k_{BA} = H k_{AA} \tag{8.23}$$

Similarly, since $P_A = k_{AB} u_B$ when $u_A = 0$,

$$P_B = H P_A = H k_{AB} u_B$$

Therefore, since $P_B = k_{BB} u_B$,

$$k_{BB} = H k_{AB} \tag{8.24}$$

In addition, the reciprocal theorem indicates that the complete matrix must be symmetrical; further, by inspection of the matrix elements, it is seen that

$$k_{AB} = k'_{BA} \tag{8.25}$$

Thus once k_{AA} is known, the three other local stiffness sub-matrices ($k_{AB} k_{BB} k_{BA}$) follow directly from eqns. (8.23) to (8.25). These stiffness transformations, and others which are developed later, can be derived more elegantly and in a more general manner by using the properties of general axis transformation, but the above procedure is simpler for presenting a simple case.

8.11 Contracted Stiffness Matrices

A stiffness matrix is not unique in that there are several ways of defining a member stiffness. In all cases however it is only possible to define a stiffness, i.e. force per unit deformation, if the deformation defined is independent (c.f. flexibility matrix condition on page 193).

In addition, a beam element has only two independent end actions so that the stiffness matrix can be reduced in size and still contain all essential information.

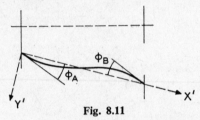

Fig. 8.11

(a) From Fig. 8.10 and eqns. (8.4) and (8.5) and using M_A and M_B only,

$$\begin{bmatrix} M_A \\ M_B \end{bmatrix} = \begin{bmatrix} 4 & 6/L & 2 & -6/L \\ 2 & 6/L & 4 & -6/L \end{bmatrix} EI/L \begin{bmatrix} \theta_A \\ \delta_A \\ \theta_B \\ \delta_B \end{bmatrix}$$

where, in this case of a uniform member, the constant EI/L has been removed from the matrix as a constant scalar multiplier.

(b) From Fig. 8.10 and the same equations, but where $\delta = (\delta_B - \delta_A)$,

$$\begin{bmatrix} M_A \\ M_B \end{bmatrix} = \begin{bmatrix} 4 & 2 & -6/L \\ 2 & 4 & -6/L \end{bmatrix} EI/L \begin{bmatrix} \theta_A \\ \theta_B \\ \delta \end{bmatrix}$$

(c) Redefining the end rotations as ϕ_A and ϕ_B in Fig. 8.11 and using eqns. (8.4) and (8.5) referred to the new axes X' and Y',

$$\begin{bmatrix} M_A \\ M_B \end{bmatrix} = \begin{bmatrix} 4 & 2 \\ 2 & 4 \end{bmatrix} EI/L \begin{bmatrix} \phi_A \\ \phi_B \end{bmatrix}$$

(*d*) As an exercise, the reader should derive the two other forms of the modified stiffness matrix using $\{M_A Q_A\}$ and $\{M_B Q_B\}$.

8.12 Structure Stiffness Matrix

The structure stiffness matrix of p. 237 is built up of the individual member stiffnesses; its elements are all expressed in a single set of structure coordinates, whereas the member stiffnesses discussed above are all more simply expressed in member coordinates. Bearing in mind this difference, it is now required to find some convenient way of building up the full K-matrix in agreement with the way in which the members of the structure are connected together.

The number of degrees of freedom of the structure corresponds to the number of elements in the R_0 set and this number of displacements is therefore sufficient to define completely the deformed shape of the structure.

Let the independent external displacements (corresponding to the constraints) be represented by the vector d and let the associated internal member deformations be u (corresponding to member end actions). Then, if d defines u in some way, it follows that

$$u = Dd \tag{8.26}$$

where D is a transformation matrix which depends only on the geometry of the structure.

From the principle of contragredience (p. 174), since the external actions R_0 correspond to d and the internal actions P_0 correspond to the deformations u, the application of forces R to remove the constraints causes internal actions P such that

$$R = D'P \tag{8.27}$$

Also, for each member, the end actions and deformations are related by the member stiffnesses

$$P_{AB} = K_{AB} u_{AB}$$

and for all members of a structure this can be written as a single large matrix

$$\begin{bmatrix} P_{AB} \\ P_{BC} \\ P_{CD} \\ \text{etc.} \end{bmatrix} = \begin{bmatrix} K_{AB} & 0 & 0 & - & - \\ 0 & K_{BC} & 0 & - & - \\ 0 & 0 & K_{CD} & - & - \\ & & \text{etc.} & & \end{bmatrix} \begin{bmatrix} u_{AB} \\ u_{BC} \\ u_{CD} \\ \text{etc.} \end{bmatrix}$$

or simply $P = K_0 u$

where K_0, called the *unassembled stiffness matrix*, is a diagonal matrix of sub-matrices K_{AB}, K_{BC}, etc. Then, using this in eqn. (8.27)

$$R = D'K_0u$$

and, using (8.26), $\qquad R = D'K_0Dd$

Therefore $\qquad\qquad R = Kd$

as on p. 237. But now $\quad K = D'K_0D$

$$= \text{assembled structure stiffness matrix}$$

Thus the final equilibrium equation for the removal of constraints (eqn. (8.20)) becomes

$$R_0 + D'K_0Dd = 0$$

and the solution follows as before. The matrix transformations

$$D'K_0D$$

have the effect of assembling the member stiffnesses, written in member coordinates, in the correct manner and transformed into the single set of general structure axes.

The matrix K_0 is very large and in practice this is not a good way of assembling the structure matrix K, but this presentation has the great advantage of conciseness and clarity.

Example 8.3
Following through the above treatment for the portal frame shown in Fig. 8.12 (*a*), there are seen to be three degrees of freedom—

$$d = \{\theta_B\theta_C\delta\} \qquad\qquad \text{Fig. 8.12}(b)$$

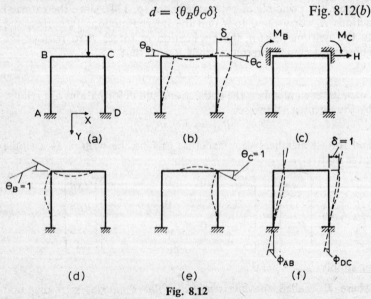

Fig. 8.12

and three corresponding constraints—

$$R_0 = \{M_B M_C H\} \qquad \text{Fig. 8.12}(c)$$

Using for each member stiffness the form

$$\begin{bmatrix} M_{AB} \\ M_{BA} \end{bmatrix} = \begin{bmatrix} 4 & 2 \\ 2 & 4 \end{bmatrix} EI/L \begin{bmatrix} \phi_A \\ \phi_B \end{bmatrix}$$

(where EI/L is constant for all members), the unassembled stiffness matrix is

$$\begin{bmatrix} M_{AB} \\ M_{BA} \\ M_{BC} \\ M_{CB} \\ M_{CD} \\ M_{DC} \end{bmatrix} = \begin{bmatrix} 4 & 2 & & & & \\ 2 & 4 & & & & \\ & & 4 & 2 & & \\ & & 2 & 4 & & \\ & & & & 4 & 2 \\ & & & & 2 & 4 \end{bmatrix} EI/L \begin{bmatrix} \phi_{AB} \\ \phi_{BA} \\ \phi_{BC} \\ \phi_{CB} \\ \phi_{CD} \\ \phi_{DC} \end{bmatrix}$$

The next step is to form, element by element, the transformation matrix D in $u = Dd$, i.e. allow $\theta_B = 1$ while $\theta_C = \delta = 0$, and from Fig. 8.12 (d)

$$\phi_{AB} = 0$$
$$\phi_{BA} = 1$$
$$\phi_{BC} = 1$$

In the same way, if $\theta_C = 1$ while $\theta_B = \delta = 0$,

$$\phi_{CB} = 1$$
$$\phi_{CD} = 1$$
$$\phi_{DC} = 0$$

Finally if $\delta = 1$ while $\theta_B = \theta_C = 0$ (Fig. 8.12 (e))

$$\phi_{AB} = -1/L$$
$$\phi_{BA} = -1/L$$
$$\phi_{BC} = 0$$
$$\phi_{CB} = 0$$
$$\phi_{CD} = -1/L$$
$$\phi_{DC} = -1/L$$

Thus

$$
\begin{bmatrix} \phi_{AB} \\ \phi_{BA} \\ \phi_{BC} \\ \phi_{CB} \\ \phi_{CD} \\ \phi_{DC} \end{bmatrix} = \begin{bmatrix} 0 & 0 & -1/L \\ 1 & 0 & -1/L \\ 1 & 0 & 0 \\ 0 & 1 & 0 \\ 0 & 1 & -1/L \\ 0 & 0 & -1/L \end{bmatrix} \begin{bmatrix} \theta_B \\ \theta_C \\ \delta \end{bmatrix}
$$

i.e. $$u = Dd$$

The assembly of the K-matrix, namely

$$K = D'K_0 D$$

is then

$$
K = \begin{bmatrix} 0 & 1 & 1 & 0 & 0 & 0 \\ 0 & 0 & 0 & 1 & 1 & 0 \\ -1/L & -1/L & 0 & 0 & -1/L & -1/L \end{bmatrix} \begin{bmatrix} 4 & 2 & & & & \\ 2 & 4 & & & & \\ & & 4 & 2 & & \\ & & 2 & 4 & & \\ & & & & 4 & 2 \\ & & & & 2 & 4 \end{bmatrix} EI/L \begin{bmatrix} 0 & 0 & -1/L \\ 1 & 0 & -1/L \\ 1 & 0 & 0 \\ 0 & 1 & 0 \\ 0 & 1 & -1/L \\ 0 & 0 & -1/L \end{bmatrix}
$$

which when multiplied out gives

$$
K = \begin{bmatrix} 8 & 2 & -6/L \\ 2 & 8 & -6/L \\ -6/L & -6/L & 24/L^2 \end{bmatrix} EI/L
$$

i.e.

$$
\begin{bmatrix} M_B \\ M_C \\ H \end{bmatrix} = \begin{bmatrix} 8 & 2 & -6/L \\ 2 & 8 & -6/L \\ -6/L & -6/L & 24/L^2 \end{bmatrix} EI/L \begin{bmatrix} \theta_B \\ \theta_C \\ \delta \end{bmatrix} \qquad (8.28)
$$

which is, of course,

$$R = Kd$$

Also, since $R_0 = \{M_B M_C H\}$ is given by $M_B = M_{BC}^F$, $M_C = M_{CB}^F$ and $H = 0$, the equation (8.20), namely

$$R_0 + Kd = 0$$

can be solved for the displacements.

8.13 Alternative Formulation of the Stiffness Matrix

As already mentioned, the unassembled stiffness matrix K_0 is sometimes very large and contains a large number of zero terms, so that when a computer is used a lot of storage space is wasted. For this reason the transformation procedure may not be the best way of assembling the matrix.

Starting from the individual member stiffnesses from p. 239, namely

$$\begin{bmatrix} P_A \\ P_B \end{bmatrix} = \begin{bmatrix} k_{AA} & k_{AB} \\ k_{BA} & k_{BB} \end{bmatrix} \begin{bmatrix} u_A \\ u_B \end{bmatrix}$$

where the matrix elements are sub-matrices, then where several members meet at a joint the structure stiffness at that joint is the sum of the stiffnesses of the members. So that, at a joint A where members AB, AC, ... AM meet, the external force on the joint for equilibrium is

$$P_A = P_{AB} + P_{AC} + P_{AD} \cdots + P_{AM}$$

where the actions P_{AB} etc. are the actions on the ends A of the members.

Thus $\quad P_A = \begin{bmatrix} k_{AA} & k_{AB} \end{bmatrix} \begin{bmatrix} u_A \\ u_B \end{bmatrix} \qquad$ for member AB

$$+ \begin{bmatrix} k_{AA} & k_{AC} \end{bmatrix} \begin{bmatrix} u_A \\ u_C \end{bmatrix} \qquad \text{for member AC}$$

$$+ \begin{bmatrix} k_{AA} & k_{AD} \end{bmatrix} \begin{bmatrix} u_A \\ u_D \end{bmatrix} \qquad \text{for member AD}$$

$+$ etc. $\qquad$ up to member AM

$$= [\textstyle\sum k_{AA} + k_{AB} + k_{AC} + \cdots k_{AM}] \begin{bmatrix} u_A \\ u_B \\ u_C \\ . \\ \text{etc.} \\ . \\ u_M \end{bmatrix}$$

In the same way at joint B,

$$P_B = [\sum k_{BB} + k_{BA} + k_{BC} + \cdots k_{BN}] \begin{bmatrix} u_B \\ u_A \\ u_C \\ \cdot \\ \text{etc.} \\ \cdot \\ u_N \end{bmatrix}$$

so that writing the complete set as a matrix of joint stiffnesses we have

$$\begin{bmatrix} P_A \\ P_B \\ P_C \\ \cdot \end{bmatrix} = \begin{bmatrix} \sum k_{AA} & k_{AB} & k_{AC} & \cdots \\ k_{BA} & \sum k_{BB} & k_{BC} & \cdots \\ k_{CA} & & \text{etc.} & \end{bmatrix} \begin{bmatrix} u_A \\ u_B \\ u_C \\ \cdot \end{bmatrix}$$

The stiffness matrix is readily built up in tabular form depending on how the members are connected, as shown in Table 8.2. Columns and rows of the table are lettered by the designation of the joints: at joint A, say, if member AB is connected, then the AAth term (i.e. row A, column A) is k_{AA} and there will be an ABth term, namely k_{AB} as shown; at joint B (row B of the Table) there will be a k_{BA} term and a k_{BB} term for member AB. If B is joined to C then member BC contributes a k_{BB} term and a k_{BC} term. In principle this process is readily continued for all joints and the complete structure stiffness matrix built up term by term. However it will be recalled that on p. 238 the member stiffness matrix, i.e. the k_{AA}, k_{BB}, etc. terms, are written in member coordinates. Since the constraints and joint displacements must be written in a single set

Table 8.2

	A	B	C	D
A	$(k_{AA})_{AB}$	k_{AB}		
B	k_{BA}	$\begin{Bmatrix}(k_{BB})_{AB}\\(k_{BB})_{BC}\end{Bmatrix}$	k_{BC}	
C				
D				

of system coordinates it is therefore necessary to transform the member stiffnesses into the system coordinates before the elements can be added into the structure stiffness matrix. In addition, and referring back to Fig. 8.3, it is seen that the signs given to end actions, e.g. Q_A and Q_B in the figure, depend on which end of the member is chosen as the origin for the member coordinates. It is more convenient now to number the member ends as *end* 1 and *end* 2, and always to use an origin at *end* 1. It is then necessary to know which end of a member is connected at any joint, i.e. the *BB*th term in Table 8.2 will become

$$k_{BB} = k_{11}\begin{Bmatrix}\text{for all members with}\\ \text{end 1 at the joint}\end{Bmatrix} + k_{22}\begin{Bmatrix}\text{for all members with}\\ \text{end 2 at the joint}\end{Bmatrix}$$

and all member stiffnesses will be written

$$\begin{bmatrix} P_1 \\ P_2 \end{bmatrix} = \begin{bmatrix} k_{11} & k_{12} \\ k_{21} & k_{22} \end{bmatrix}\begin{bmatrix} u_1 \\ u_2 \end{bmatrix}$$

8.14 Member Transformations

A set of forces at a point, e.g. on a member end, is defined as the set P related to a set of member axes X' and Y' rotated by an angle α from the system coordinates X and Y (Fig. 8.13). Considering a plane frame where

$$P = \begin{bmatrix} N \\ Q \\ M \end{bmatrix}$$

i.e. the axis X' defines the member axis.

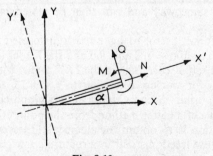

Fig. 8.13

These actions are resolved along the system axes to give the $\bar{P}$ set, i.e.

$$\bar{P} = \begin{bmatrix} P_x \\ P_y \\ M' \end{bmatrix}$$

where $P_x = N \cos \alpha - Q \sin \alpha$
$\quad\quad\; P_y = N \sin \alpha + Q \cos \alpha$
$\quad\quad\; M' = M$

Writing this out in matrix form,

$$\bar{P} = \begin{bmatrix} P_x \\ P_y \\ M' \end{bmatrix} = \begin{bmatrix} \cos & -\sin & 0 \\ \sin & \cos & 0 \\ 0 & 0 & 1 \end{bmatrix} \begin{bmatrix} N \\ Q \\ M \end{bmatrix}$$

i.e. $\quad\quad\quad\quad \bar{P} = TP$

where T is a transformation matrix.

Then, by the Principle of Contragredience,

$$d = T'\bar{d}$$

where d and $\bar{d}$ are sets of displacements corresponding to the P and $\bar{P}$ sets of forces, respectively. If the member stiffness in member coordinates is

$$P_1 = k_{11}d_1 \cdots \text{etc.}$$

This then becomes, in system coordinates,

$$TP_1 = Tk_{11}T'\bar{d}$$
or $\quad\quad\quad\quad \bar{P}_1 = Tk_{11}T'\bar{d}$

and all the sub-matrices of the member stiffness matrix will transform in the same way and can then be added into the structure stiffness matrix.

It is sometimes useful to have the transformed element stiffness matrix in full as above in Table 8.3 and the reader should check this result for himself by carrying out the TKT' transformation on the full 6×6 matrix for an element similar to Fig. 8.13. End 1 is the end nearer to the origin, S and C represent $\sin \alpha$ and $\cos \alpha$ respectively.

An example of a better method from the point of view of systematic computation is to obtain the member stiffnesses directly in terms of the structure freedoms and in structure coordinates, and then to build up the structure stiffness matrix term by term.

Table 8.3

$$
\begin{bmatrix} P_{x1} \\ P_{y1} \\ M_1 \\ P_{x2} \\ P_{y2} \\ M_2 \end{bmatrix} =
\begin{bmatrix}
(C^2A + S^2 12I/L^2) & SC(A - 12I/L^2) & -S6I/L & (S^2 12I/L^2 - C^2A) & -SC(12I/L^2 - A) & -S6I/L \\
 & (S^2A + C^2 12I/L^2) & -C6I/L & -SC(A - 12I/L^2) & -(S^2A + C^2 12I/L^2) & C6I/L \\
 & & 4I & S6I/L & -C6I/L & 2I \\
 & \text{Symmetric} & & C^2A + S^2 12I/L^2 & SC(A - 12I/L^2) & S6I/L \\
 & & & & (S^2A + C^2 12I/L^2) & -C6I/L \\
 & & & & & 4I
\end{bmatrix}
\begin{bmatrix} U_{x1} \\ U_{y1} \\ \theta_1 \\ U_{x2} \\ U_{y2} \\ \theta_2 \end{bmatrix}
$$

The structure stiffness matrix from p. 237 is

$$\begin{bmatrix} R_1 \\ R_2 \\ R_3 \end{bmatrix} = \begin{bmatrix} k_{11} & k_{12} & k_{13} \\ k_{21} & k_{22} & k_{23} \\ k_{31} & k_{32} & k_{33} \end{bmatrix} \begin{bmatrix} d_1 \\ d_2 \\ d_3 \end{bmatrix}$$

where the individual elements (which may themselves be matrices) are identified by subscripts referring to the "freedom numbers." Then, if the freedoms are numbered consecutively as in the example in Fig. 8.12,

$$\theta_B = d_1$$
$$\theta_C = d_2$$
$$\delta = d_3$$

each member stiffness can be written directly in terms of these freedoms with the rows and columns of the matrices numbered accordingly.

For member AB (using para. (*d*), p. 241),

$$\begin{matrix} M_{BA} \\ Q_{BA} \end{matrix} = \begin{bmatrix} 4 & -6/L \\ -6/L & 12/L^2 \end{bmatrix} \begin{bmatrix} d_1 \\ d_3 \end{bmatrix} EI/L$$

where the rows and columns of K_{AB} can be numbered as follows

$$K_{AB} = \begin{matrix} 1 \\ 3 \end{matrix} \begin{matrix} 1 & \quad 3 \\ \begin{bmatrix} 4 & -6/L \\ -6/L & 12/L^2 \end{bmatrix} \end{matrix}$$

Similarly for member BC,

$$\begin{matrix} M_{BC} \\ M_{CB} \end{matrix} = \begin{matrix} 1 \\ 2 \end{matrix} \begin{matrix} 1 & 2 \\ \begin{bmatrix} 4 & 2 \\ 2 & 4 \end{bmatrix} \end{matrix} \begin{bmatrix} d_1 \\ d_2 \end{bmatrix} EI/L$$

In this case the member stiffness used before is already in terms of freedom displacements.

Finally, for member CD,

$$\begin{matrix} M_{CD} \\ Q_{DC} \end{matrix} = \begin{matrix} 2 \\ 3 \end{matrix} \begin{matrix} 2 & \quad 3 \\ \begin{bmatrix} 4 & -6/L \\ -6/L & 12/L^2 \end{bmatrix} \end{matrix} \begin{bmatrix} d_2 \\ d_3 \end{bmatrix} EI/L$$

The structure stiffness matrix is then assembled by selecting the correct elements in turn, e.g.

Element $k_{11} = (\text{term}_{11} \text{ from member } K_{AB}) + (\text{term}_{11} \text{ from } K_{BC})$
$$= 4 + 4 = 8$$

Thus the structural stiffness matrix becomes

$$K = \begin{matrix} & 1 & 2 & 3 \\ 1 \\ 2 \\ 3 \end{matrix} \begin{bmatrix} (4+4) & 2 & -6/L \\ 2 & (4+4) & -6/L \\ -6/L & -6/L & (12/L^2 + 12/L^2) \end{bmatrix}$$

This method lends itself well to systematic programming for a computer and, while the above example could be written down directly, it will usually be necessary to transform the member stiffness by calculation.

Demonstrating this for member AB in Fig. 8.12, the initial member stiffness used is

$$\begin{matrix} M_{AB} \\ M_{BA} \end{matrix} = \begin{bmatrix} 4 & 2 \\ 2 & 4 \end{bmatrix} \begin{bmatrix} \phi_A \\ \phi_B \end{bmatrix} EI/L$$

but for this member $u_{AB} = Dd$

or

$$\begin{matrix} \phi_A \\ \phi_B \end{matrix} = \begin{bmatrix} 0 & 0 & -1/L \\ 1 & 0 & -1/L \end{bmatrix} \begin{bmatrix} d_1 \\ d_2 \\ d_3 \end{bmatrix}$$

and using contragradience, as on p. 242, we get

$$R = D'P = D'K_{AB}Dd$$

Multiplying out, using matrix multiplication,

$$D'P = \begin{bmatrix} M_{BA} \\ \theta \\ \frac{1}{2}(M_{BA} + M_{AB})/L \end{bmatrix}$$

and

$$D'K_{AB}D = \begin{bmatrix} 4 & 0 & -6/L \\ 0 & 0 & 0 \\ -6/L & 0 & 12/L^2 \end{bmatrix} \begin{bmatrix} d_1 \\ d_2 \\ d_3 \end{bmatrix} EI/L$$

so that, if the zero rows and columns are omitted, we get—

$$\begin{matrix} M_{BA} \\ Q_{BA} \end{matrix} = \begin{bmatrix} 4 & -6/L \\ -6/L & 12/L^2 \end{bmatrix} \begin{bmatrix} d_1 \\ d_3 \end{bmatrix} EI/L$$

This is, of course, the same matrix that was already used, but by carrying out the transformations on each member singly the large computer storage space used for the $D'K_0D$ multiplication on p. 243 is avoided and the techniques required to assemble Table 8.2 are simplified.

A useful standard procedure for automatically 'coding' the element stiffnesses and collecting together the structural stiffness matrix is set out below using Fig. 8.12 as an example.

Joint numbering

The joints are numbered in such a way as to reduce the numerical difference between the joint numbers at the ends of each member. This helps to preserve the 'banded' nature of the structural stiffness matrix. In the case of long narrow structures this is relatively easy by numbering consecutively across the narrow dimension, e.g. B, C, A and D become 1, 2, 3 and 4 respectively.

Restraint Array

Restrained joints are listed in the following table—

Joint	Restraint		
	x	y	θ
3	1	1	1
4	1	1	1

A 'marker' digit is then used to indicate the restraint present. In the case of no restraint a zero is inserted. In the case of Fig. 8.12 both support joints are fully restrained so that a marker appears in all columns of the table.

Node Freedom Array

All joints are listed in the table below then, taking each in turn, the Restraint Array is searched. If the joint is recorded and a restraint

Node	Freedom No.		
	x	y	θ
1	1	0	2
2	1	0	3
3	0	0	0
4	0	0	0

marked, i.e. no freedom, a zero is inserted in the corresponding column. If the joint is not recorded or if there is a 'zero' restraint the node freedom is given a consecutive number.

Development of the Node Freedom Array is particularly valuable and allows a wide variety of structural possibilities to be catered for.

For example, in this case, where the change in length of 1–2 is neglected the x displacements of nodes 1 and 2 are the same. They are therefore given the same freedom number.

Element Freedom Array

For each element an array (1 × 6) is defined in the table below.

Element	End 1			End 2		
1–3	1	0	2	0	0	0
1–2	1	0	3	1	0	4
2–4	1	0	4	0	0	0

At end '1' of element 1–3 the node number is 1 so that the freedom numbers are found from node 1 in the Node Freedom Array and entered. At end '2' the node number is 3 and again the freedom numbers are found and entered.

These arrays now number correctly the rows and columns of an element stiffness matrix which must have been already transformed into the overall structural axes system.

Structural Stiffness

From the above the coefficient 1,3 in the stiffness matrix for structural element 1–3 carries the freedom numbers 1,2 and therefore has to be added to the coefficient k_{12} of the structure stiffness matrix. This automatic process is readily programmed for computer use.

8.16 Member Loads

The procedures so far have treated structures loaded only at joints. Where concentrated loads are applied within a member length it is possible to consider the load point to be a joint. This allows the procedures to be used unaltered but increases the size of the problem.

Normally superposition of solutions would be used, e.g. Fig. 8.14 (*a*) shows such a structure. In Fig. 8.14 (*b*) restraints are added to remove all degrees of freedom and the restraint actions calculated by orthodox methods. The restraints are then removed, Fig. 8.14 (*c*), and present the standard 'joint loads only' problem.

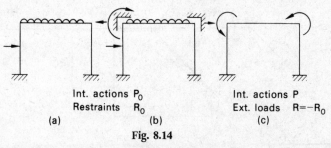

	Int. actions P_0	Int. actions P
	Restraints R_0	Ext. loads $R = -R_0$
(a)	(b)	(c)

Fig. 8.14

The complete solution is then Fig. 8.14 (*b*) superposed on Fig. 8.14 (*c*).

Then as in 8.8

$$R = Kd$$

or $$-R_0 = Kd$$

i.e. $$d = -K^{-1}R_0$$

then: $$P = ku = kDd \text{ (as before)}$$

final ext. actions $R_0 + R = 0$

final int. actions $= P_0 + P$

8.17 Examples for Practice

1. Sketch the bending moment diagram and calculate the support reactions for the uniform portal frames shown in Figs. 8.15 and 8.16.

In the case of Fig. 8.15 the free body for the sway equation is obtained as described on p. 232 using sections 1–1 and 2–2 but, due to the unequal column lengths and taking moments about A, say, the equilibrium equation is

$$M_{BA} + M_{AB} + M_{CD} + M_{DC} + H_D a + Wb = 0$$

(*Ans.* $M_{AB} = -10$ kN-m; $M_{DC} = -8.6$ kN-m.)

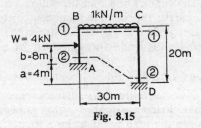

Fig. 8.15

Figure 8.16 shows a more complex type of sway problem; observing the distorted shape it is seen that member BC also rotates. The inset figure shows the geometrical relations between δ_{BA} and δ_{BC}; the effects at joint C are similar. Since joint C is pinned, θ_C is not relevant and only two degrees of freedom remain viz, θ_B and δ_1.

(*Ans.* $M_{DC} = -9.05$ kN-m.)

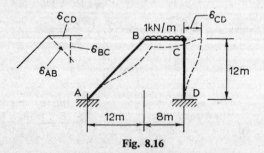

Fig. 8.16

Note: Dealing with this example without modifying the matrix procedures presented it is necessary to define two additional freedoms θ_{CB} and θ_{CD} to allow for the two separate member end slopes at the pin joint C, i.e. the Node Freedom Array would be:

Node	Freedom No.		
	x	y	θ
B	1	2	3
C (of CB)	1	0	4
C (of CD)	1	0	5
etc.			

The 6×6 element matrices are then used as before. In computer programming a standard approach such as this is desirable while, for hand calculations, it would be easier to use reduced element stiffnesses (page 241 for example) for elements BC and CD.

2. The frame shown in Fig. 8.17 carries distributed loading on the beam and a side load at D. Calculate the secondary bending moments in the members if

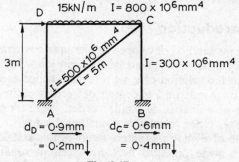

Fig. 8.17

all joints are fabricated rigid. In this case, due to the lateral load on the beam, the slope–deflexion equations for the beam will contain the fixed end moment, i.e.

$$M_{AB} = (2EI/L)(2\theta_A + \theta_B - 3\delta/L) + M_{AB}^F$$

(*Ans.* $M_{CA} = +18\cdot53$ kN-m.)

Note: Use 'Secondary Stress' methods of 8·7 for this problem.

3. Set up the stiffness matrix for the continuous beam shown in Fig. 8.18, where the stiffness of the spring support at B is k_B tonf/in.

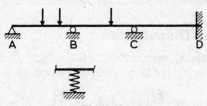

Fig. 8.18

9

Successive Approximations

9.1 Introduction

The stiffness method developed in the previous chapter provides a convenient procedure for analysing rigid jointed frames, requiring at the end the solution of a set of simultaneous equations equal in number to the independent degrees of freedom of the structure or the inversion of a matrix of similar order. Manual calculations are therefore restricted to relatively small problems.

The use of digital computers removes this restriction, but there remains a wide practical field in which personal calculation is desirable and the method of moment distribution provides one of the most useful methods of analysis. Moment distribution is a method of successive approximations forming a special case of general relaxation methods and was developed much earlier. This chapter develops the method of moment distribution and also a systematic procedure for successive approximation to the solution of the beam-column differential equation (p. 110 para. 4.2). In all cases the structures are such that axial and shear deformations of members can again be neglected.

9.2 Moment Distribution

Considering a typical member BC of the rigid frame ABCD of Fig. 9.1 (a), and using the slope–deflexion equation (8.4) (p. 224),

$$M_{BC} = \frac{2EI}{L}\left(2\theta_B + \theta_C - \frac{3\delta}{L}\right) + M_{BC}^F$$

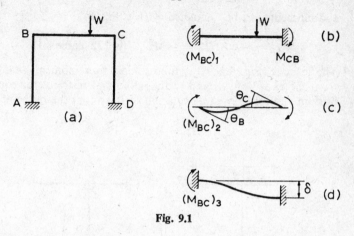

Fig. 9.1

This can be conveniently split into three parts and each part dealt with separately—

(a) $(M_{BC})_1 = M_{BC}^F =$ the effect of lateral loads when all movements of the joints are prevented; i.e. the fixed end moment, (Fig. 9.1 (b)).

(b) $(M_{BC})_2 = \dfrac{2EI}{L}(2\theta_B + \theta_C) =$ the effect of the rotations of the joints at the member ends, i.e. no lateral loads and $\delta = 0$ (Fig. 9.1 (c)).

(c) $(M_{BC})_3 = \dfrac{2EI}{L}\left(-\dfrac{3\delta}{L}\right) =$ the effect of sway (rotation) of the member only i.e. no lateral loads and $\theta = 0$ (Fig. 9.1 (d)).

In the first instance, the method of moment distribution will be used to deal separately with (a) and (b), and the sway (c) will be eliminated. This is best done by a specific demonstration as in Example 9.1 (p. 261).

9.3 Moment Distribution Procedure

(a) Following the stiffness procedure (Chap. 8 p. 235), all degrees of freedom of the joints are restrained. In Fig. 9.2 the only free joint is B, which cannot translate if change in length of the members is negligible, so that the only possible displacement is a rotation of B. Fixing joint B isolates each member and the fixed end moments can be calculated as (Fig. 9.2 (b))

$$M_{AB}^F = -wL_1^2/12 = -M_{BA}^F \quad \text{(p. 222 Table 8.1)}$$

(b) The joint B (Fig. 9.2 (c)) is thus subject to a moment $-M_{AB}^F$ from the end B of AB and to the additional restraint for equilibrium and for preventing rotation of B, so that the restraint is

$$R = +M_{BA}^F$$

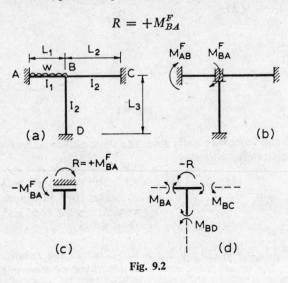

(a)   (b)   (c)   (d)

Fig. 9.2

The restraint R is called the *out of balance moment* on joint B; if others of the members meeting at the joint carried lateral loads,

$$R = \sum M_{(joint\ B)}^F$$

(c) The constraint R is artificial and must finally be released, i.e. joint B must be allowed to rotate. A balancing moment equal and opposite to R will reduce the constraint to zero and this is now applied (Fig. 9.2 (d)) and will cause rotation of B and moments in all members meeting at the joint. Thus the balancing moment is distributed among all the members meeting at the joint such that

$$-R + \bar{M}_{BA} + \bar{M}_{BC} + \bar{M}_{BD} = 0 \qquad (9.1)$$

for equilibrium, where $\bar{M}$ indicates the moment *on the joint*, i.e.

$$\bar{M}_{BA} = -M_{BA}, \text{ etc.}$$

If the rotation of B is θ and the far ends of all members remain fixed, using the original notation (p. 224),

$$\left.\begin{array}{l} M_{BA} = 4EI_1\theta/L_1 \text{ (eqn. (8.4) for the moment} \\ M_{BC} = 4EI_2\theta/L_2 \qquad\qquad\quad \text{\textit{on the member})} \\ M_{BD} = 4EI_3\theta/L_3 \end{array}\right\} \quad (9.2)$$

Therefore, from (9.1),

$$-R = (4EI_1/L_1 + 4EI_2/L_2 + 4EI_3/L_3)\theta$$

From (9.2),

$$M_{BA} = \frac{-R}{4EI_1/L_1 + 4EI_2/L_2 + 4EI_3/L_3}(4EI_1/L_1)$$

From Chap. 8

$$4EI/L = \text{direct end stiffness of a member} = k$$

Therefore

$$M_{BA} = -R\frac{k_{BA}}{\sum_B k}$$

Thus the balancing moment is distributed to the members meeting at a joint in proportion to their stiffnesses, and in the usual case where E is constant it is sufficient to take

$$k \equiv I/L$$

Therefore

$$M_{BA} = -R\frac{(I/L)_{BA}}{\sum (I/L)}$$

The joint B has now been freed, allowed to rotate, and the member end moments at B are now the sum of the above and the original fixed end moments.

(d) Due to the rotation of B, moments will be developed at the far ends of members, i.e. " carried over,"

$$M_{AB} = (2EI_1/L)(\theta_B) \text{ from eqn. (8.4)}$$
$$= \tfrac{1}{2}M_{BA} \text{ from eqn. (9.2)}$$

Thus one-half of the balancing moment is carried over to the far ends of all members.

(e) Where more than one joint restraint has been applied, this procedure is applied to each one in turn, and the cycle repeated until the out-of-balance moments are as small as desired, remembering from paragraph (c) that each time a joint is balanced (released) all other joints are held fixed.

Example 9.1. Numerical Evaluation

(a) *Member stiffnesses.* In this case (Fig. 9.2 (a))

$$k_1 = I_1/L_1 = 3$$
$$k_2 = I_2/L_2 = 2$$
$$k_3 = I_3/L_3 = 1$$

(b) *Distribution factors.* The proportions of the balancing moment which is carried by each member at a joint is given by

$$M_1 = R \frac{k_1}{\sum k} = R \times \text{(distribution factor)} = R \times D$$
$$D_1 = \tfrac{3}{6} = \tfrac{1}{2}$$
$$D_2 = \tfrac{2}{6} = \tfrac{1}{3}$$
$$D_3 = \tfrac{1}{6}$$

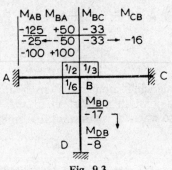

Fig. 9.3

(Note that for equilibrium $D_1 + D_2 + D_3 = 1 \cdot 0$.) It is useful to write this factor on a diagram of the structure (Fig. 9.3).

(c) *Fixed-end moments* (Table 8.1, p. 222)

$$M_{BA}^F = + wL_1^2/12 = 100, \text{ say}$$
$$M_{AB}^F = - wL_1^2/12 = -100$$

(d) For simple structures it is convenient to record the calculations on top of a diagram of the structure; thus in Fig. 9.3 at A write M_{AB}^F and at end B of AB write M_{BA}^F.

(e) Examine each joint for the out-of-balance moment; thus at B (the only free joint) the out-of-balance is $+100$.

(f) Apply a balancing moment of -100 at B and distribute this to the members at B, giving

$$\left. \begin{array}{l} M_{BA} = \tfrac{1}{2}(-100) = -50 \\ M_{BC} = \tfrac{1}{3}(-100) = -33 \\ M_{BD} = \tfrac{1}{6}(-100) = -17 \end{array} \right\} \sum M = -100$$

These figures are recorded in columns and a line drawn to show that joint B has been released.

(g) Carry over one-half of the balancing moments to the far ends of all members and again record on the diagram.

(h) There are no other joints to be released, so that this solution is complete and the final end moments are obtained by summing up all the columns as shown. It will have been observed that in this example all figures have been rounded off to zero decimal places, but that always, at every joint to be balanced,

$$\sum M = 0$$

If additional figures are retained in the calculation it will be seen that this rounding-off has led to errors of about

4 per cent in the small moments M_{DB} and M_{CB}

2 per cent in the moment M_{BD}

These errors can be made as small as desired by retaining the necessary accuracy in the arithmetic.

(i) The final bending moment diagram can now be obtained by drawing, first, the end moment (or reactant moment) diagram, e.g. AabB for member AB (Fig. 9.4 (a)) and adding to it the simple beam moments due to lateral loads (Fig. 9.4 (b)) to give

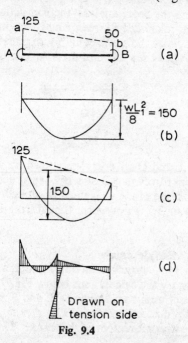

Drawn on
tension side

Fig. 9.4

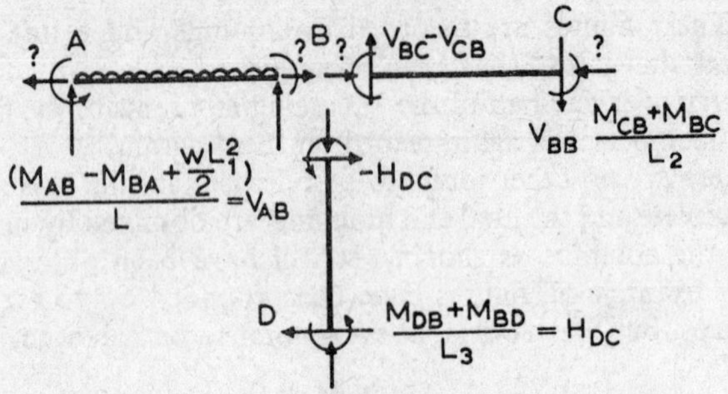

$V_{BC} - V_{CB}$

$\dfrac{(M_{AB} - M_{BA} + \dfrac{wL^2}{2})}{L} = V_{AB}$

$-H_{DC}$

V_{BB} $\dfrac{M_{CB} + M_{BC}}{L_2}$

$\dfrac{M_{DB} + M_{BD}}{L_3} = H_{DC}$

Fig. 9.5

the final diagrams Figs. 9.4 (c) and (d). It must be remembered that the sign convention developed for the slope deflexion equations (p. 221) is different from that used to draw a bending moment diagram (i.e. a stress resultant diagram) in Chap. 2. In this case since M_{AB} is negative, i.e. anticlockwise, the upper fibres of AB are in tension at end A and the ordinate has been drawn on this side. This convention has been used for all diagrams in Fig. 9.4; it is quite arbitrary and should be stated on all such sketches.

(j) To complete the force analysis the end actions on all members can now be calculated by removing them one at a time as free bodies and using the static equilibrium equations as in Fig. 9.5. No account has been taken of the axial changes in length and as

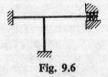

Fig. 9.6

both ends A and C are completely fixed no values can be found for the axial forces in AB and BC. In order to specify this the condition at one of the ends must be altered, e.g. in Fig. 9.6, the axial force in BC is now zero, though rotational fixity is maintained.

Example 9.2. Multiple Joints

The extension to a structure having more than one joint free to rotate is shown by the calculations below Fig. 9.7. The rigid frame ABCDEF is continuous and joints D, E and F are fully continuous; thus the short pin-ended member FG serves no purpose other than to prevent sideways movement of the frame. The supports A, B and

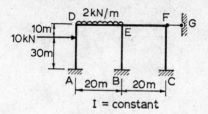

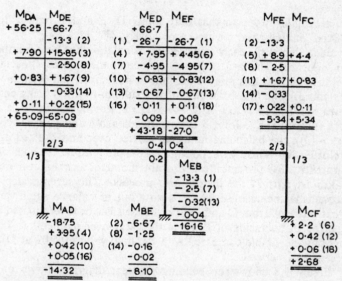

1. *Fixing moments*

$$M_{AD}^F = \frac{-10 \times 30 \times 10^2}{40^2} = -18\cdot75 \text{ kN-m}$$

$$M_{DA}^F = \frac{10 \times 30^2 \times 10}{40^2} = +56\cdot25 \text{ kN-m}$$

$$M_{DE}^F = \frac{-2 \times 20^2}{12} = -66\cdot7 \text{ kN-m} = -M_{ED}^F$$

2. *Stiffness factors*

$$(I/L)_{BE} = (I/L)_{EF} = I/20$$
$$(I/L)_{columns} = I/40$$

3. *Balancing factors*

Joint D $D_{DA} = \dfrac{I/40}{I/40 + I/20}$

$\qquad = \tfrac{1}{3} \; (= D_{FO})$

Joint E $D_{ED} = \dfrac{I/20}{I/20 + I/20 + I/40}$

$\qquad = 2/5 \; (= D_{EF})$

Fig. 9.7

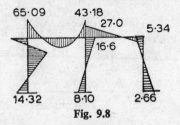

65·09   43·18
              27·0   5·34
                16·6

14·32   8·10   2·66

Fig. 9.8

C are rigidly fixed so that only joints D, E and F require to be "fixed" initially and released successively.

Each step in the early stages of the calculation has been numbered in sequence. The largest out-of-balance moment ($+66\cdot7$) occurs at joint E, so that this joint is the first to be relaxed and the moments balanced. The first balancing of E results in ($-13\cdot3$) being carried over to D, ($-13\cdot3$) to F and ($-6\cdot67$) to B, leaving an out-of-balance of ($-23\cdot75$) at D. Joint E is now locked again and joint D relaxed (step (3)), the balancing moments carried over and D fixed again. Joint F is now dealt with, noting that the pin joints at the ends of the short strut FG prevent it carrying any moments and that it therefore takes no part in the distribution process. The seventh step is a second balancing of E which is now out of balance again due to "carry overs" from D and F. The process has been terminated after the fourth balancing of E, since the carry-over moments ($-0\cdot045$) are now negligible compared with the final values ($65\cdot09$ at D), i.e. less than 1 per cent.

Figure 9.8 shows the bending moment diagram drawn on the "tension" side of members.

In subsequent problems it will be necessary to calculate the force in the short prop FG; this is done in Fig. 9.9. Particular short cuts

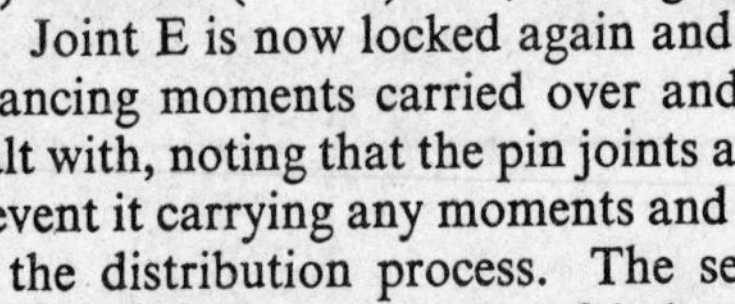

$H_1 \dashv$ $H_2 \dashv$ $\dashv H_F$
                                $\dashv H_3$
10kN

$$H_1 = \frac{(10 \times 30) + 65\cdot09 - 14\cdot32}{40} = 8\cdot77$$

$$H_2 = \frac{16\cdot6 + 8\cdot10}{40} = 0\cdot62$$

$$H_3 = \frac{5\cdot34 + 2\cdot66}{40} = 0\cdot20$$

$$H_F = 8\cdot77 - 0\cdot62 + 0\cdot20 = 8\cdot35 \text{ kN}$$

Fig. 9.9

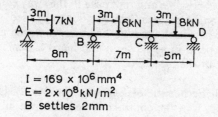

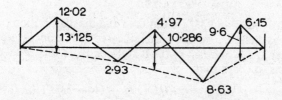

1. Fixed end moments

$$M_{AB}^{F} = \frac{-3 \times 5^{2} \times 3}{8^{2}} - \frac{6 \times 2 \times 10^{8} \times 169}{\phantom{8^{2}}} \times 10^{6} \times 2 \times 10^{-3}$$
$$= -8 \cdot 2 - 6 \cdot 35 = -14 \cdot 55 \text{ kN-m}$$

$$M_{BA}^{F} = \frac{+3 \times 5 \times 3^{2}}{8^{2}} - 6 \cdot 35 = -1 \cdot 43 \text{ kN-m}$$

$$M_{BC}^{F} = \frac{-6 \times 4^{2} \times 3}{7^{2}} + 6 \cdot 35 \left(\frac{8}{7}\right)^{2}$$
$$= -5 \cdot 88 + 8 \cdot 28 = 2 \cdot 40 \text{ kN-m}$$

$$M_{CB}^{F} = \frac{+6 \times 4 \times 3^{2}}{7^{2}} + 8 \cdot 28 = 12 \cdot 69 \text{ kN-m}$$

$$M_{CD}^{F} = \frac{-8 \times 2^{2} \times 3}{5^{2}} = -3 \cdot 54 \text{ kN-m}$$

$$M_{DC}^{F} = \frac{+8 \times 2 \times 3^{2}}{5^{2}} = +5 \cdot 76 \text{ kN-m}$$

2. Stiffness factors

Member BA $\frac{3}{4}I/8$
Member BC $I/7$
Member CD $\frac{3}{4}I/5$

3. Distribution factors

Joint B $D_{BA} = \dfrac{3/32}{3/32 + 1/7}$
$= 0 \cdot 396$

Joint C $D_{CD} = \dfrac{3/20}{3/20 + 1/7}$
$= 0 \cdot 512$

Fig. 9.10

are sometimes possible and reference to the sway equation on p. 232 will show a rapid treatment which can then be used. But the method of Fig. 9.8, where each member has been removed and analysed on its own, is general in application and helps to remove ambiguity in the direction of the force FG.

Example 9.3. Settlement of Supports and Pin Ends

If the support B of the continuous beam shown in Fig. 9.10 settles under load by 2 mm., moment distribution methods can be applied readily. The calculations of fixed end moments are carried out as usual and the settlement allowed to take place on first application of the loads with the joints restrained. Thus

$$M_{AB}^F = M_{BA}^F = -6\frac{EI}{L^2}\delta \text{ (due to settlement only)}$$

$$= -\frac{6 \times 2 \times 10^8 \times 169 \times 10^{-6} \times 2 \times 10^{-3}}{8^2}$$

$$= -6{\cdot}35 \text{ kN-m}$$

Similarly M_{BC}^F and M_{CB}^F are altered and the distribution procedure can be as before. In this case, since ends A and D are simply-supported, $M_{AB} = M_{DC} = 0$. It follows, from the slope–deflexion equations (p. 224), that

$$M_{AB} = (2EI/L)(2\theta_A + \theta_B - 3\delta/L)$$

Since $M_{AB} = 0$, we get

$$4EI\theta_A = -2EI\theta_B + 6EI\delta/L$$

and substituting in eqn. (8.4), namely

$$M_{BA} = (2EI/L)(2\theta_B + \theta_A - 3\delta/L)$$

we get $\qquad M_{BA} = (3EI/L)(\theta_B - \delta/L)$

Thus, in Fig. 9.11 (*a*), the stiffness of the beam with far end fixed is,

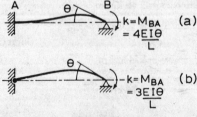

Fig. 9.11

from eqn. (8.4),

$$k = M_{BA} = 4EI/L, \qquad (\delta/L = 0)$$

and in Fig. 9.11 (*b*), if the far end is a pin joint as above

$$k' = M_{BA} = 3EI/L, \qquad (\delta/L = 0)$$

i.e. $\qquad k' = \tfrac{3}{4}k$

and this can be allowed for in the distribution procedure by adjusting the distribution factors, i.e. for joint B—

$$D_{BA} = \frac{\tfrac{3}{4}k_{BA}}{\sum k} = \frac{\tfrac{3}{4}I/8}{\tfrac{3}{4}I/8 + I/7} = 0.396$$

D_{CD} is obtained in the same way.

The sequence of operations is shown in Fig. 9.10 and the first steps are seen to be the relaxing of A and D and carry-overs to B and C respectively. Subsequent cycles of balancing at B and C have already allowed for the fact that A and D are pins, so that there are no carry-overs back to these points.

9.4 Symmetry and Skew Symmetry

In a symmetrical structure symmetrically loaded as in Fig. 9.12, symmetrical joints rotate the same amount but in opposite directions. Making use of this fact, the end stiffness of a member such as AB in

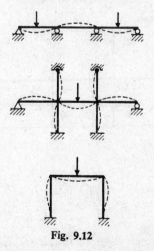

Fig. 9.12

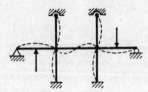

Fig. 9.13

Fig. 9.13 is

$$M_{AB} = (2EI/L)(2\theta_A + \theta_B) \tag{9.3}$$

But $\theta_A = -\theta_B$

Therefore $M_{AB} = 2EI\theta_A/L$

or $k = 2EI/L$

Comparing this with Fig. 9.11 (*a*),

$$k'' = \tfrac{1}{2}k$$

Thus if the distribution factors at the ends of the members common to each half of the structure are adjusted the distribution procedure need only be carried out on one half and there will be no carry-over moments across the axis of symmetry.

A symmetrical structure with skew-symmetrical loading is shown in Fig. 9.14, where symmetrical joints rotate the same amount but, this time, in the same direction. Thus (Fig. 9.15) there is a point of contraflexure at mid-span, i.e.

$$\theta_A = \theta_B$$

and from eqn. (9.3),

$$M_{AB} = 6EI\theta/L$$

or $k''' = 6EI/L$

Fig. 9.14

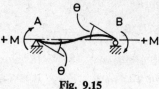

Fig. 9.15

so, comparing with Fig. 9.11 (a),

$$k''' = \tfrac{3}{2}k$$

(Alternatively, consider one half as a pin-ended member, $L/2$ long, then $k''' = \tfrac{3}{4}I/(\tfrac{1}{2}L) = \tfrac{3}{2}k$ as above.) Again, if the distribution factors at the ends of the common members are adjusted, there is no carry-over across the axis of symmetry and only half the structure need be analysed.

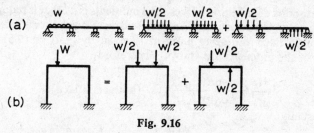

Fig. 9.16

Since any load system can be broken down into two systems, one symmetrical and one skew symmetrical, these devices are so useful that they cannot be classed as "dodges." Two examples are given in Fig. 9.16.

9.5   Moment Distribution with Sway

So far the structures dealt with have been such that sway did not occur or was prevented (Fig. 9.7). A typical problem involving side sway is shown in Fig. 9.17 (a). Here

$$M_{AB} = (2EI/L)(2\theta_A + \theta_B - 3\delta/L) \qquad \text{(from eqn. (8.4) p. 224)}$$

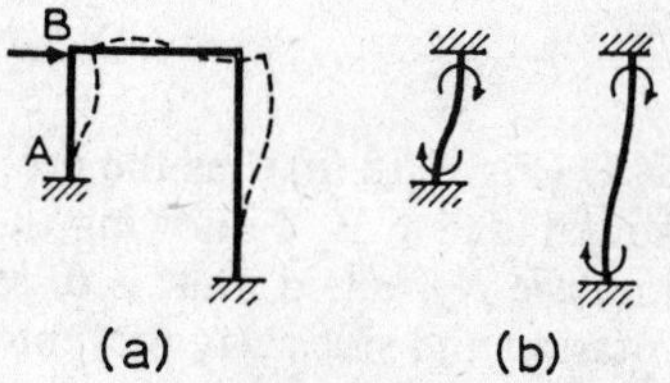

Fig. 9.17

But $\theta_A = 0$. Therefore

$$M_{AB} = \qquad (2EI/L)\theta_B \qquad - \qquad \frac{6EI\delta}{L^2}$$

$$\begin{Bmatrix} \text{Joint rotations of} \\ \text{previous analysis} \end{Bmatrix} \qquad \begin{Bmatrix} \text{Sway or translation} \\ \text{previously neglected} \end{Bmatrix}$$

If the magnitude of the sway (value of δ) is known, the problem is identical to Example 9.3, the free joints are locked against rotation and all the sway allowed to occur. The member end moments (fixed end moments) are then

$$M = -(6EI\delta/L^2) \qquad \text{for fixed ends} \tag{9.4}$$

$$M = -(3EI\delta/L^2) \qquad \text{for one end pinned} \tag{9.5}$$

and orthodox distribution of these end moments is all that is required.

In general the sway is not known, and may not be required, but since each fixed end sway moment is given by (9.4) or (9.5), the ratios of the moments are known. Therefore

$$\frac{M_1}{M_2} = \frac{-6EI_1\delta_1/L_1^2}{-6EI_2\delta_2/L_2^2} \qquad \text{(both ends fixed)}$$

and usually

$$\delta_1 = \delta_2 \qquad \text{for simple cases}$$

Therefore

$$\frac{M_1}{M_2} = \frac{I_1/L_1^2}{I_2/L_2^2}$$

Two methods are available—
 (i) an arbitrary value for δ (Fig. 9.17) is assumed, or
 (ii) arbitrary values for the fixed end sway moments, in the correct ratios, are assumed.

Then all the member end moments are obtained by orthodox moment distribution, i.e. allowing no more sway but releasing the joint rotations. Finally any necessary corrections are made, as in the following example.

Example 9.4

The rigid frame ABCD (Fig. 9.18 (*a*)) has the proportions given and carries a 2 kN horizontal load at B. Following the usual procedure, all degrees of freedom are restrained, that is to say, joints B and C are locked against rotation and side sway prevented by adding a side support CE (Fig. 9.18 (*b*)).

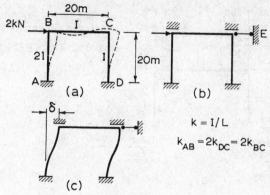

Fig. 9.18

Fixed end moments should now be calculated. There are no lateral loads on members, hence there are no fixed end moments and therefore, for equilibrium, the force in CE must be 2 kN compression. Since there are no end moments there is no out-of-balance moment at any joint and no moments to distribute, and the joint rotational restraints at B and C have nothing to do.

Knowing that the structure does in fact sway, retain the rotational constraints but allow some sway (δ) to take place (Fig. 9.18 (c)). Therefore

$$M_{AB} = M_{BA} = -(6EI\delta/L^2)_{AB} = -(6Ek\delta/L)_{AB}$$
$$M_{DC} = M_{CD} = -(6EI\delta/L^2)_{DC} = -(6Ek\delta/L)_{DC}$$

or
$$\frac{M_{BA}}{M_{CD}} = \frac{k_{AB}}{k_{DC}} = \frac{2}{1}$$

(since $L_{AB} = L_{DC}$ and $\delta_{AB} = \delta_{DL}$).

Assume that $\quad M_{BA} = M_{AB} = +120$
Therefore $\quad M_{DC} = M_{CD} = +60$

The fixed-end moments (due to sway δ) are now as shown in Fig. 9.19 (a) as the first entries in the respective columns. Relaxing the rotational constraints while retaining the translation constraint gives the standard moment distribution process shown. Balancing commences with joint B and has been stopped after a second balance at B.

Isolating each member as a free body (Fig.9.20(a)), for equilibrium the force in the prop CE is found to be 11·0 kN compression, i.e. the moments are the internal actions of the frame ABCD loaded as in Fig. 9.20 (b) and the solution for the frame in Fig. 9.20 (c) is also known (zero moments in this case). The solution is required for the

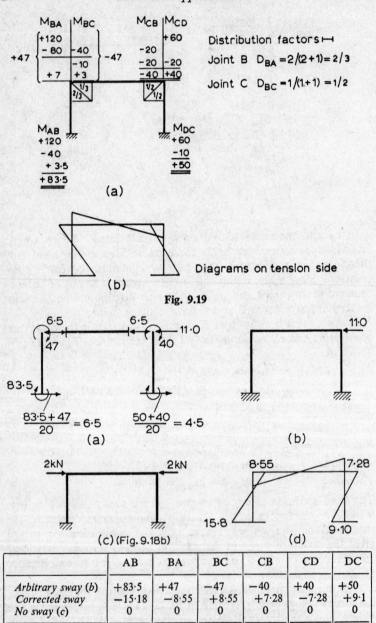

Distribution factors ⊢—⊣

Joint B $D_{BA} = 2/(2+1) = 2/3$

Joint C $D_{BC} = 1/(1+1) = 1/2$

Diagrams on tension side

(b)

Fig. 9.19

$\frac{83.5+47}{20} = 6.5$

$\frac{50+40}{20} = 4.5$

(a)

(b)

(c) (Fig. 9.18b)

(d)

	AB	BA	BC	CB	CD	DC
Arbitrary sway (b)	+83·5	+47	−47	−40	+40	+50
Corrected sway	−15·18	−8·55	+8·55	+7·28	−7·28	+9·1
No sway (c)	0	0	0	0	0	0
Total	−15·18	−8·55	+8·55	+7·28	−7·28	+9·10

Fig. 9.20

loading of Fig. 9.18 (*a*), that is all forces and moments of Fig. 9.20 (*b*) multiplied by a correction factor

$$c = -2/11 \cdot 0$$

and combined with Fig. 9.20 (*c*) give the moments and forces of Fig. 9.20 (*d*), which is the solution of the original problem.

Summarizing, the procedure is as follows—

(1) Lock all joint rotations and provide support to prevent sway,
(2) Carry out normal moment distribution (none in this case),
(3) Calculate the force R_0 to prevent sway,
(4) Allow an arbitrary sway (or moments in proportion) with zero joint rotations,
(5) Balance and distribute sway moments,
(6) Calculate a new force R_1 to prevent sway,
(7) Taking the correction factor $c = -R_0/R_1$,
(8) Multiply all moments in (5) by c and add to the moments in (2) as collected in the table in Fig. 9.20.

In order to keep the two separate moment distributions to the same accuracy it is desirable to choose arbitrary sway moments of as near to the correct values as possible (correction factor $c = 1$); this, of course, is difficult and should be regarded by the student as a refinement for later consideration.

Example 9.5

Consider the rigid frame in Fig. 9.21 (*a*). Add a horizontal restraint R_0 at C to prevent sway (Fig. 9.21 (*b*)) and lock all joints against rotation. Calculation of the distribution factors and fixed end moments gives the values shown in rows 1 and 2 of the table and the results of the first (no sway) distribution in row 3.

Separation of the members and equilibrium calculations as shown at (*c*) give the force R_0 equal to 71·6 kN compression.

Allowing an arbitrary sway to the right, with no joint rotation, produces the fixed end sway moments of row 4 in the table; balancing and distribution of these gives row 5. Analysis of the equilibrium of each member in (*d*) shows the restraint R_1 to be 1,184 kN tension.

Then, for no lateral restraint,

$$R_0 + cR_1 = 0$$

or

$$c = -R_0/R_1 = \frac{71}{1,532} = 0 \cdot 046$$

and the final two rows of the table are completed.

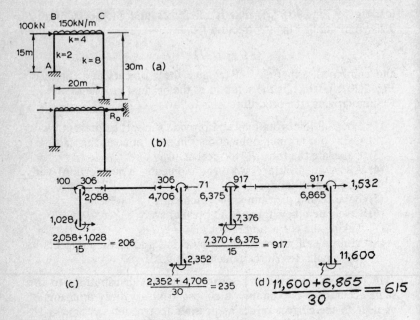

Fig. 9.21

		AB	BA	BC	CB	CD	DC	R_0
1	*Dist. factor*		$\frac{1}{3}$	$\frac{2}{3}$	$\frac{1}{8}$	$\frac{2}{3}$		
2	M^F kN-m			−5,000	+5,000			
3	*No sway*	+1,028	+2,058	−2,058	+4,706	−4,706	−2,352	−71
4	*Pure sway*	−8,350	−8,350	—	—	−16,700	−16,700	
5	*Balanced sway*	−7,370	−6,375	+6,375	+6,865	−6,865	−11,600	+1,532
6	*Corrected sway*	−341	−295	+295	+318	−318	−537	+71
7	*Total* (3) + (6)	+687	+1,763	−1,763	+4,388	−4,388	−2,889	0

9.6 Sloping Members

In frames with sloping members the sway is a little more complicated but is dealt with in the same way as for the slope-deflexion equations, p. 255. A typical example is shown in Fig. 9.22 where the single degree of freedom is restrained by a single horizontal prop. Allowing an arbitrary sway, shown dotted in the figure, will then produce a rotation of the horizontal member as well as the columns and there is a simple geometrical relationship between the rotations.

Evaluation of the restraint due to the arbitrary sway, as shown in the free-body diagrams, requires a little extra care since the beam shears will now affect the equilibrium of the sloping member DE.

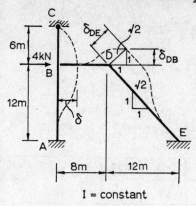

I = constant

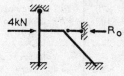

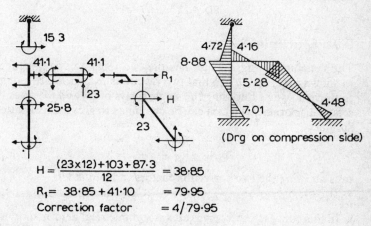

(Drg on compression side)

$$H = \frac{(23 \times 12) + 103 + 87 \cdot 3}{12} = 38 \cdot 85$$

$$R_1 = 38 \cdot 85 + 41 \cdot 10 = 79 \cdot 95$$

Correction factor $= 4/79 \cdot 95$

Distribution factors

Joint B $\quad D_{BC} = \dfrac{3/4\ I/6}{3/4\ I/6 + I/12 + I/8} = \dfrac{3}{8}$

$$D_{BA} = \dfrac{I/12}{3/4\ I/6 + I/12 + I/8} = \dfrac{1}{4}$$

Joint D $\quad D_{DB} = \dfrac{I/8}{I/8 + I/12\sqrt{2}} = 0 \cdot 68$

Fig. 9.22

Sway moments

$$M_{BA} = M_{AB} = \frac{-6EI\delta}{12^2} = \frac{-EI\delta}{24}$$

$$M_{BC} = \frac{+3EI\delta}{6^2} = \frac{+EI\delta}{12} = -2M_{BA}$$

$$M_{BD} = M_{DB} = \frac{+6EI\delta}{8} = -2 \cdot 25 M_{BA}$$

$$M_{DE} = M_{ED} = \frac{-6EI\delta\sqrt{2}}{(12\sqrt{2})^2} = 0 \cdot 707 M_{BA}$$

	AB	0·25	0·375	0·375	0·68	0·32	ED	
		BA	BC	BD	DB	DE		
1	−100	−100	+200	+225	+225	+70·7	−70·7	*Pure sway*
2	−136·5	−173	+92	+81	+103	−103	−87·3	*Balanced sway*
3	−7·01	−8·88	+4·72	+4·16	+5·28	−5·28	−4·48	*Corrected sway moments*

Fig. 9.22 (continued)

9.7 Multiple Sway

The structure of Fig. 9.23 (a) has the two separate sway freedoms, as shown in (c) and (d), so that two restraints R_1 and R_2 are required to prevent sway. Imposing the single sways of (c) and (d) gives two sets of displacements which can be combined to give the real pattern, so that

$$R_{10} + \alpha R_{11} + \beta R_{12} = 0$$
$$R_{20} + \alpha R_{21} + \beta R_{22} = 0$$

thus allowing the two correction factors to be found as

$$\alpha = 1 \cdot 69; \qquad \beta = 1 \cdot 93$$

In this case each sway gives an anti-symmetrical deformation with a point of contraflexure at the centre of each beam and only half of the frame need be analysed. Thus the effective stiffness of BE is

$$k_{BE} = \tfrac{3}{4}I/\tfrac{1}{2}L$$

or the distribution factor for BE is

$$D_{BE} = \frac{\tfrac{3}{4}I/\tfrac{1}{2}L}{\tfrac{3}{4}I/\tfrac{1}{2}L + I/L + I/L}$$
$$= \tfrac{3}{7}$$

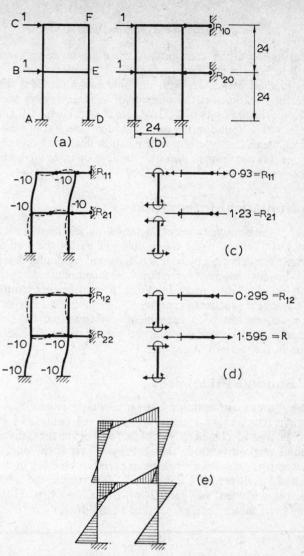

Fig. 9.23

(Linear dimensions in metres)

	AB	2/7 BA	4/7 BE	2/7 BC	2/5 CB	3/5 CF
Sway 1: (c)	+1·19	+2·37	+3·50	−5·87	−5·31	+5·31
Sway 2: (d)	−8·53	−7·06	+4·41	+2·66	+0·88	−0·88
α × (c)	+2·02	+4·02	+5·95	−9·97	−9·02	+9·02
β × (d)	−16·45	−13·60	+8·50	+5·12	+1·70	−1·70
Total	−14·43	−9·58	+14·45	−4·85	−7·32	+7·32

and during the distribution process there are no carry-overs from joints B to E or from C to F.

This method of dealing with multiple sways provides the same number of equations as the number of arbitrary sways and is, of course, not the only way of dealing with the problem. Other special methods will be found in the technical literature, but in view of the ease with which this type of problem can be dealt with on the digital computer it is not thought necessary to develop these methods in the present context.

9.8 Numerical Integration

In Chap. 3 the bending moment internal stress resultant in a laterally-loaded beam is seen to be the double integral of the loading, and in Chap. 4 the deflexion of beams is obtained by double integration of the bending moment variation. For complicated patterns of loading or of bending moments these integrations are troublesome and of frequent practical occurrence. The numerical integration of such equations has been conveniently systematized by Newmark and is of such value that a condensed presentation of this treatment is given in the rest of this chapter.

9.9 General Principles

We use the sign convention for stress resultants from Chap. 3, that is, a right-hand set of axes having the X-axis coincident with the centroidal axis of a beam (Fig. 9.24 (*a*)) and define the signs of stress resultants as positive along the coordinate axis at the positive face of an element: thus a positive moment produces tension in the upper fibres and a positive load is downwards. Referring to the equilibrium of the beam element in Fig. 9.24 (*b*) (*see* Chap. 4, p. 110), it has already been shown that, for vertical equilibrium,

$$\frac{dQ}{dx} = -w$$

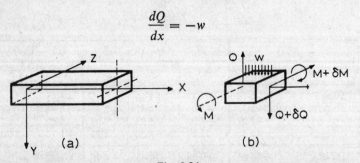

(a)   (b)

Fig. 9.24

Integrating,

$$Q = Q_0 - \int w \, dx \tag{9.6}$$

where $Q = Q_0$ at $x = 0$. For moment equilibrium, omitting second-order terms in w,

$$\frac{dM}{dx} = -Q$$

Integrating,

$$M = M_0 - \int Q \, dx \tag{9.7}$$

								Multiplier
(i)	0	4	1	2	3	0	0	Loads W	
(ii)		5	1	0	-2	-5	-5	Shear = $Q_0 - \Sigma W$	
(iii)	0	-5	-6	-6	-4	+1	+6	Moments ($M_0 - Qdx$)	λ
(iv)	0	-1	-2	-3	-4	-5	-6	$M_g \# 6$ ie. $_aQ_b \# 5$ (Linear correction)	
(v)	0	-6	-8	-9	-8	-4	0	Final moments	λ

Bending moment diagram

Fig. 9.25

where $M = M_0$ at $x = 0$. Thus, substituting for Q,

$$M = M_0 - \int Q_0 \, dx + \iint w \, dx \, dx$$

The relations are summarized as

$$w = -\frac{dQ}{dx} = \frac{d^2M}{dx^2}$$

and, provided the boundary conditions Q_0 and M_0 are known, M can be found by the double integration of the load w.

Thus any methods developed for the calculation of shears and moments in beams are, in fact, general numerical procedures for double integration and can be used as such.

Dealing, in the first instance, with uniformly-spaced concentrated loads, this procedure is conveniently set out in Fig. 9.25. The integrations now become simple summations and it is convenient to proceed from load point to load point along the beam. Thus

$$Q = Q_0 - \int w \, dx$$

becomes

$$_bQ_c = {}_aQ_b - W_b \text{ etc.}$$

and assuming $_aQ_b$ equal to $+5$, the shear in each element between load points can be written down directly. Line (i) of Fig. 9.25 records the concentrated loads and the shears are recorded in the respective intervals of line (ii).

Integration of the average shears to get the bending moments is also dealt with element by element. Thus

eqn. (9.7) $$M = M_0 - \int Q \, dx$$

becomes $$M_b = M_a - {}_aQ_b\lambda$$

and so on. Then assuming that $M_a = 0$,

$$M_b = 0 - 5\lambda = -5\lambda$$
$$M_c = (-5\lambda) - (1\lambda) = -6\lambda$$

and finally $$M_g = (+\lambda) - (-5\lambda) = +6\lambda \qquad \text{(line (iii))}$$

This value for M_g does not satisfy the known end condition of zero moment and indicates an error in the assumed end condition $_aQ_b$, that is to say, a wrong value for the first constant of integration. Thus, after the second integration the error is a linear function of x and a linear correction is necessary.

At g the correction $= -6$

that is   at f the correction $= (-\frac{6}{6})5 = -5$

at e the correction $= (-\frac{6}{6})4 = -4$, etc.

The corrections are entered in line (iv). Line (v) records the final bending moments at the load points, (the sum of lines (iii) and (iv)), and the bending moment diagram is shown below.

9.10 Treatment of Distributed Loads

With distributed loading such as Fig. 9.26 (*a*), numerical integration is carried out by breaking the loading up into finite elements, that is an equivalent set of concentrated loads which will produce the correct shears and moments at selected points along the beam.

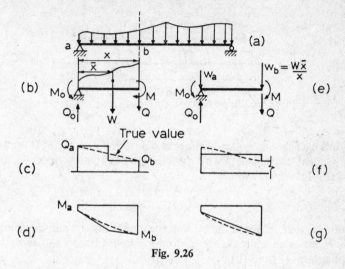

Fig. 9.26

Between these points the values will not be correct, but the points may be selected as close together as required to give the desired accuracy.

The change of shear force between any two sections is given by

$$\delta Q = -\int w \, dx$$

$$= -(\text{area of loading diagram})$$

Thus any set of concentrated loads W will do, provided that, up to any point,

$$\int w \, dx = [\text{area of load diagram}] = \sum W$$

For the bending moments, the bending moment at any point (Fig. 9.26 (b)) is

$$M = M_0 - Q_0 x + W(x - \bar{x})$$

where W is the concentrated load equal to the area of the loading diagram ($\int w \, dx$), and $\bar{x}$ is the distance to the C.G. of the distributed loading.

The first moments, then, of the equivalent concentrated loads about a section must be the same as for the distributed loading, and this can be achieved by placing equivalent loads at the centres of gravity of the elements of the distributed loading which they replace. The result of this would be to give the correct values for Q_b and M_b as shown, but in the length a–b there would be a discontinuity in each

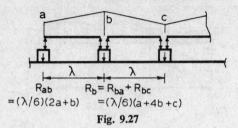

$$R_{ab}$$
$$=(\lambda/6)(2a+b)$$

$$R_b = R_{ba} + R_{bc}$$
$$=(\lambda/6)(a+4b+c)$$

Fig. 9.27

diagram as shown by the full lines in diagrams (c) and (d). It is more convenient to replace W by the statically-equivalent pair W_a and W_b of Fig. 9.26 (e) where

$$W = W_a + W_b$$

and
$$W_a\bar{x} + W_b(x - \bar{x}) = 0$$

(taking moments about the C.G. of the load) to give again the correct values but with a simple linear variation in the length a–b as shown in diagrams (f) and (g).

This is equivalent to the assumption that the distributed loading is carried on short, simply-supported stringers and the equivalent concentrated loads are then the reactions from these stringers.

Linear variation in loading is easily dealt with numerically and the reader should verify the values given in Fig. 9.27. One way of dealing with a general distribution of loading (as long as it is not discontinuous) is shown in Fig. 9.28, where the loaded length is divided up into a number of equal lengths and a second-degree parabola assumed for the load-variation over each pair of intervals. Taking the origin at a,

$$w = a + \alpha x + \beta x^2$$

and α and β are arranged to give

$$w = b \quad \text{at} \quad x = \lambda$$
$$w = c \quad \text{at} \quad x = 2\lambda$$

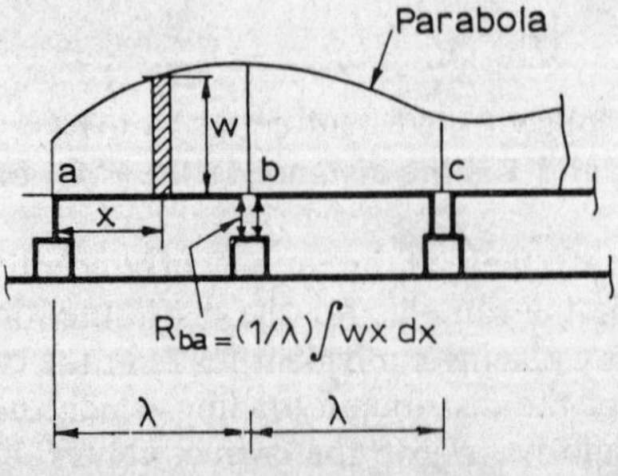

Parabola

$$R_{ba} = (1/\lambda)\int wx \, dx$$

Fig. 9.28

Then, for the length a to b,

$$R_{ba} = (1/\lambda)\int wx\,dx$$

$$= (\lambda/24)(3a + 10b - c)$$

The algebra is tedious but simple. In the same way

$$R_{ab} = \int w\,dx - R_{ba} = \frac{\lambda}{24}(7a + 6b - c)$$

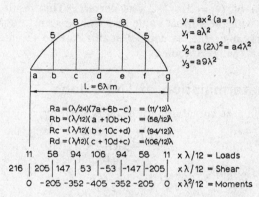

$y = ax^2\ (a = 1)$
$y_1 = a\lambda^2$
$y_2 = a(2\lambda)^2 = a4\lambda^2$
$y_3 = a9\lambda^2$

$L = 6\lambda$ m

$R_a = (\lambda/24)(7a + 6b - c) = (11/12)\lambda$
$R_b = (\lambda/12)(a + 10b + c) = (58/12)\lambda$
$R_c = (\lambda/12)(b + 10c + d) = (94/12)\lambda$
$R_d = (\lambda/12)(c + 10d + c) = (106/12)\lambda$

11	58	94	106	94	58	11	$\times \lambda/12$ = Loads
216	205	147	53	-53	-147	-205	$\times \lambda/12$ = Shear
0	-205	-352	-405	-352	-205	0	$\times \lambda^2/12$ = Moments

Fig. 9.29

Thus, combining results for a–b and b–c we get;

$$R_b = \frac{1}{12}\lambda(a + 10b + c)$$

There are other methods of selecting equivalent concentrated loads, but for most purposes the assumption of a parabolic load variation is adequate. For a deeper study the reader should refer to Newmark's paper (ref. 1, p. 291) and the references given there. An example using a distributed loading is shown in Fig. 9.29. This beam carries a parabolically-distributed loading of maximum intensity 9 kN/m. The equivalent concentrated loadings are obtained as shown and recorded in the first line of the calculations. Assuming $Q_0 = 216\lambda/12$, the shears in each length are recorded in the second line—

a–b Shear $= Q = Q_0 - W_1 = 216 - 11 = 205$

b–c Shear $= Q = 205 - 58 = 147$ etc.

In the third line, assuming $M_a = M_0 = 0$,

$$M_b = M_0 - \int Q\, dx = M_0 - Q\lambda$$

$$= 0 - 205 = -205\,(\times \lambda^2/12)$$

$$M_c = M_b - Q\lambda$$

$$= -205 - 147 = -352\,(\times \lambda^2/12)$$

and so to M_g, which is seen to satisfy the end-condition of zero moment, so that we have a correct solution; in other words, the assumption of $Q_0 = 216\lambda/12$ was correct. This result is readily compared with the correct value and will be seen to agree exactly at all points a–g, due, of course, in this case to the fact that the loading is actually parabolic.

9.11 Determination of Deflexions

From p. 109, the curvature of a beam is given by

$$\phi = \frac{d^2y}{dx^2} = \frac{M}{EI}$$

Therefore the slope is

$$\frac{dy}{dx} = A + \int \frac{M}{EI}\, dx$$

or

$$\theta = \theta_0 + \int \phi\, dx$$

where θ_0 = slope at $x = 0$.

Again, deflexion is

$$y = B + \int A\, dx + \iint \frac{M}{EI}\, dx$$

$$= y_0 + \int \theta_0\, dx + \iint \phi\, dx$$

or, simply

$$y = y_0 + \int \theta\, dx$$

where A and B are constants of integration. These equations are identical with eqns. (9.6) and (9.7), except for a change in sign, so that once the values of curvature ($M/EI = \phi$) are known, all techniques developed for the calculation of shears and moments are directly applicable to the calculation of slopes and deflexions. The continuously variable curvature ($\phi = M/EI$) is replaced by a *concentrated angle change* Φ, in the same way as the varying load was previously replaced by equivalent concentrated loads, and instead of the average shear between two sections we have the average slope.

It is convenient, in computation, to omit constant multipliers as much as possible, and this has been done in the example of Fig. 9.30. The beam is uniformly loaded so that the bending moment diagram is parabolic, that is, the curvature ϕ varies parabolically.

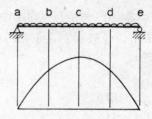

0	-3	-4	-3	0	× $wL^2/32$	= Bending moment
0	-3	-4	-3	0	× $wL^2/32EI$	= Curvature ϕ
-7	-34	-46	-34	-7	× $wL^2/32EI$ × $\lambda/12$	= Conc. angle chg.
57	23	-23	-57		× $wL^3\lambda/384EI$	= Av. slope
0	57	80	57	0	× $wL^2\lambda^2/384EI$	= Deflexion

Fig. 9.30

From Fig. 9.30, the concentrated angle change at point a is

$$\Phi_a = (\lambda/24)(7\phi_a + 6\phi_b - \phi_c)$$
$$= (\lambda/24)(0 + 6 \times 3 - 4) = (7\lambda/12)\left(\frac{wL^2}{32EI}\right)$$

and at point b

$$\Phi_b = (\lambda/12)(\phi_a + 10\phi_b + \phi_c)$$
$$= (\lambda/12)(0 + 30 + L) = (34\lambda/12)\left(\frac{wL^2}{32EI}\right)$$

and in the same way the values at c, d and e are obtained. Assuming the slope (equivalent to shear) in the length a–b is 57, we get the slope

$$_b\theta_c = {_a\theta_b} + \Phi_b$$
$$= 57 - 34 = 23$$
$$_c\theta_d = {_b\theta_c} + \Phi_c$$
$$= -23$$

Assuming zero deflexion at end a, the deflexion at the other points follows directly—

$$y = y_0 + \int \theta \, dx$$

i.e.
$$y_b = y_a + {}_a\theta_b\lambda$$
$$= 0 + 57\lambda = 57\lambda$$
$$y_c = y_b + {}_b\theta_c\lambda = 80\lambda \qquad \text{and so on.}$$

In this case the value obtained for y_e does satisfy the end condition of zero deflexion, so that the assumption of $\left(57\dfrac{wL^2}{32EI}\dfrac{\lambda}{12}\right)$ for the average slope between a and b is correct.

$$\text{Central deflexion} = 80\frac{wL^2}{32EI}\frac{\lambda^2}{12}$$

$$= \frac{5}{384}\frac{wL^4}{EI}$$

(since $\lambda = L/4$). This is the correct answer for this case.

9.12 Non-uniform Section

The curvatures of a beam are given by M/EI, so that it is only necessary to re-draw the bending moment diagram in the form of the curvature diagram and proceed as before. Some ingenuity can be exercised to keep the figures simple by a suitable choice of common multiplier, as in the example shown in Fig. 9.31. In cases such as this where an abrupt change in section occurs, the division of the curvature diagram into a number of equal lengths may be difficult but, of course, the equivalent "loadings" of Fig. 9.28 can be adjusted accordingly. Cases of non-linear elasticity, or partially plastic beams

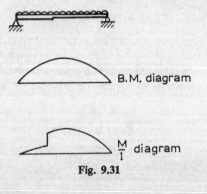

B.M. diagram

$\dfrac{M}{I}$ diagram

Fig. 9.31

are dealt with in the same way provided the moment-curvature relationship is known.

9.13 Axial Loads on Beams

Where a beam is subject to an axial load as well as to lateral loadings (Fig. 9.32) the moment at any cross-section contains a term dependent on the deflexion. Here

$$M_x = -Q_0 x + W_1 x + W_2(x - a) - Py$$

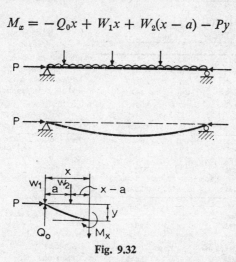

Fig. 9.32

where y is not known. The methods of numerical integration are extremely valuable in such calculations, but the reader should omit the rest of this chapter until the chapter on columns (Chap. 12) has been read.

The technique is as follows—

(a) Assume a deflected shape (this provides values for the Py term above).

(b) Draw the bending moment diagram and follow through the normal procedure of numerical integration to obtain the deflexions.

(c) Compare the calculated shape with the assumed shape; if they are identical then the assumed shape is the correct one.

(d) If the derived shape is different, repeat the calculations using this new shape or a revised assumption (in most cases use of the derived shape as a second approximation will yield a third and closer approximation).

(e) Repeat the cycle (a), (b), (c) as often as necessary.

It is true to say that in most cases this procedure will converge to the correct solution rapidly. However, this is not always so and special techniques become necessary.

Figure 9.33 shows this method applied to the problem from p. 347 of Chap. 12. The beam is divided into six segments, and the answer

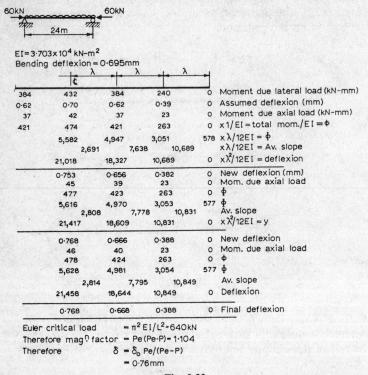

Fig. 9.33

is seen to be obtained after three trials, starting with the deflexion due to the lateral loads only. Due to symmetry only half of the span is recorded. Needless to say a good initial assumption of deformed shape aids convergence and a ridiculously bad one may lead to divergence instead of convergence of the solution or to oscillation about the correct solution.

9.14 Column Buckling Loads

Referring to Chap. 12, the critical load or elastic buckling load of a member may be defined as that load which is just sufficient to

maintain the column in a deflected shape. The difficulty is to determine the deformed shape and which of the possible forms gives the
lowest value for the initial load. It is shown in Chap. 12 that if a
reasonably correct form can be found, a close approximation to the
critical load is obtained. In this chapter the methods of numerical
integration are used to work back from an assumed shape to the
required axial load, the only real difficulty being again that of making

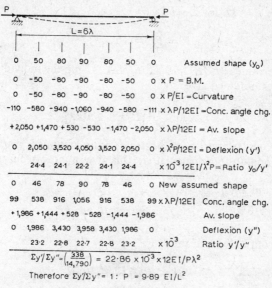

0	50	80	90	80	50	0	Assumed shape (y_0)
0	-50	-80	-90	-80	-50	0	x P = B.M.
0	-50	-80	-90	-80	-50	0	x P/EI = Curvature
-110	-580	-940	-1,060	-940	-580	-111	x λP/12EI = Conc. angle chg.
+2,050	+1,470	+530	-530	-1,470	-2,050		x λP/12EI = Av. slope
0	2,050	3,520	4,050	3,520	2,050	0	x λ^2P/12EI = Deflexion (y')
	24·4	24·1	22·2	24·1	24·4		x 10^3 12EI/λ^2P = Ratio y_0/y'
0	46	78	90	78	46	0	New assumed shape
99	538	916	1,056	916	538	99	x λP/12EI Conc. angle chg.
+1,986	+1,444	+528	-528	-1,444	-1,986		Av. slope
0	1,986	3,430	3,958	3,430	1,986	0	Deflexion (y'')
	23·2	22·8	22·7	22·8	23·2		x 10^3 Ratio y'/y''

$$\Sigma y'/\Sigma y'' = \left(\frac{338}{14,790}\right) = 22\cdot86 \times 10^3 \times 12EI/P\lambda^2$$

Therefore $\Sigma y'/\Sigma y'' = 1: P = 9\cdot89\ EI/L^2$

Fig. 9.34

sure that the final shape is in fact that which gives the lowest critical
load, that is, the fundamental buckling mode.

The problem of the simple, uniform, pin-ended strut as set out
in Fig. 9.34 is straightforward and serves to illustrate the procedure.
An initial shape (y_0) is chosen, and the ordinates recorded (parabolic,
as a first guess). The bending moment at each section is calculated,
followed by the curvatures, and from the curvatures the concentrated
angle changes as before. Assuming the average slope at each side
of the centre line to be $+530$ and -530, i.e. using the symmetry
of the problem, the remaining average slopes and deflexions
(y') follow.

Final deflexions all contain a constant multiplier $\lambda^2 P/12EI$. If a
value of P can be found which makes the derived deflexions identical
with the original assumption, then this value of P will maintain
equilibrium in this mode, that is, P is the critical load corresponding

to this mode of deformation. The ratios y_0/y' are not constant, so that no value of P will make $y_0 \equiv y'$, but if we take the ratio of the sums of deflexions an approximation is obtained.

$$\Sigma\, y_0/\Sigma\, y' = 23{\cdot}06 \frac{12EI}{P\lambda^2}\, 10^{-3}$$

and since $\lambda = \frac{1}{6}L$ for $\Sigma\, y_0/\Sigma\, y = 1$,

$$P = P_{cr} = 10{\cdot}5EI/L^2$$

(accurate value $\pi^2 EI/L^2 = 9{\cdot}870EI/L^2$). Using the individual ratios, the values obtained give

$$9{\cdot}60 < P_{cr}L^2/EI < 10{\cdot}55$$

A second trial is now possible using assumed displacements proportional to those calculated; this gives a more accurate answer.

(*a*) From $\Sigma\, y_0/\Sigma\, y$

$$P_{cr} = 9{\cdot}89EI/L^2$$

(*b*) From individual ratios

$$9{\cdot}81 < PL^2/EI < 10{\cdot}02$$

It is apparent that the closer the original assumption is to the correct buckling mode the easier an accurate solution will be obtained. This is the essential feature of all approximate treatments of the buckling problem and in unskilled hands will cause failure of the procedure to converge to the desired solution. In fact, convergence is difficult to achieve unless the initial assumption contains a major component of the correct solution. Further development of this aspect is beyond the scope of this text.

9.15 Further Reading

1. NEWMARK, N. M., "Numerical procedure for computing deflexions, moments and buckling loads," *Trans. Amer. Soc. Civil Eng.*, **108**, 1943.
2. GODDEN, W. G., *Numerical analysis of beam and column structures* (New Jersey, Prentice-Hall, 1965).

9.16 Examples for Practice

Determine the moments and reactions at the base of the columns in the structures shown in Figs. 9.35 to 9.38.

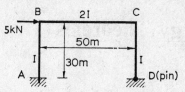

Fig. 9.35

(*Ans.* $M_A = 66.0$ kN-m, $H_D = 1.15$ kN, $V_D = 1.68$ kN, $H_A = 3.85$ kN.)

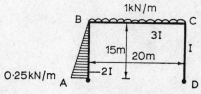

Fig. 9.36

(*Ans.* $M_A = M_D = 0$, $M_B = 10.7$ kN-m, $M_C = 20.2$ kN-m, $H_D = 1.35$ kN, $H_A = 0.52$ kN, $V_D = 10.48$ kN, $V_A = 9.52$ kN.)

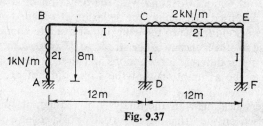

Fig. 9.37

(*Ans.* $M_A = 162$ kN-m, $M_D = 28$ kN-m, $M_F = 117$ kN-m, $H_F = 3.22$ kN, $H_D = 1.33$ kN, $H_A = 6.12$ kN.)

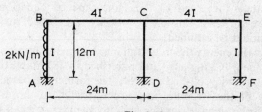

Fig. 9.38

(*Ans.* $M_A = 51.0$ kN-m, $M_D = 24.7$ kN-m, $M_F = 23.0$ kN-m, $H_A = 16.5$ kN.)

10

The Energy Theorems

10.1 Introduction

In the section of Chap. 4 dealing with the deflexion of structures, strain energy was used to calculate the deflexion at a loaded point by equating the work done due to the deflexion of a single applied load to the increase of strain energy of the structure. Expressions for this strain energy under different force actions had already been obtained in Chap. 3 on the assumption that the stress did not exceed the elastic limit.

The chapters dealing with the analysis of redundant structures have been based on the Principle of Virtual Work, using either the flexibility or the stiffness methods of formulating the equations. There are however other approaches to these types of problems and whilst ultimately they lead to the same type of equations it is felt that these different approaches—which are based on energy theorems—should be outlined.

Historically the energy theorems fall into two main divisions. The earlier were due to Castigliano and published in 1879, and the later due to Engesser and published in 1889. The former are based on the strain energy of the structure and are restricted to those having linear load-displacement diagrams with no member stressed beyond the elastic limit. Engesser's Theorems are based on the Complementary Energy (q.v.) and do not specify that the structure shall have a linear load–displacement diagram.

10.2 Castigliano's First Theorem

The body shown in Fig. 10.1 (a) has the linear load–displacement diagram shown in Fig. 10.1 (b). As the load increases from W to $(W + \delta W)$ with a corresponding increase in the deflexion by $\delta\Delta$, the work done is given by $(W + \tfrac{1}{2}\delta W)\delta\Delta$ which, neglecting second-order infinitesimals, is $W\,\delta\Delta$. This work is by definition equal to the increase in strain energy; hence the total increase in strain energy U

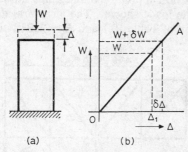

Fig. 10.1

as the load increases from 0 to W_1 is given by

$$U = \int_0^{\Delta_1} W \, d\Delta \qquad (10.1)$$

where Δ_1 is the deflexion corresponding to W_1. This integral is equal to the area under the curve OA. It follows from eqn. (10.1) that

$$\frac{\partial U}{\partial \Delta} = W \qquad (10.2)$$

This relationship is sometimes referred to as Castigliano's Theorem Part I. It is one way of expressing his first theorem, which deals with the relationship between strain energy, load and deflexion.

Since OA in Fig. 10.1 (b) is a straight line,

$$\int_0^{\Delta_1} W \, d\Delta = \int_0^{W_1} \Delta \, dW$$

hence it follows that

$$\frac{\partial U}{\partial W} = \Delta \qquad (10.3)$$

This alternative way of expressing the relationship between strain energy, load and deflexion is Part II of Castigliano's First Theorem.

The form (10.3) is particularly useful in obtaining the deflexions at points in a structure, dummy-loads being used at unloaded points.

Expressed in words it states that if the total strain-energy be partially differentiated with respect to an applied load the result gives the displacement of that load in its line of action.

This method will be applied to the example given in p. 150, which was solved using virtual work methods (the details are repeated to save reference back).

Example 10.1
Calculate the deflexion at point Y of the frame shown in Fig. 10.2 (*a*) when it carries the loads shown. The structural data are given in the second column of Table 10.1 and $E = 230 \text{ kN/mm}^2$.

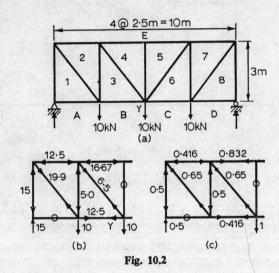

Fig. 10.2

Table 10.1

Member	L/A mm^{-1}	p_0 kN	$\dfrac{\partial p_0}{\partial W}$	$\dfrac{p_0 L}{A} \cdot \dfrac{\partial p_0}{\partial W}$
E1	12	−15	−0·50	+90
E2	6	−12·5	−0·416	+31·2
E4	6	−16·67	−0·832	+83·2
B3	6	12·5	+0·416	+31·2
12	12	19·9	+0·65	+152·1
23	12	−5·0	−0·50	+30·0
34	12	6·5	+0·65	+50·7
			Σ	468·4

$$\Delta_y = (2 \times 468\cdot4)/230 = 4\cdot05 \text{ mm}$$

SOLUTION

Since the members of the frame are stressed within the elastic limit, eqn. (10.3) will apply, giving

$$d_y = \frac{\partial U}{\partial W_Y}$$

where d_y is used in place of Δ_y for displacements corresponding to W_Y.

The members are all subject to direct load only, hence

$$U = \frac{p_0{}^2 L}{2AE}$$

where p_0 = stress resultant and

L/A = length/cross-sectional area.

For the whole frame

$$U = \sum \frac{p_0{}^2 L}{2AE}$$

and

$$d_y = \frac{\partial U}{\partial W_Y} = \sum \frac{p_0 L}{AE} \cdot \frac{\partial p_0}{\partial W_Y}$$

The forces p_0 are determined in the usual way for statically-determinate structures. Their values are given on the diagram in Fig. 10.2 (*b*). The partial differential $\partial p_0 / \partial W_Y$ is obtained by putting a unit load at Y and calculating the forces in the members due to this. These are given in Fig. 10.2 (*c*). The values are, of course, the values p_1 as defined on p. 149 of Chap. 5, thus giving again the displacement

$$d_1 = \sum \frac{p_1 p_0 L}{AE} \quad \text{or} \quad \sum (p_1 \times \text{change in length of member})$$

The summation is obtained as shown in Table 10.1. The frame is symmetrical; consequently only half of the members are given and the resulting sum multiplied by 2.

In this example the changes in length have resulted from applied loads. They could have resulted from temperature changes, errors in length, etc. If λ = magnitude of such changes then the deflexion due to them is given by $d = p_1 \lambda$, where the sign of λ must correspond to the sign convention adopted for stress resultants. Usually this is positive for tensions and therefore for elongations.

10.3 Castigliano's Second Theorem or Castigliano's Theorem of Compatibility

This theorem is concerned with the relationship between the strain energy and force action in a statically-indeterminate frame.

Consider the braced frame shown in Fig. 10.3 (*a*), which is simply-supported at A and B but redundant in that it contains too many members for analysis by the application of the equations of static

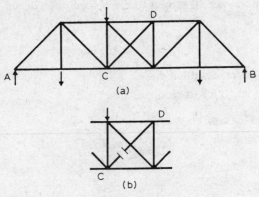

(a)

(b)

Fig. 10.3

equilibrium. Let CD be taken as the redundant member and let the tensile force in this member be R.

If this member is cut as shown in Fig. 10.3 (*b*) then the structure becomes statically determinate and the forces in the members can be expressed in terms of the external loads and R.

The amount Δ_R by which C and D *approach* one another is given from eqn. (10.3) by

$$\Delta_R = \partial U' / \partial R$$

where U' is the strain energy of the frame excluding the bar CD.

The extension e_R of the bar CD due to the force R is given by

$$e_R = \partial u / \partial R$$

where u is the strain energy of bar CD alone.

If in the unloaded statically-determinate state (Fig. 10.3 (*b*)) the points C and D were distant L apart and the bar CD which is to be inserted in position had an initial length $(L + \lambda)$, then in order for the strained bar to fit into the deflected positions of C and D,

$$\Delta_R = -e_R - \lambda$$

or
$$\Delta_R + e_R = -\lambda$$

or
$$\frac{\partial U'}{\partial R} + \frac{\partial u}{\partial R} = -\lambda$$

Since the structure has a linear load-displacement diagram the principle of superposition applies, and

$$U' + u = U$$

where U is the strain energy of the whole frame including the bar CD. Hence

$$\partial U / \partial R = -\lambda \qquad (10.4)$$

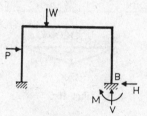

Fig. 10.4

This equation is the second theorem of Castigliano, which, expressed in words, states that if in a redundant structure the total strain energy is partially differentiated with respect to the load in a redundant member the result is the initial lack of fit of that member, due attention being paid to signs.

In the special case where there is no initial lack of fit, i.e. $\lambda = 0$, eqn. (10.4) becomes

$$\partial U / \partial R = 0 \qquad (10.5)$$

This is a condition for minimum value of the strain energy. The increase in strain energy is, however, equal to the work done by the deflexion of the applied loads. Thus the relationship given in eqn. (10.5) also expresses a condition for the minimum value of the work done. This relationship is therefore known as the Principle of Least Work.

Statically-indeterminate structures can be analysed by this principle, which in some cases, e.g. the portal frame shown in Fig. 10.4, is an application of the First Theorem, Part II.

In Fig. 10.4 the redundancies are H, V and M, the reactions at the support B. Since this support is fully-fixed, i.e. there is no horizontal or vertical movement or rotation, by Part II of Theorem I

$$\frac{\partial U}{\partial H} = \frac{\partial U}{\partial V} = \frac{\partial U}{\partial M} = 0$$

Example 7.7 on p. 207, which was solved using the principle of virtual work, will now be solved by the principle of least work.

Example 10.2
The king-posted beams AB and BC shown in Fig. 10.5 (*a*) are simply-supported at A and C but continuous over support B. The second moment of area of ABC is 5×10^7 mm^4 units and the area

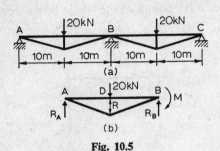

Fig. 10.5

of each of the four ties is 500 mm^2. The king-posts can be taken as rigid. The loads are 20 kN at the mid-points of each of the two beams.

If E is constant throughout, determine the load in the king-posts.

SOLUTION
This problem has two redundancies, the moment M over the support B due to continuity, and the load in the king-posts which will be taken as R.

Since the beams and the loading are symmetrical, the structure can be dealt with as two propped cantilevers and there is no rotation of the beam at B; hence $\partial U/\partial M = 0$.

Since there is no initial lack of fit in the posts, $\partial U/\partial R = 0$.

Consider first the redundant moment M. The beam AB is shown in Fig. 10.5 (*b*) together with the force actions. Then, if $W = 20$ kN load,

$$M_x = \frac{x}{20} M - \tfrac{1}{2}(W - R)x + [(W - R)(x - 10)]$$

$$U = \int \frac{M_x^2 \, dx}{2EI}$$

$$\frac{\partial U}{\partial M} = \int \frac{M_x}{EI} \cdot \frac{\partial M_x}{\partial M} \cdot dx \qquad\qquad \frac{\partial M_x}{\partial M} = \frac{x}{20}$$

Hence

$$\frac{\partial U}{\partial M} = \frac{1}{EI} \int_0^{20} \frac{x}{20} \left\{ \frac{Mx}{20} - \frac{(W-R)x}{2} \right\} dx$$

$$+ \frac{1}{EI} \int_{10}^{20} \frac{x}{20} (W-R)(x-10) \, dx$$

$$= \frac{1}{EI.20^2} \left[\frac{Mx^3}{3} - \frac{20(W-R)x^3}{6} \right]_0^{20} + \frac{1}{EI.20} \left[(W-R) \left(\frac{x^3}{3} - \frac{10x^2}{2} \right) \right]_{10}^{20}$$

Since $\partial U/\partial M = 0$, the term $1/EI$ cancels out, leaving

$$\frac{20M}{3} - \frac{400(W-R)}{6} + \frac{1}{20} \left\{ (W-R) \left(\frac{8,000 - 1,000}{3} - \frac{4,000 - 1,000}{2} \right) \right\} = 0$$

This leads to

$$\frac{20M}{3} = (W-R) \cdot \frac{150}{6}$$

or

$$M = \frac{15}{4}(W-R)$$

Substituting for W gives

$$M = \frac{150}{2} - \frac{15R}{4}$$

The reaction at A is then

$$10 - \tfrac{1}{2}R - \frac{1}{20} \left(\frac{150}{2} - \frac{15R}{4} \right) = \frac{50}{8} - \frac{5}{16} R$$

Dealing now with the second redundancy R, the Principle of Least Work gives

$$\frac{\partial U}{\partial R} = 0$$

For the beam,

$$U = \int \frac{M_x^2 \, ds}{2EI}$$

and for the ties,

$$U = \sum \frac{p^2 L}{2AE}$$

Since the structure is symmetrical and the final partial differential is to be equated to zero, only half the structure need be considered.

For the beam from A to D,

$$M_x = \left(-\frac{50}{8} + \frac{5R}{16}\right)x$$

$$\frac{\partial M_x}{\partial R} = \frac{5x}{16}$$

and

$$\left[\frac{\partial U}{\partial R}\right]_{AD} = \frac{1}{EI} \cdot \frac{5}{16} \int_0^{10} \left(\frac{5R}{16} x^2 - \frac{50}{8} x^2\right) dx$$

$$= \frac{25}{256EI}\left[\frac{Rx^3}{3} - \frac{20x^3}{3}\right]_0^{10}$$

$$= \frac{25,000(R - 20)}{3 \times 256EI}$$

For the beam from D to B, measuring x from D,

$$M_x = -\frac{50}{8}(x + 10) + 20x + \frac{5R}{16}(x + 10) - Rx$$

$$= -\frac{500}{8} + \frac{110}{8} x + R\left(\frac{50}{16} - \frac{11x}{16}\right)$$

$$\frac{\partial M_x}{\partial R} = \left(\frac{50}{16} - \frac{11x}{16}\right)$$

$$\left[\frac{\partial U}{\partial R}\right]_{DB} = \frac{1}{256EI} \int_0^{10} (50 - 11x)\{(220x - 1,000) + R(50 - 11x)\} dx$$

$$= \frac{1}{256EI}\left[\frac{22,000x^2}{2} - \frac{2,420x^3}{3} - 50,000x\right.$$

$$\left. + \left(\frac{121x^3}{3} - \frac{1,100x^2}{2} + 2,500x\right)R\right]_0^{10}$$

$$= \frac{1}{3 \times 256EI}(31,000R - 620,000)$$

For the whole beam,

$$\frac{\partial U}{\partial R} = \frac{1}{3 \times 256EI}(56,000R - 1,120,000)$$

For the ties, $p_0 = \frac{1}{2}\sqrt{17}R$, hence $\dfrac{\partial p_0}{\partial R} = \frac{1}{2}\sqrt{17}$.

The length of each tie $= 10\cdot31$ m. Hence for the two ties,

$$\left[\frac{\partial U}{\partial R}\right] = R \times \frac{17R}{4} \cdot \frac{10\cdot31}{AE} \times 2.$$

For the beam and ties together,

$$\frac{\partial U}{\partial R} = \frac{1}{3 \times 256EI}(56{,}000R - 1{,}120{,}000) + \frac{87\cdot3R}{AE} = 0$$

The lengths have been measured in metre units, hence when substituting for I and A metre units must be used, giving as the final equation—

$$56{,}000R - 1{,}120{,}000 + 87\cdot3 \times 3 \times 256 \times 10^{-1}\times R = 0$$

leading to $R = 1{,}120{,}000/62{,}710 = 17\cdot85$ kN.

10.4 First Theorem of Complementary Energy

Figure 10.6 shows the load-deformation relationship of a body similar to that shown in Fig. 10.1 (*a*). Figure 10.6 differs however from Fig. 10.1 (*b*) in that the load-deformation relationship is not linear.

The area to the right of curve OA is expressed mathematically as $\int W\, d\Delta$ and is the strain energy of the structure.

The area to the left of the curve in Fig. 10.6, i.e. the area between the curve and the W-axis, is expressed mathematically as $\int \Delta\, dW$ and is known as the *Complementary Energy*. It is denoted by the symbol C and has the same dimensions as strain energy.

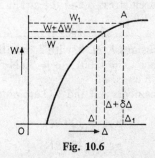

Fig. 10.6

Since $C = \int \Delta\, dW$,

$$\frac{\partial C}{\partial W} = \Delta \qquad\qquad (10.6)$$

This relationship expressed in words states that if the total complementary energy be partially differentiated with respect to an applied load the result is the displacement of that load in its line of action. This is the *First Theorem of Complementary Energy*.

Castigliano's First Theorem Part II is a special case of (10.6) which applies when the load-deflexion diagram is linear. Under this condition

$$\int \Delta\, dW = \int W\, d\Delta = U = C$$

Example 10.3

The frame shown in Fig. 10.7 (*a*) has the members AB and BC made from a material having the stress–strain relationship shown in Fig. 10.7 (*b*). Calculate the deflexion under the load at B if the areas of AB and BC are 2×10^4 and 3×10^4 mm² respectively.

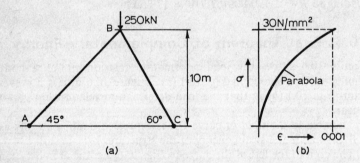

(a)   (b)

Fig. 10.7

SOLUTION

The relationship between stress σ and the strain ϵ is given by

$$\epsilon = k\sigma^2$$

where k has the dimensions $F^{-2}L^{+4}$, with F denoting force units and L length units. When $\epsilon = 0.001$, $\sigma = 30$ N/mm². Hence $k = 1.11 \times 10^{-6}$ N⁻² mm⁺⁴ units.

The forces in the members AB and AC are obtained from statics, giving

$$p_{AB} = 129 \text{ kN}$$

and

$$p_{BC} = 183 \text{ kN}$$

The complementary energy C for each member is given by

$$C = \int_0^{p_1} \Delta \, dp$$

But $\Delta = \epsilon L$, where L is the length of the member, and $p = \sigma A$, where A is the cross-sectional area. Hence

$$C = \int_0^{\sigma_1} AL\epsilon \, d\sigma = ALk \int_0^{\sigma_1} \sigma^2 \cdot d\sigma$$

$$= \tfrac{1}{3} AkL\sigma^3$$

Total complementary energy

$$C = \sum (C_{AB} + C_{BC}) = \sum \tfrac{1}{3} ALk\sigma_1^3$$

The deflexion Δ is given by

$$\Delta = \frac{\partial C}{\partial W} = \sum \frac{\partial C}{\partial \sigma} \cdot \frac{\partial \sigma}{\partial W} = \sum ALk\sigma^2 \cdot \frac{1}{A} \frac{\partial p}{\partial W}$$

$$= \sum Lk\sigma^2 \, \partial p / \partial W$$

It is helpful at this stage to check the dimensions of this expression

$$\Delta = L^{+1}$$

$$Lk\sigma^2 = L^{+1}, F^{-2}L^{+4}, F^{+2}L^{-4} = L^{+1}$$

Thus there is a dimensional check.

For member AB, $\sigma = 6 \cdot 45 \text{ N/mm}^2$.

For member BC, $\sigma = 6 \cdot 10 \text{ N/mm}^2$.

$$\left[\frac{\partial C}{\partial \sigma} \cdot \frac{\partial \sigma}{\partial W} \right]_{AB} = 10{,}000\sqrt{2} \times 1 \cdot 11 \times 10^{-6} \times 6 \cdot 45^2 \times 129/250$$

$$= 3 \cdot 38 \times 10^{-1}$$

$$\left[\frac{\partial C}{\partial \sigma} \cdot \frac{\partial \sigma}{\partial W} \right]_{BC} = \frac{20{,}000}{\sqrt{3}} \times 1 \cdot 11 \times 10^{-6} \times 6 \cdot 10^2 \times 183/250 = 3 \cdot 50 \times 10^{-1}$$

Hence

$$\Delta = 6 \cdot 88 \times 10^{-1} \text{ mm}$$

10.5 Second Theorem of Complementary Energy or Engesser's Theorem of Compatibility

If the members of the frame shown in Fig. 10.3 (*a*) have a load-deflexion relationship as shown in Fig. 10.6, then the amount Δ_R by which C and D approach one another is, by eqn. (10.6),

$$\Delta_R = \frac{\partial C'}{\partial R}$$

where C' is the complementary energy of the frame excluding the bar CD.

The extension e_R of the bar CD is given by

$$e_R = \frac{\partial c}{\partial R}$$

where c is the complementary energy of bar CD.

Again joining C and D by a bar initially λ too long gives

$$\frac{\partial C}{\partial R} = -\lambda$$

where C is the complementary energy of the whole frame.

This states that if the total complementary energy of a structure be partially differentiated with respect to the load in a redundant member the result is the initial lack of fit of that member.

10.6 Theorem of Minimum Energy

Castigliano's Theorem Part I is expressed mathematically by eqn. (10.2), namely $\partial U/\partial \Delta = W$. At all unloaded points on the frame this gives $\partial U/\partial \Delta = 0$. This is a condition for a minimum value of U and the physical interpretation of this is that the deflected form of

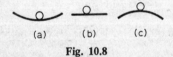

(a)   (b)   (c)

Fig. 10.8

the structure is such as to make the strain energy a minimum. (This is similar to the result obtained from the Second Theorem of Castigliano, where in the case of no initial lack of fit the load in the redundant member is such as to give a minimum value for the strain energy.)

The fact that the potential energy of a body has a stationary value when that body is in neutral equilibrium and a minimum value when it is in stable equilibrium is demonstrated by Figs. 10.8 (a) (b) and (c). These show a body which is in (a) stable, (b) neutral and (c) unstable equilibrium. It is clear from these figures that the potential energy has a minimum value at (a), a constant value at (b) and a maximum at (c).

If the potential energy is denoted by V, then the condition that this shall be a minimum when small displacements Δ are given about the equilibrium position is that

$$\frac{\partial V}{\partial \Delta} = 0 \tag{10.7}$$

Combining this equation with the expression for minimum strain energy $\partial U/\partial \Delta = 0$ gives

$$\frac{\partial (U + V)}{\partial \Delta} = 0 \tag{10.8}$$

$(U + V)$ is the sum of the strain and potential energies, i.e. the total energy, or *total potential*, of the body.

Equation (10.8) expresses the relationship for a minimum value of the total energy and is particularly useful in dealing with stability problems as shown in Chap. 12.

11

The Use of Models in Structural Analysis

11.1 Introduction

There are two distinct aspects of the use of structural models, *direct* and *indirect*.

A *direct* model requires dynamical similarity with the prototype structure, so that all effects such as stresses, strains, deformations, etc. are known, scaled replicas of those in the prototype.

Indirect methods follow from the general reciprocal theorem and only a limited geometrical similarity is required, so that only specific features of the model, e.g. displacements, bear known relations to those of the prototype. These methods provide an experimental means for the rapid determination of influence-line diagrams.

11.2 Indirect Models (Displacement Models)

(*a*) *Influence-line diagram for deflexion.*
From the development of the reciprocal theorem in Chap. 5,

$$W^{\mathrm{I}}\Delta^{\mathrm{II}} = W^{\mathrm{II}}\Delta^{\mathrm{I}} \tag{11.1}$$

the superscripts I and II referring to two separate load-displacement systems.

This method will be applied to the two systems shown in Fig. 11.1. Equation (11.1) is tabulated below; the forces and displacements

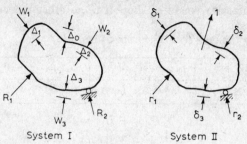

System I   System II

Fig. 11.1

of system I are written in the first row, whilst the second row contains the displacements and forces of system II, i.e. the order is reversed.

System I	W_1	W_2	W_3	R_1	R_2	Δ_1	Δ_2	Δ_3	Δ_0	0	0
System II	δ_1	δ_2	δ_3	0	0	0	0	0	1	r_1	r_2

The products are found by multiplying each term by the one immediately below it, giving

$$W_1\delta_1 + W_2\delta_2 + \cdots = 1 \times \Delta_0$$

or
$$\Delta_0 = \sum W\delta \qquad (11.2)$$

Thus the ordinates to the deflected shape of system II when multiplied by the corresponding loads of system I give the value of the displacement Δ_0, i.e. the deformed shape of system II is the influence-line diagram for the displacement Δ_0.

(b) Influence-line diagram for stress resultant.
Instead of applying a known (or unit) load to system II, a known displacement can be introduced as shown in Fig. 11.2, where δ_0 is

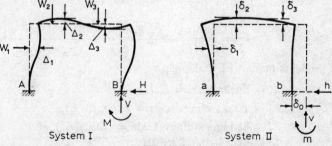

System I   System II

Fig. 11.2

the horizontal movement of support b, no other movements there being allowed.

Applying the reciprocal theorem as before

System I	W_1	W_2	$\cdots$	H	M	V	Δ_1	Δ_2	$\cdots$	0	0	0
System II	δ_1	δ_2	$\cdots$	δ_0	0	0	0	0	$\cdots$	h	m	v

$$\sum W_1\delta_1 + H\delta_0 = 0 \qquad (11.3)$$

i.e.
$$H = -\sum W(\delta/\delta_0)$$

Thus the deformed shape of system II when correctly scaled, i.e. divided by δ_0, is the influence-line diagram for the action in system I corresponding to δ_0.

This is the theorem of Muller Breslau, which states that the ordinates to the influence line for a redundant force action in a structure are equal to those of the deflected form, suitably scaled, obtained when unit load replaces the redundant force action.

11.3   Scale Factors

The method shown in Fig. 11.2 uses two load systems on the same structure and is of little practical value unless a small-scale model can be used for system II. This must be made so that its deformed shape is similar to that of the prototype. That is to say, if δ is any displacement and L is any dimensional characteristic, then

$$(\delta/L)_m = (\delta/L)_p \qquad (11.4)$$

where the suffixes m and p denote model and prototype respectively.

From Chap. 5 the displacement component at some point of a structure due to axial changes in length and flexure of the members is given by

$$\delta = \sum \frac{n_1 n_0 l}{AE} + \sum \int \frac{m_1 m_0}{EI}\, ds$$

i.e.

$$\frac{\delta}{L} = \sum \frac{n_1 n_0 l}{AEL} + \sum \int \frac{m_1 m_0}{EIL}\, ds \qquad (11.5)$$

If a model is made with scale factors,

$$k_1 = \text{linear scale} = L_m/L_p$$
$$k_2 = \text{axial stiffness scale} = (AE)_m/(AE)_p$$
$$k_3 = \text{bending stiffness scale} = (EI)_m/(EI)_p$$
$$k_4 = \text{load scale} = W_m/W_p$$

the ratio $(\delta/L)_m$ is, in the same way,

$$\left(\frac{\delta}{L}\right)_m = \Sigma \left(\frac{n_1 n_0 l}{AEL}\right)_m + \Sigma \int \left(\frac{m_1 m_0}{EIL} ds\right)_m$$

Since n_1 and m_1 are stress resultants due to unit loads and are unaffected by a load scale change, the scale factors can be introduced to give

$$\left(\frac{\delta}{L}\right)_m = \Sigma \frac{n_1(n_0 k_4)(l k_1)}{(EA)k_2(L k_1)} + \Sigma \int \frac{(m_1 k_1)(k_4 m_0 k_1)}{(EI)k_3(L k_1)} (ds k_1)$$

$$= k_4 \Sigma \frac{n_0 n_1 l}{EALk_2} + k_1{}^2 k_4 \Sigma \int \frac{m_1 m_0}{EILk_3} ds \qquad (11.6)$$

In order to satisfy the similarity condition, eqn. (11.4), the right-hand side of eqns. (11.5) and (11.6) must be equal for all values of n_0 and m_0, i.e. k_2 and k_3 must be constants. If this is so, then eqn. (11.6) can be written

$$\left(\frac{\delta}{L}\right)_m = \frac{k_4}{k_2} \Sigma \frac{n_0 n_1 l}{EAL} + \frac{k_1{}^2 k_4}{k_3} \Sigma \int \frac{m_1 m_0}{EIL} ds$$

Continued equality of the right-hand sides of eqns. (11.5) and (11.6) is satisfied if the scale factors satisfy the relationship

$$\frac{k_4}{k_2} = \frac{k_1{}^2 k_4}{k_3} = 1 \qquad (11.7)$$

or $$k_3 = k_1{}^2 k_2$$

It is seen that the load scale does not affect the actual form of the deformed shapes, but by eqn. (11.7) once two of the scale factors have been chosen the other one is fixed. Using the relationship of eqn. (11.7) in (11.6) and removing the dimension L gives

$$\delta_m = \frac{k_4 k_1}{k_2} \Sigma \frac{n_1 n_0 l}{EA} + \frac{k_1{}^3 k_4}{k_3} \Sigma \int \frac{m_1 m_0}{EI} ds$$

$$= k_1 \Sigma \frac{n_1 n_0 l}{EA} + k_1 \Sigma \int \frac{m_1 m_0}{EI} ds$$

$$= k_1 \delta_p$$

Thus the model displacements are also to the linear scale k_1.

In the important group of structures where axial changes in length of the members is not significant, only the second group (the bending terms) of eqns. (11.5) and (11.6) need be considered, so that in eqn. (11.7) there remains only the condition

$$\frac{k_1{}^2 k_2}{k_3} = 1$$

Both the linear scale k_1 and the stiffness scale k_3 can therefore be chosen arbitrarily, the only essential feature being that k_3 is constant at all cross-sections, though its actual value need not be known. Usually models are cut from sheet material of constant thickness and variations in EI_p are obtained by varying the width of the model members d_m so that $d_m \propto \sqrt[3]{(EI_p)}$.

11.4   Practical Applications of the Indirect Method

In structures where bending flexibility predominates, the essentials of the procedure are to make a model to some arbitrary linear scale and with the ratio EI_m/EI_p constant throughout.

A known displacement is then given corresponding to any stress resultant and the deformed shape (when correctly scaled) is the influence-line diagram for the stress resultant.

It is convenient to alter the definition of linear scale and let

$$s = \text{scale ratio} = L_p/L_m = 1/k_1$$

Example 11.1
Figure 11.3 I shows the prototype and Fig. 11.3 II the scale model of system I. A known displacement δ_0 is applied to the model in the direction H which causes model displacements δ_1, etc. These must all be multiplied by the scale factor s to give the full-size system I displacement.

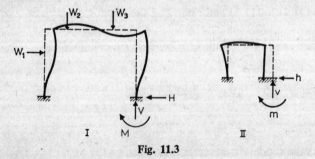

Fig. 11.3

Using the reciprocal theorem gives

System I	W_1	W_2	$\cdots$	H	M	V	Δ_1	Δ_2	$\cdots$	0	0	0
System II	$s\delta_1$	$s\delta_2$	$\cdots$	$s\delta_0$	0	0	0	0	$\cdots$	h	m	v

where the system II forces h, m and v are unknown but could be measured and scaled up if necessary.

Carrying out the multiplications,

$$\sum Ws\delta + Hs\delta_0 = 0$$
$$H = -\sum W(\delta/\delta_0) \quad \text{as before}$$

In this case of concentrated loads only and an influence-line diagram for a force, the scale factor does not appear.

Example 11.2

In order to determine the influence-line diagram for a restraining moment M (Fig. 11.3I) it is necessary to give a rotation corresponding to M at the same section of the model. Fig. 11.4 shows

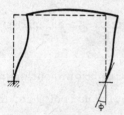

Fig. 11.4

the displaced model. The tabulation of the reciprocal equation is then as given below, but note here that the angular rotation is unaffected by the linear scale factor s.

System I	W_1	W_2	$\cdots$	H	M	V	Δ_1	$\cdots$	0	0	0
System II	$s\delta_1$	$-s\delta_2$	$\cdots$	0	ϕ	0	0	$\cdots$	h	m	v

Multiplying gives

$$\sum Ws\delta + M\phi = 0$$

or $$M = -\sum W(s\delta/\phi) \qquad (11.8)$$

In this case the scale factor *does* appear. It should also be noted that the sign of $s\delta_2$ is negative because the displacement shown in Fig. 11.4 is in the opposite direction to the corresponding force. The definition of corresponding displacement, p. 157 must be carefully adhered to in problems on models.

With a uniformly-distributed load as shown in Fig. 11.5 it is the area under the influence line which is required and the (scale)^2 should therefore appear, but care is required with the units. The loading shown in Fig. 11.5 can be dealt with as a series of concentrated loads of magnitude $w\delta l$ and the tabulation then becomes

System I	$w\delta l$	$\cdots$
System II	$s\delta_1$	$\cdots$

Multiplying out gives

$$\sum w\delta l s \delta_1 + \cdots \qquad (11.9)$$

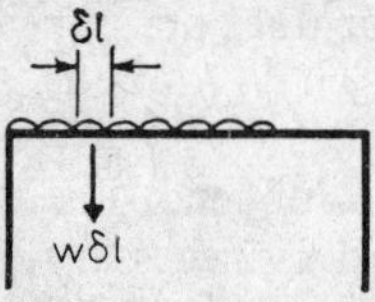

Fig. 11.5

Writing $s\delta x$ for δl, where δx is a distance on the model, gives eqn. (11.6) as

$$\sum w s^2 \delta_1\, \delta x + \cdots$$

i.e. ws^2 [area below model outline]

Care must be taken that the units of w and the area are consistent.

11.5 Experimental Procedure in the Indirect Method

As already stated, the indirect method of model analysis consists of determining the deflected form of a model of the structure due to a known displacement imposed in the direction of the force action. This implies measurement of the deflexion by some means or another and the main subdivision of this method of analysis depends on the the procedure used for measuring the deflexions. One method is to use an ordinary scale divided into millimetres. This, in order to give results to, say, two significant figures, means that the displacement at the redundancy must be of the order of twenty millimetres and leads to what is known as the *large-displacement method*.

As an alternative to using a scale, the displacements can be measured by a micrometer microscope. Then there is no need for a large displacement since quite small quantities can be accurately measured: hence the *small-displacement method*. Begg's Apparatus is an example of this method. It relies on the use of cylindrical and rectangular plugs located in Vee notches, but is difficult to manufacture with sufficient accuracy. A simplified alternative by Magnel is shown in Fig. 11.6.

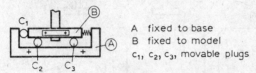

A   fixed to base
B   fixed to model
c_1, c_2, c_3, movable plugs

Fig. 11.6

The large-displacement method is much simpler to operate, but the question must arise as to the errors it introduces. For example, a horizontal displacement of 25 mm in a model which only has a span of 250 mm corresponds to a 10 per cent alteration in span. The errors so introduced can largely be eliminated by displacing in both directions at the redundancy. Referring back to Fig. 11.3 where a model method is used to determine the influence line for horizontal thrust at the support B, an inward displacement is given as shown and the movement of W_2, which will be in an upward direction, is measured. Then an outward displacement is given at B of the same magnitude as the initial inward displacement. The corresponding displacement at W_2, this time in a downward direction, is measured and a second value calculated for the horizontal thrust at B due to unit load at W_2. These two values will be found to differ slightly and an average is taken to get a reasonably accurate value.

It has to be remembered that the displacement must be in the direction of the redundancy. When, for example, the horizontal thrust at the supports is required, only horizontal movement must be given at the supports and no vertical movement or rotation must take place.

In a portal frame with built-in ends there are three redundancies at the support, the horizontal thrust H, the vertical reaction V, and the fixing moment M. Each of these can be obtained in turn by large (or small) displacement model analysis.

Consider the rigid portal frame shown in Fig. 11.7 (*a*), which has the end A firmly clamped in position. A unit horizontal shift Δ_H is given to support B (Fig. 11.7 (*b*)); during this operation no vertical movement or rotation takes place. Measurement of the displacement at X gives the horizontal thrust at B due to a load at X.

Then a unit vertical displacement Δ_V is given to B (Fig. 11.7 (*c*)); during this displacement no horizontal movement or rotation of the

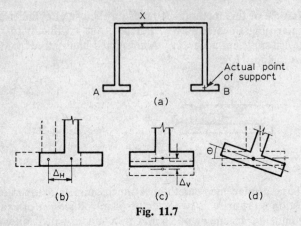

Fig. 11.7

support takes place. Again, measurement at X gives the vertical force at B.

Finally a rotation is given, at B (Fig. 11.7 (*d*)), without any vertical or horizontal movement. The displacement at X is measured and the ratio between this displacement ($\times s$) and the angular rotation of support A gives the fixing moment at B due to unit load at X.

Figures 11.7 (*b*) (*c*) and (*d*) show the displacement in one direction only, but as already pointed out equal and opposite displacements should be given and the average value taken for the redundancy if the large-displacement method is being used.

As an alternative to analysing the frame by determining the redundant external reactions, an analysis by models can be made to find the internal forces at any point in the frame. Consider, for example, the portal frame of Fig. 11.8 (*a*), where it is required to

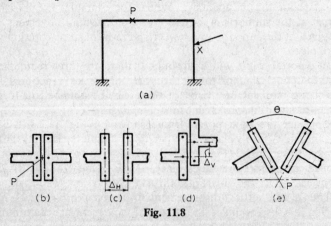

Fig. 11.8

find the internal force action at any point P by drawing the influence lines for the thrust, shear and moment at that point. A model of the portal is made which is divided into two parts at P and the two ends meeting at this point thickened out as shown in Fig. 11.8 (*b*). Then the internal force actions at P due to a unit load applied at X are obtained by giving known thrust, shear and moment displacements at P (*see* (*c*), (*d*) and (*e*)) and measuring the displacement produced at X.

One disadvantage of the *internal reaction* method is the thickening up required in the model on either side of the cut. This upsets the scale for flexural rigidity at that point and will introduce errors in the results.

The material used in indirect model analysis is in general in some sheet form. It should be easily cut and readily deflected, with a linear load-displacement diagram. Three main classes of material are possible, metals, plastics and cardboard. Brass is possibly the only metal which is used in this method. It has all the properties referred to, but in order to get measurable deflexions thin sections must be used and these tend to buckle under load.

Perspex is the most commonly used plastic of the considerable number that are available, as it satisfies most of the requirements. Its stress–strain diagram, however, is not linear and this may introduce errors if the stresses in the model are high. In the fixed-displacement technique employed in the indirect method of model analysis, creep has no effect on the deflected form.

Cardboard has doubtful elastic properties, but despite this it is a very useful simple material for use in the preliminary analysis of a structure and good results can be obtained with it.

The following examples for external and internal redundancies show the application of the method.

Example 11.3
A model was used for the analysis of the structure loaded as in Fig. 11.9 (*a*). The following data were obtained from the model—

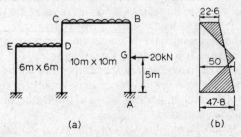

(a)   (b)

Fig. 11.9

Movement A	Corresponding movement at G	Corresponding area under load BC	Corresponding area under load DE
Horizontal 50 mm to right	−37·5 mm	1,935 mm²	645 mm²
Rotational $\pi/6$ clockwise	−75 mm	500 mm²	195 mm²

Full-scale/model = 16:1; u.d. load = 3 kN/m run.

Sketch the bending moment diagram for column AGB giving the critical values.   (*Glasgow*)

SOLUTION

In order to draw the bending moment diagram for AGB it is necessary to determine the horizontal thrust and the fixing moment at A. Let H_A be the horizontal thrust at A and let it act towards the left.

For the point load at G,

$$H_A = -20 \times \left(-\frac{37\cdot5}{50}\right) = 15 \text{ kN}$$

(The negative sign in front of the movement at G indicates that it is towards the right, i.e. opposite the load at G.)

For the uniformly-distributed loads on BC and DE,

$$H_A = -\frac{1,935}{50} \times \frac{3\cdot0 \times 16}{10^3} - \frac{645}{50} \times \frac{3\cdot0 \times 16}{10^3}$$

$$= -1\cdot86 - 0\cdot62 = -2\cdot48 \text{ kN}$$

leading to a resultant $H_A = 12\cdot52$ kN.

In dealing with uniformly-distributed loads it is the area under the influence line which is important, and the ratio of the area to the displacement is affected by the scale. If the linear scale is $s:1$, then the ratio of the area under the influence line to the displacement in the actual structure is s times its value in the model. Hence the insertion of 16 in the numerator in the terms given above. The units employed in the example are, however, in millimetres; consequently the intensity of loading must be given in kN/mm run, thus resulting in a 10^3 in the denominator.

This explanation having been given, the point load and uniform loads will be dealt with, simultaneously, for the fixing moment,

$$M_A = \frac{20}{1 . \pi/6} \times 75 \cdot 0 \times 16 - \frac{500}{1 . \pi/6} \times \frac{3 \cdot 0 \times 16 \times 16}{10^3}$$

$$- \frac{195}{1 . \pi/6} \times \frac{3 \cdot 0 \times 16 \times 16}{10^3}$$

$$= 48,800 - 732 - 286$$

$$= 47,780 \text{ kN-mm} = 47 \cdot 78 \text{ kN-m}$$

Here the scale has to be introduced into the calculations for point load, since in the actual structure the ratio of displacement to angular rotation will be s times what it is in the model. This accounts for the 16 in the numerator.

The bending moment at B is

$$M_B = M_A - 10 H_A + 20 \times 5$$

$$= 47 \cdot 78 - 125 \cdot 2 + 100$$

$$= 22 \cdot 6 \text{ kN-m}$$

The bending moment diagram for the member is shown in Fig. 11.9 (*b*).

The two displacements given in this example are not sufficient to solve the frame completely, since there are six redundancies. Similar displacements at the feet of the other legs would be required. Once H and M at each support have been determined the frame can be solved.

Example 11.4

A model test carried out on a $\frac{1}{24}$th scale model of the frame shown in Fig. 11.10 gave the movements of the points X and Y on the right-hand column as tabulated in the figure. It was also noticed that a 2·5 mm separation of the cut had occurred due to shear movement at C. Draw the bending moment diagram for the structure loaded as shown.

SOLUTION

The loads on the brackets are converted into a horizontal load and a bending moment. On the right-hand side the moment is $70 \times 1 = 70$ kN-m and the horizontal load is 5·0 kN.

In order to get the displacement corresponding to the moment, the rotation of the portion XY for the different displacements must be obtained. The result is given below (clockwise rotations being reckoned positive).

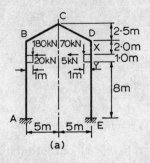

Displacement at C	(a) 50mm ←2·5mm	(b) 50mm	(c) 7½°
Horizontal at X	+16mm	−21mm	+2·5mm
Horizontal at Y (inwards +)	+13mm	−16mm	+4·5mm

(b)

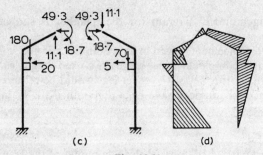

Fig. 11.10

	Model displacement	Rotation (rad)
	(a)	$(24 \times 3 \cdot 0)/10^3 = -0 \cdot 072$
	(b)	$(24 \times 5 \cdot 0)/10^3 = +0 \cdot 120$
	(c)	$(24 \times 2 \cdot 0)/10^3 = +0 \cdot 048$

Using the reciprocal theorem as in Section 11.4 (r.h. loads only) gives

System I	H	V	M		5·0	−70 × 10³	Δ_1	Δ_2	⋯	0	0	0
System II(a)	2·5s	50s	0		16s	−0·072	0	0	⋯	h	v	m
System II(b)	50s	0	0		−21s	+0·120	0	0	⋯	h	v	m
System II(c)	0	0	0·262		2·5s	+0·048	0	0	⋯	h	v	m

where s = scale factor. Multiplying out gives

$$2 \cdot 5H + 50V = -80 - \frac{0 \cdot 072}{24}(70 \times 10^3) \qquad (11.10)$$

$$50H = 105 + \frac{0 \cdot 120}{24}(70 \times 10^3) \qquad (11.11)$$

$$0 \cdot 262M = -5 \cdot 0 \times 2 \cdot 5 \times 24 + 70 \times 10^3 \times 0 \cdot 048 \quad (11.12)$$

giving

$$M = 11 \cdot 7 \text{ kN-m}$$

$$H = 9 \cdot 1 \text{ kN}$$

$$V = -6 \cdot 26 \text{ kN}$$

On the left-hand side the direct force corresponding to X is -20 kN and the moment $180 \times 1 = 180$ kN-m.

Using the reciprocal theorem as before and dealing with this load and moment gives—

System I	H	V	M	-20	180×10^3	Δ_1	Δ_2	$\cdots$	0	0	0
System II(a)	$2 \cdot 5s$	$50s$	0	$16s$	$+0 \cdot 072$	0	0	$\cdots$	h	v	m
System II(b)	$50s$	0	0	$-21s$	$-0 \cdot 120$	0	0	$\cdots$	h	v	m
System II(c)	0	0	$0 \cdot 262$	$2 \cdot 5s$	$-0 \cdot 048$	0	0	$\cdots$	h	v	m

Multiplying out gives

$$2 \cdot 5H + 50V = 320 - (180 \times 72)/24 \qquad (11.13)$$

$$50H = -21 \times 20 + (120 \times 180)/24 \qquad (11.14)$$

$$0 \cdot 262M = 20 \times 2 \cdot 5 \times 24 + 180 \times 48 \qquad (11.15)$$

giving
$$M = 37 \cdot 6 \text{ kN-m}$$

$$H = 9 \cdot 6 \text{ kN}$$

$$V = -4 \cdot 88 \text{ kN}$$

Combining the two loads gives the forces at the cut as

$$H = 9 \cdot 1 + 9 \cdot 6 = 18 \cdot 7 \text{ kN}$$

$$V = -6 \cdot 26 - 4 \cdot 88 = 11 \cdot 14 \text{ kN}$$

$$M = 11 \cdot 7 + 37 \cdot 6 = 49 \cdot 3 \text{ kN-m}$$

The free-body diagrams for each half of the structure are shown in Fig. 11.10 (c).

Considering first the right-hand portion and working in kN-m units gives

$$M_D = 49{\cdot}3 + 11{\cdot}1 \times 5 - 18{\cdot}7 \times 2{\cdot}5$$
$$= 58{\cdot}0$$
$$M_{XR} = 49{\cdot}3 + 11{\cdot}1 \times 5 - 18{\cdot}7 \times 4{\cdot}5$$
$$= 20{\cdot}7$$
$$M_E = 49{\cdot}3 + 11{\cdot}1 \times 5 - 18{\cdot}7 \times 13{\cdot}5 + 5 \times 9 + 70 \times 1$$
$$= -32{\cdot}7$$

Considering the left-hand side and working in kN-m units gives

$$M_B = 49{\cdot}3 - 11{\cdot}1 \times 5 - 18{\cdot}7 \times 2{\cdot}5 = -53{\cdot}0$$
$$M_{XL} = 49{\cdot}3 - 11{\cdot}1 \times 5 - 18{\cdot}7 \times 4{\cdot}5 = -90{\cdot}3$$
$$M_A = 49{\cdot}3 - 11{\cdot}1 \times 5 - 18{\cdot}7 \times 13{\cdot}5 + 180 \times 1 - 20 \times 9$$
$$= -258{\cdot}7$$

The bending moment diagram is given in Fig. 11.10 (*d*).

11.6 Direct Methods of Model Analysis

The requirements of a direct model are that all features of its behaviour must bear a known relation to those of the prototype. These requirements are described as *dynamical similarity* and, if satisfied, all measurements made on the model directly and quantitatively predict the behaviour of the prototype.

It is not always necessary to achieve true dynamical similarity. When using the indirect method only specific similarities are required, and these can be obtained from the theories of engineering mechanics. For more general cases it is better to use the methods of dimensional analysis.

Consider the problem of the strains in a structural member which depend on the member-end actions, member sizes and material properties. When the structure is loaded with concentrated loads the factors affecting the strains and their units of measurement M, L and T (mass, length and time) are—

$$\text{Load } W = \text{force} = M^{+1}L^{+1}T^{-2}$$
$$\text{Size } L = \text{length} = L^{+1}$$
$$\text{2nd moment of area } I = L^{+4}$$
$$\text{Young's modulus } E = M^{+1}L^{-1}T^{-2}$$
$$\text{Density } w = M^{+1}L^{-2}T^{-2}$$
$$\text{Cross-sectional area } A = L^{+2}$$
$$\text{Poisson's ratio} = \nu$$

The strain ϵ is given by some unknown function of all the variables, i.e.

$$\epsilon = \phi\{W^a, L^b, I^c, E^d, w^e, A^f, v^g\} \qquad (11.16)$$

The dimensions on each side of the equation must be the same, hence

$$M^0 L^0 T^0 = \phi\left\{\left(\frac{ML}{T^2}\right)^a, L^b, (L^4)^c, \left(\frac{M}{LT^2}\right)^d, \left(\frac{M}{L^2T^2}\right)^e (L^2)^f v^g\right\} \qquad (11.17)$$

Equating the indices of mass, length and time gives the following three equations—

$$a + d + e = 0$$
$$a + b + 4c - d - 2e + 2f = 0$$
$$-2a - 2d - 2e = 0$$

The first and third equations are identical, thus leaving two equations from which a and b are obtained in terms of the other quantities as

$$a = -d - e$$
$$b = -4c + 2d + 3e - 2f$$

Substituting for a and b in the function ϕ and collecting together terms with the same index gives

$$\epsilon = \phi\left\{\left(\frac{I}{L^4}\right)^c \left(\frac{EL^2}{W}\right)^d \left(\frac{wL^3}{W}\right)^e \left(\frac{A}{L^2}\right)^f v^g\right\} \qquad (11.18)$$

For the strains ϵ to be constant for a series of models each dimensionless group of variables must be a constant.

For a model to a linear scale $s = L_p/L_m$ the groups I/L^4 and A/L^2 do remain constant, giving geometrical similarity, but it is also necessary that

$$\frac{E_m L_m{}^2}{W_m} = \frac{E_p L_p{}^2}{W_p} \qquad (11.19)$$

or for a model of the same material,

$$\frac{W_m}{W_p} = \frac{L_m{}^2}{L_p{}^2} = \frac{1}{s^2}$$

i.e. concentrated loads on the model must be reduced by the (scale)².

This argument would apply only when the external force consists of a concentrated load. A uniformly-distributed load of q force/unit length leads to the similarity condition

$$\frac{q_m}{q_p} = \frac{E_m}{E_p} \cdot \frac{L_m}{L_p}$$

and for the same material $q_m/q_p = 1/s$ as distinct from $1/s^2$ for concentrated loads.

On the other hand, if the external force action is a bending moment then it can be shown that $M_m/M_p = 1/s^3$.

The last group (v^9) shows that Poisson's ratio remains a single term. Hence v for the model must be the same as v for the prototype.

The difficult group of terms is that containing the self or dead weight w; for similarity the dimensionless group of terms (wL^3/W) must remain constant
i.e.

$$\frac{w_m L_m{}^3}{W_m} = \frac{w_p L_p{}^3}{W_p} \qquad (11.20)$$

For a model with a linear scale s and a load scale of s^2,

$$\frac{w_m}{w_p} = \frac{W_m}{W_p} \cdot \frac{L_p{}^3}{L_m{}^3} = \frac{1}{s^2} \times s^3 = s$$

This means that the density of the model material must be increased by the linear scale, or alternatively, if the same material is used extra weights equal to $[(s-1) \times$ model weight$]$ must be added.

This form of dimensional analysis can be carried out for all effects and to include any variables which may affect the problem and if done for the deformed shape, i.e. the ratio δ/L as for indirect models, the same dimensionless groups are obtained. It is possible to use models which are not dynamically similar to the prototype, but great care is required not to confuse the differing scale factors in transferring results from the model to the prototype. One obvious difficulty is the fact that the dead load cannot be scaled correctly without special provisions. But these special provisions are often simple or, alternatively, dead load stresses, not differing greatly in form from live load stresses, can be estimated directly once the latter have been determined.

The size of a model is governed by conflicting factors; the general requirements are—

1. It must fit into the space available.
2. Deflexions, etc. must be measured accurately.
3. The required loadings must be convenient.

For example, in a model of an arch dam to a scale of 1/50th the total distributed loading should be (Prototype Load)/50^2, but if fluid loading is used the volume is scaled down by $(1/50)^3$ so that the same fluid will not give dynamical similarity.

If mercury (S.G. = 13·6) is used as the loading fluid in the model then

$$\text{Model load} = 13\text{·}6 \times (\text{water load prototype}) \div 50^3$$

Therefore

$$\text{Equivalent full-scale loading} = 50^2 \times \text{model load}$$
$$= 0\text{·}272 \times \text{true load}$$

The measured stresses in the model must therefore be multiplied by 1/0·272 to get the stresses in the prototype, or, in other words, the model stresses are about a quarter of those in the prototype. Such stresses could be measured quite accurately, whereas those which would have resulted from water as a loading fluid in the model would have been much too small to measure accurately.

Similarly in a 1/20th scale model of a shell roof the dead load stresses in the model would only be 1/20th of those in the prototype, and very small. Dead load stresses are difficult to measure anyway, due to the necessity for attaching gauges while the structure is unstrained. Several methods have been devised for obtaining them, but the techniques are specialized and call for a deeper study than is possible in this book.

11.7 Examples for Practice

1. A portal structure ABCDEF is rigidly fixed at A and F (Fig. 11.11). It supports uniform loads of 15 kN/m on both beams BC and DE together with a horizontal load of 40 kN applied at the mid-point of EF. The force analysis is carried out by using a flexible model having a scale factor of 12.

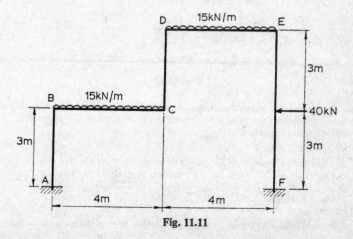

Fig. 11.11

The results from the three separate movements given to the constraint at F are contained in the following table—

Movement at F	Movement at 40 kN load	Net area on DE	Net area on BC
←50 mm	←38 mm	1,935 mm² ↑	1,870 mm² ↑
50 mm ↑	←8 mm	10,000 mm² ↑	2,710 mm² ↑
π/6 clockwise	48 mm→	5,420 mm² ↓	2,710 mm² ↓

Determine the reactions at F and A and sketch the form of the bending moment diagram for the member CD.

(*Ans.* $H_A = 23\cdot2$ kN; $V_A = 80\cdot2$ kN; $M_A = -89\cdot4$ kN-m; $H_F = 16\cdot8$ kN; $V_F = 39\cdot8$ kN; $M_F = 10\cdot4$ kN-m.)

2. State and prove Maxwell's Theorem of Reciprocal Displacements and show its application to redundant structures.

   A beam ABCD is continuous over three spans (7·5 m, 5·0 m, 6·0 m) and is simply-supported at A and D. It carries a live load of uniform intensity of 30 kN/m run having a length greater than the total span. The experimental figures from a model (scale 1/40) are given in Fig. 11.12.

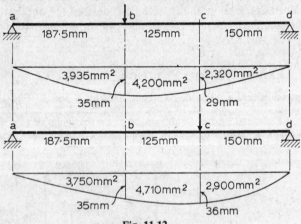

Fig. 11.12

Determine the maximum reaction on an interior support and sketch the bending moment diagram for the beam with the load in the position to give this maximum reaction.

(*Ans.* $R_B = 233$ kN.)

3. A continuous beam ABCD on three equal 10 m spans has the following ordinates to the influence line for reaction at A at 2·5 m intervals: 1, 0·70, 0·40, 0·19, 0, −0·070, −0·073, −0·045, 0, 0·027, 0·030, 0·018, 0.

   Make a scale sketch of the influence line for bending moment at the centre of span BC.

   (*Ans.* Maximum ordinates +0·35 kN-m and −1·77 kN-m.)

4. The two-bay portal frame of Fig. 11.13 is of varying section; all columns
   are of equal stiffness I, but the beam varies uniformly from I at D and F to
   $2I$ at E. In order to carry out a model analysis a model 3 mm thick was
   made to a scale of 1 to 20 and the model columns made 12 mm × 3 mm.
   What is the model beam cross-section at E?

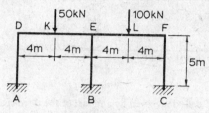

Fig. 11.13

Due to defects in the deformeter, true horizontal and vertical displacements
were not obtained but the readings were as tabulated.

Determine all the reactions at A.

Movement at A of model	Vertical deflexion of model	
	K (mm)	L (mm)
$\frac{1}{6}$ rad clockwise	2·0 up	0·2 down
50 mm up 6 mm left	29·0 up	0·6 down
12 mm up 50 mm left	9·0 down	0·2 up

(*Ans.* $M_A = 24\cdot0$ kN-m; $H_A = 15\cdot7$ kN; $V_A = 29\cdot6$ kN.)

5. A continuous beam ADBEC is built-in at A, continuous over B and pinned
   at C. AB = 10 m and BC = 15 m. D and E are the mid-points of AB and
   BC respectively.
   In order to analyse the beam a model is made to a scale of 1 to 25.
   The supports B and C are pinned, A is moved and the deflexions of points
   D and E noted as in the table below.
   Determine the reactions at B and C when loads of 10 kN and 15 kN act
   at D and E respectively.

Movement of A	D	E
Rotation of 10° clockwise	2·5 mm down	1·0 mm up
Displacement of 25 mm down	10·0 mm down	2·0 mm up

(*Ans.* $R_B = 16\cdot2$ kN; $R_C = 6\cdot0$ kN.)

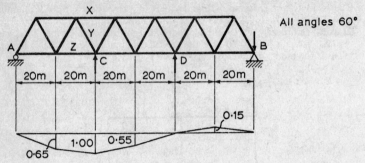

Fig. 11.14

6. The ratio of deflexions given by a model of the three-span braced girder with support C removed are shown in Fig. 11.14. Calculate the reaction at C when a uniformly-distributed load of intensity 25 kN/m run and 40 m long covers the second and third panels from the left. Determine the load in the members X, Y, Z for this condition. All angles are 60°.

   (*Ans.* $R_c = 800$ kN; $W_X = 100/\sqrt{3}$; $W_Y = -300/\sqrt{3}$; $W_Z = 50/\sqrt{3}$ kN.)

12

Compression Members and Stability Problems

12.1 Introduction

If a 6 mm diameter steel rod 1 m long is placed in a testing machine and subjected to a pull, as shown in Fig. 12.1 (*a*), it will be found to carry a load of about 7,000 N before failure occurs. If on the other hand this same rod had been subjected to compression, as shown in Fig. 12.1 (*b*), then the maximum load which would have been carried would have been about 35 N—a very big difference.

Failure in the first test occurs by the fracture of the member; in the second it is due to bending out of the line of action of the load, as indicated by the dotted line in Fig. 12.1 (*b*).

Since the load-carrying capacity of a member in compression is

(a)   (b)

Fig. 12.1

very different from that of a similar member in tension, the former requires special treatment.

It is seen that failure takes place by bending. This cannot occur unless a moment acts on the member and this moment results from a number of effects which make an apparently axial load act eccentrically. The causes are—

(1) The fact that no member can be made perfectly straight.
(2) Imperfections in manufacture leaving some part of the member with slightly different mechanical properties from the remainder.
(3) Inability to ensure that the load actually acts along the centre of area of the cross-section.

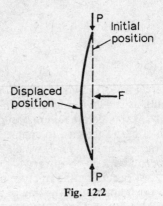

Fig. 12.2

Because of these imperfections, compression members in a structure are all subject to bending as well as direct thrust; in determining the load-carrying capacity of the member this bending must be taken into account.

It should be clear that the amount of buckling will be less in a heavy cross-section strut than it would be in a light cross-section of the same length. Consequently the strength of a strut will be governed to some extent by the bending flexibility l/EI or l/EAk^2 where $k = $ radius of gyration $= \sqrt{(I/A)}$. For a strut of given material and cross-section the ratio l/k becomes the governing factor. This ratio is known as the *slenderness ratio*. The problem of stability has been referred to in Chap. 10 Section 6 and the three states of equilibrium defined.

A perfectly straight strut subject to axial load can be displaced from its initial position by a disturbing force F at mid-height acting as shown in Fig. 12.2. The state of equilibrium of the strut depends on its behaviour after the disturbing force is removed.

If the strut returns to its position prior to the application of F, then it is in stable equilibrium. If on removal of F it remains in the deflected position, it is in neutral equilibrium. But if it continues to deflect, it is in unstable equilibrium.

It will be found that for low values of P the equilibrium is stable, but that as P is increased a load value is obtained which causes the strut to be in a state of neutral equilibrium. This load value is known as the *critical* or *buckling load* of a strut.

In practice it is found that struts fail by buckling without the application of the disturbing force F. This is due to the fact that the imperfections, etc. previously mentioned cause the applied load P to act eccentrically and this in itself constitutes a disturbing force on the strut.

Engineering structures do not contain single isolated struts similar to the problems analysed below. A strut generally forms part of a structure within which all the members interact to affect the stability of the structure.

What follows must be regarded only as the necessary basic introduction to a study of the problem. An adequate treatment of such problems is considered to be outside the scope of the present text and the reader is directed to reference 7 for a concise introduction.

12.2   Critical or Buckling Load of a Pin-ended Strut

Consider the strut AB of length l shown in Fig. 12.3. If a disturbing force acting in the y-direction is applied at O, then for the neutral equilibrium condition the strut remains in the disturbed state after

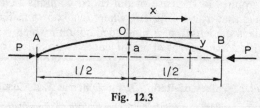

Fig. 12.3

this force has been removed. Fig. 12.3 shows the final deflected form of the strut, with a central deflexion of a and a deflexion at a distance x from the origin of $(a - y)$.

The differential equation of bending gives

$$EI\frac{d^2y}{dx^2} = M$$
$$= P(a - y) \qquad (12.1)$$

Putting $\mu^2 = P/EI$ gives

$$\frac{d^2y}{dx^2} + \mu^2(y - a) = 0 \qquad (12.2)$$

Back substitution shows that a solution of this is

$$y = A \sin \mu x + B \cos \mu x + a \qquad (12.3)$$

where A and B are constants of integration. When $x = 0$, $y = 0$, hence $B = -a$; also when $x = 0$, $dy/dx = 0$, hence $A = 0$. The solution is therefore

$$y = a(1 - \cos \mu x) \qquad (12.4)$$

when $x = \pm \frac{1}{2}l$, $y = a$ and substitution gives

$$a \cos \tfrac{1}{2}\mu l = 0$$

But the problem implies $a \neq 0$; hence $\cos \frac{1}{2}\mu l = 0$, or $\frac{1}{2}\mu l = \frac{1}{2}\pi$ or

$$\frac{Pl^2}{EI} = \pi^2$$

Hence the smallest value of P to produce buckling is

$$P = \pi^2 EI/l^2 \qquad (12.5)$$

This load is known as the *Euler critical load*. At this load the column is in a state of neutral equilibrium and at any slightly increased load it becomes unstable.

The load $\pi^2 EI/l^2$ is symbolized by P_E and the resulting stress σ_E is given by

$$\sigma_E = \pi^2 Ek^2/l^2$$

This value applies in the case of pin-ended columns. Generally it may be stated that $\sigma_E = \pi^2 Ek^2/l_e^2$, where l_e is known as the *effective length*. This is the length of a pin-ended column which would have the same critical load as that of one of length l with the end conditions specified.

Table 12.1 **Values of Ratio l_e/l for Different End Conditions**

Ends	Both pinned	Both built-in	One built-in other free	One built-in other pin
l_e/l	1·0	0·5	2·0	1·435

We have shown that σ_E depends only on the elastic modulus of the material and the slenderness ratio l/k. The relationship between σ_E and l/k for mild steel is drawn in Fig. 12.4 and it is seen from this that for values of l/k less than, say, 100 the Euler stress exceeds the yield stress in the material. It is clear that such a state of affairs is not possible and the Euler buckling load can only represent column behaviour at higher values of the slenderness ratio. For lower values an alternative approach is required. The alternatives fall into two main classes, theoretical and empirical. Some examples of each class will be given: to the former belong the Reduced Modulus and the Tangent Modulus Methods, to the latter the Perry–Robertson formula.

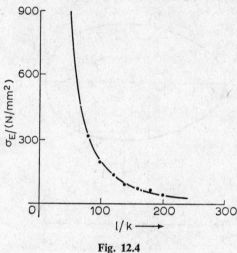

Fig. 12.4

12.3 The Reduced Modulus Theory

The Euler analysis assumes that the critical stress is still below the elastic limit when unstable equilibrium occurs. As Fig. 12.4 shows, this is only true of columns with a high slenderness ratio. With lower values the elastic limit will be reached before buckling takes place and the value to be taken for E will depend on the stress.

The assumptions in the reduced modulus theory are—
(1) Displacements are small relative to the cross-section.
(2) Plane sections remain plane and normal to centre-line after bending.
(3) The relationship between stress and strain in any longitudinal fibre is given by the stress–strain curve for the material.
(4) The plane of bending is a plane of symmetry of the column section.

Consider the column shown in section in Fig. 12.5 (*a*), which has an axis of symmetry YY. Let this be subject to an axial load P such that $\sigma = P/A$ is greater than the elastic limit. Apply a disturbing force to the column which produces a slight deflexion. This will increase the stress on the concave side and reduce it on the convex, but there will be an axis NN where the stress remains unchanged. On the concave side the rate of increase will be proportional to $d\sigma/d\epsilon = E_t$, where E_t is the tangent modulus at the stress σ; but on the convex side the reduction relieves only the elastic portion of the strain and the normal value of E will apply. The stress diagrams in Fig. 12.5 (*b*) are different for the two sides.

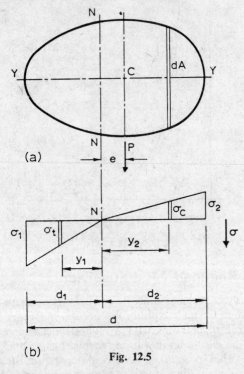

Fig. 12.5

Since the axial load is unaltered by the deformation,

$$\int_0^{d_1} \sigma_t \, dA = \int_0^{d_2} \sigma_c \, dA \qquad (12.6)$$

If the applied load acts at C, the centroid of the end section, then if NN is distant e from C,

$$\int_0^{d_1} \sigma_t(y_1 + e) \, dA + \int_0^{d_2} \sigma_c(y_2 - e) \, dA = M = Py \qquad (12.7)$$

$$\sigma_t = \frac{y_1}{d_1} \sigma_1 \text{ and } \sigma_c = \frac{y_2}{d_2} \sigma_2$$

The relative rotation of two vertical sections distance dx apart is shown diagrammatically in Fig. 12.6 and it is seen that $\Delta\,dx = d_1\,d\phi$. Since $\Delta\,dx = \sigma_t\,dx/E$,

$$\frac{d\phi}{dx} = \frac{\sigma_t}{E d_1} = \frac{\sigma_c}{E_t d_2} \tag{12.8}$$

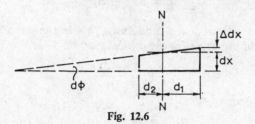

Fig. 12.6

For small deformations $d\phi/dx = d^2y/dx^2$; hence

$$\sigma_1 = E d_1 \frac{d^2y}{dx^2} \quad \text{and} \quad \sigma_2 = E_t d_2 \frac{d^2y}{dx^2}$$

Thus eqn. (12.6) becomes

$$E \frac{d^2y}{dx^2} \int_0^{d_1} y_1\, dA - E_t \frac{d^2y}{dx^2} \int_0^{d_2} y_2\, dA = 0 \tag{12.9}$$

or

$$E S_1 - E_t S_2 = 0 \tag{12.10}$$

where S_1 and S_2 are the first moments of area of the portions to the left and right of NN about NN.

This equation, together with the relationship $d_1 + d_2 = d$, enables the position of NN to be determined. Equation (12.7) gives

$$\frac{d^2y}{dx^2}\left(E \int_0^{d_1} y_1{}^2\, dA + E_t \int_0^{d_2} y_2{}^2\, dA \right)$$
$$+ e \frac{d^2y}{dx^2}\left(E \int_0^{d_1} y_1\, dA - E_t \int_0^{d_2} y_2\, dA \right) = Py \tag{12.11}$$

But the second term on the left-hand side has been shown in eqn. (12.9) to be equal to zero. Hence

$$\frac{d^2y}{dx^2} (E I_1 + E_t I_2) = Py \tag{12.12}$$

where I_1 and I_2 are the second moments of the areas to the left and right of NN taken about NN.

Putting $$\bar{E}I = EI_1 + E_tI_2 \qquad (12.13)$$

gives

$$\bar{E}I \frac{d^2y}{dx^2} + Py = 0$$

where

$$\bar{E} = E\frac{I_1}{I} + E_t\frac{I_2}{I} \qquad (12.14)$$

is known as the reduced modulus. This is similar to the *Euler equation* and the value for the critical load P_r is

$$P_r = \frac{\pi^2 \bar{E} I}{l^2} \qquad (12.15)$$

and the maximum stress

$$\sigma_r = \frac{\pi^2 \bar{E} k^2}{l^2}$$

12.4 The Tangent Modulus Theory

The reduced modulus theory assumed that a strain reversal took place on the convex side and that such strain reversal relieved only the elastic portion of the stress. The tangent modulus theory assumes that no strain reversal takes place and that the tangent modulus E_t applies over the whole cross-section. The analysis is then the same as the standard Euler approach with the substitution of E_t for E, giving the critical load P_t from

$$P_t = \pi^2 E_t I / l^2$$

and the stress

$$\sigma_t = \pi^2 E_t k^2 / l^2 \qquad (12.16)$$

Since I_1 and I_2 in eqn. (12.14) are together greater than I and since E is greater than E_t, it follows that $\bar{E}$ is greater than E_t: consequently the critical load from the reduced modulus approach P_r is greater than that derived from the tangent modulus theory.

In the study of compression members much use has to be made of experimental results. Many carefully controlled experiments show that the loads carried agree more closely with those predicted by the tangent modulus theory than those predicted by the more refined reduced modulus theory. The explanation of such behaviour is due to Shanley.

12.5 Shanley's "Hinged-Column" Explanation

The tangent modulus load is lower than the reduced modulus load, but the latter theory does not allow for the possibility that, after reaching the tangent modulus load, bending deflexion begins and that this bending increases as the load increases. Shanley showed that this was the case and that there is a continuous spectrum of deflected forms satisfying the condition of stable equilibrium as the load increases from P_t to P_r, with the deflexion increasing from zero at the former to infinity at the latter.

Shanley's idealized column is shown in Fig. 12.7. It consists of two rigid legs AC and BC connected at C by an elastic-plastic hinge.

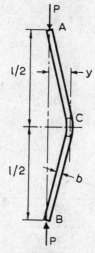

Fig. 12.7

He shows that the relationship between the load P $(>P_t)$ and the deflexion y is given by

$$P = P_t\left(1 + \frac{1}{b/2y + (1 + \tau)/(1 - \tau)}\right)$$

where $\tau = E_t/E$ and is assumed constant.

The equations giving the variation of strain as a function of the ratio $R = P/P_t$ are, according to Shanley,
Concave side

$$\Delta\epsilon_1/\epsilon_t = \frac{2(1/\tau - R)}{(1 - \tau)/(R - 1) - (1 + \tau)}$$

Convex side

$$\Delta\epsilon_2/\epsilon_t = \frac{2(R - 1)}{(1 - \tau)/(R - 1) - (1 + \tau)}$$

where ϵ_t is strain corresponding to P_t.

If a value is assumed for τ of 0·75, the resulting relationship between change of strain and load is shown in Fig. 12.8. It is seen from this that after reaching the tangent modulus load the strain on the concave side shows a very rapid increase and on the convex side a slow decrease. The very rapid increase on the concave side shows that the column will have deflected a considerable amount soon after passing the tangent modulus load. This means that it will only

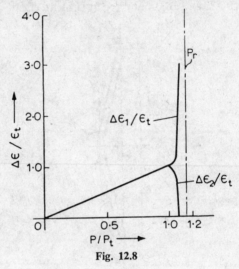

Fig. 12.8

be possible to exceed the tangent modulus load by a small amount and explains why experiments show that the load-carrying capacity of the column is given by the tangent modulus theory.

12.6 Perry-Robertson Formula for Axially-Loaded Struts

The theoretical analysis given in Section 12.2 was based on the fact that the deflected position shown in Fig. 12.3 had been produced by the application of a disturbing force. In the laboratory, however, struts fail without any application of a disturbing force due to the eccentricity which is present from the various causes mentioned earlier in this chapter. Engineers have therefore tended to produce column formulae to fit experimental results. There are many such formulae in existence, but it is only proposed to give that due to Perry and Robertson since this forms the basis of the permissible stresses for columns given in the British Standard Code of Practice for Structural Steelwork.

The Perry approach is based on the assumption that the strut has an initial curvature which is given by the equation

$$y_0 = c_0 \cos \pi x / l \tag{12.17}$$

where c_0 is the initial departure from straightness at the centre O (*see* Fig. 12.9). Under a load P the deflexion is increased by an amount y, the differential equation of bending being

$$EI \frac{d^2y}{dx^2} = M = -P[y + c_0 \cos (\pi x / l)] \tag{12.18}$$

or
$$\frac{d^2y}{dx^2} + \mu^2(y + c_0 \cos (\pi x / l)) = 0 \tag{12.19}$$

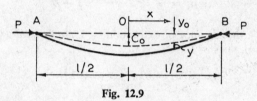

Fig. 12.9

where $\mu^2 = P/EI$. A solution of this equation is shown by back substitution to be

$$y = A \sin \mu x + B \cos \mu x + \frac{\mu^2 c_0 \cos (\pi x / l)}{\pi^2 / l^2 - \mu^2} \tag{12.20}$$

where A and B are constants of integration.

When $x = \pm l/2$, $y = 0$, hence $A = B = 0$, and

$$y = \frac{\mu^2 c_0 \cos (\pi x / l)}{\pi^2 / l^2 - \mu^2} \tag{12.21}$$

If $P_E = \pi^2 EI / l^2$, the Euler critical load, then

$$y = \frac{P c_0 \cos (\pi x / l)}{P_E - P} \tag{12.22}$$

If $P/A = \sigma$, and $P_E/A = \sigma_E$, then

$$y = \frac{\sigma}{\sigma_E - \sigma} c_0 \cos (\pi x / l) \tag{12.23}$$

The total deflexion at any point is

$$y_0 + y = \left(\frac{\sigma}{\sigma_E - \sigma} + 1 \right) c_0 \cos (\pi x / l)$$

$$= \left(\frac{\sigma_E}{\sigma_E - \sigma} \right) c_0 \cos (\pi x / l)$$

The maximum deflexion occurs when $x = 0$, and is given by y_{max} where

$$y_{max} = \frac{\sigma_E}{\sigma_E - \sigma} \cdot c_0 \qquad (12.24)$$

The maximum bending moment is

$$Pc_0\left(\frac{\sigma_E}{\sigma_E - \sigma}\right)$$

The maximum compressive stress σ_2 occurs on the concave side of the strut and is given by

$$\sigma_2 = \frac{d_2}{I} Pc_0\left(\frac{\sigma_E}{\sigma_E - \sigma}\right) + \sigma \qquad (12.25)$$

where d_2 is the distance of extreme fibre from neutral axis. Putting $c_0 d_2/k^2 = \eta$ gives

$$\sigma_2 = \sigma\left(\frac{\eta \sigma_E}{\sigma_E - \sigma} + 1\right) \qquad (12.26)$$

It is further assumed that failure will take place when the maximum fibre stress reaches the yield stress σ_y. Substituting this value for σ_2 and solving for σ gives

$$\sigma_p = \tfrac{1}{2}[\sigma_y + (\eta + 1)\sigma_E] - \sqrt{[\{\tfrac{1}{2}\sigma_y + \tfrac{1}{2}(\eta + 1)\sigma_E\}^2 - \sigma_y\sigma_E]}$$

$$(12.27)$$

This is the Perry formula for the intensity of end loading which will cause the strut to yield.

Robertson's part in the development of this formula was the determination of the dimensionless constant η. As a result of a number of experiments on circular sections he concluded that for mild steel $\eta = 0.003l/k$.

12.7 Load Factor for Struts

The value given in eqn. (12.27) for the intensity of end loading which will cause the fibre stress to reach yield point and the stresses σ_E, σ_r, σ_t and σ_p given in eqns. (12.5), (12.15), (12.16) and (12.27) can obviously not be used as working stresses since they represent the state of stress at the ultimate load-carrying capacity of the member. The conversion from ultimate stress to working stress is normally done by introducing either (1) a factor of safety or (2) a load factor.

The factor of safety (F.S.) is based on a stress ratio and defined as the ratio of ultimate or yield stress to working stress.

The load factor (N) is defined as the ratio of ultimate load to working load and is therefore based on the load-carrying capacity of the member. It is this latter value which is the important property of the member when considering the safety of a strut.

If it is assumed that the yield stress in mild steel is 230 N/mm² then using a factor of safety of 2·0 gives a working stress of 115 N/mm² whatever the value of the l/k ratio. Reference to Fig. 12.4 shows that for values of l/k greater than, say, 130 this stress is greater than σ_E. Consequently buckling will take place before the load on the column reaches that calculated with an average stress of 115 N/mm².

The only method for use with struts in order to ensure that the member is safe is the load factor. The intensity of end loading, σ_E, σ_r, σ_t or σ_p as the case may be, is divided by the load factor to get the working stress on the section.

The value generally taken for N is 2·0.

12.8 Experimental Determination of Critical Load: The Southwell Plot

The value of experiments in the study of struts has already been mentioned and in this connexion a method has been devised by Sir Richard Southwell whereby the critical load can be determined experimentally before failure is actually reached. The general solution of eqn. (12.5) is

$$P = n^2\pi^2 EI/l^2$$

where n is any integer. Because buckling must occur at the lowest value of the critical load, the Euler value was taken with $n = 1$. It is, however, important when studying the fundamentals of the Southwell plot to remember that n can be any integer.

The strut is assumed, as in the Perry approach, to have an initial departure from the straight of y_0 and the differential equation for the strut is

$$\frac{d^2y}{dx^2} + \mu^2 y = \frac{d^2 y_0}{dx^2}$$

where $\mu^2 = P/EI$ and in this case $y =$ total deflexion, and the origin is at the end of the strut. A solution of this equation may be obtained by expressing y and y_0 in a Fourier series. If

$$y = \sum_{n=1}^{\infty} [w_n \sin n\pi x/l] \tag{12.28}$$

and

$$y_0 = \sum_{n=1}^{\infty} [\bar{w}_n \sin n\pi x/l] \tag{12.29}$$

then it can be shown in a similar manner to that of Section 12.6 that

$$w_n = \frac{\bar{w}_n}{1 - \mu^2 l^2 / n^2 \pi^2} \qquad (12.30)$$

If P_n is the critical load corresponding to the substitution of n^2 in eqn. (12.5) then eqn. (12.30) can be written

$$\frac{w_n}{\bar{w}_n} = \frac{1}{1 - P/P_n} \qquad (12.31)$$

This equation gives the ratio in which the component $\bar{w}_n \sin n\pi x / l$ of the series representing the initial deflexion is magnified by the end thrust P. As P approaches P_1 (or P_E) it is seen that $\bar{w}_1$ becomes very largely magnified but that the higher harmonics are not appreciably increased. Thus it is the first harmonic which is of vital importance.

The deflexion at the centre of the strut δ can be written as

$$\delta = w_1 + w_3 + w_5 + \cdots$$

the even harmonics making no contribution. If $P \simeq P_E$ then the first harmonic becomes critical and

$$\delta \simeq w_1 = \frac{\bar{w}_1}{1 - P/P_E} = \bar{w}_1 \frac{P_E}{P_E - P} \qquad (12.32)$$

Putting $\Delta = \delta - \delta_0$ and using only the first harmonic of the series for the latter gives

$$\Delta = \frac{\bar{w}_1}{(P_E/P) - 1}$$

or

$$(P_E - P)\Delta = \bar{w}_1 P \qquad (12.33)$$

Dividing both sides of this equation by P gives

$$P_E \cdot \frac{\Delta}{P} - \Delta = \bar{w}_1 \qquad (12.34)$$

This is a linear equation between Δ/P and Δ. Thus the plot of the experimental results of Δ/P against Δ should lie on a straight line, the slope of which is $1/P_E$ and the intercept on the Δ/P axis is $\bar{w}_1$. The quantity Δ is the horizontal deflexion at the mid-height of the strut at a given load P. Thus if a run of experimental readings of P and Δ are taken, P_E can be obtained from this data before the strut has actually collapsed.

The following is an example showing the use of the Southwell Plot.

Example 12.1

The following observations were made in a test of a pin-jointed steel tubular strut of length 1·76 m.

Load (kN)	0·2	2·22	4·45	6·67	8·90	9·78
Central deflexion from initial position	—	0·25	2·75	4·75	6·75	8·25

Load (kN)	10·69	11·12	11·54	11·94	12·20	12·37
Central deflexion from initial position	10·25	14·00	14·75	22·5	28·5	75·0

Estimate from these observations the critical load of the strut and deduce the flexural rigidity EI.

Why is it not necessary to specify the units in which the deflexions were measured? *(Oxford, 1935)*

SOLUTION

In order to determine the critical load it is necessary to plot Δ/P against Δ. Values obtained from the test figures are given in the table below.

Δ	0·25	2·75	4·75	6·75	8·25	10·25	14·0	14·75	22·5	28·5	75
$\Delta/P \times 10^1$	1·12	6·18	7·11	7·60	8·45	9·61	12·6	12·76	18·85	22·37	60·7

The plot of Δ/P against Δ is given in Fig. 12.10. Assessing the best straight line through these points is not easy, but the one shown in the figure has a slope of $7 \cdot 5 \times 10^{-2}$, giving $P_E = 13 \cdot 35$ kN. The critical load is $\pi^2 EI/l^2$. Hence the flexural rigidity is

$$EI = P_E l^2/\pi^2$$
$$= \frac{13 \cdot 35 \times 1 \cdot 76^2 \times 10^6}{\pi^2}$$
$$= 4 \cdot 2 \times 10^6 \text{ kN-mm}^2 \text{ units}$$

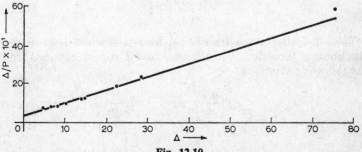

Fig. 12.10

The reason why the units in which Δ is measured do not have to be specified is because the plot is of P/Δ against Δ and the slope of the line is therefore independent of Δ.

12.9 Energy Methods of Calculating Critical Loads

Section 6 of Chap. 10 showed that for a structure in equilibrium the rate of change of total energy is zero, i.e. $d(U + V)/d\Delta = 0$. Thus in the case of a column displaced by a small amount the gain in strain energy δU equals the loss in potential energy δV. Note carefully that the external loads remain constant during the displacement.

Internal strain energy

$$\delta U = \int_0^l \frac{M^2\,dx}{2EI}$$

or, since $M/EI = d^2y/dx^2$,

$$\delta U = \frac{1}{2}\int_0^l EI\left(\frac{d^2y}{dx^2}\right)^2 dx \qquad (12.35)$$

Loss in potential energy

$$\delta V = P \times \text{change in length of the strut}$$
$$= P\,\delta l$$

Change in length $= \int_0^l (ds - dx)$

$$= \int_0^l \left[\left\{1 + \left(\frac{dy}{dx}\right)^2\right\}^{1/2} - 1\right] dx$$

$$= \frac{1}{2}\int_0^l \left\{\left(\frac{dy}{dx}\right)^2 + \text{higher terms}\right\} dx$$

$$\simeq \frac{1}{2}\int_0^l \left(\frac{dy}{dx}\right)^2 dx \qquad (12.36)$$

Since the strut when the critical load is reached is in neutral equilibrium, the value of this critical load P can be obtained from the following expression—

$$P = \frac{\int_0^l EI\left(\frac{d^2y}{dx^2}\right)^2 dx}{\int_0^l \left(\frac{dy}{dx}\right)^2 dx} \quad \text{or} \quad \frac{\int_0^l \frac{M^2}{EI}\,dx}{\int_0^l \left(\frac{dy}{dx}\right)^2 dx} \qquad (12.37)$$

Here y represents the *characteristic mode* of distortion. If this is known, then the integrals can be evaluated and the corresponding critical load deduced. In cases where the deflected form is not known, then y is assumed as a function of x which satisfies the conditions at $x = 0$ and $x = l$.

Suppose, for example, it is assumed that

$$y = a_1 \sin \frac{\pi x}{l} + a_2 \sin \frac{2\pi x}{l} + \cdots + a_n \sin \frac{n\pi x}{l}$$

$$= a_1 y_1 + a_2 y_2 + \cdots + a_n y_n$$

Then

$$\frac{dy}{dx} = a_1 \frac{dy_1}{dx} + a_2 \frac{dy_2}{dx} + \cdots + a_n \frac{dy_n}{dx}$$

and

$$\frac{d^2 y}{dx^2} = a_1 \frac{d^2 y_1}{dx^2} + a_2 \frac{d^2 y_2}{dx^2} + \cdots + a_n \frac{d^2 y_n}{dx^2}$$

It can be shown that the expression for P becomes

$$P = \frac{P_1 a_1{}^2 I_1 + P_2 a_2{}^2 I_2 \cdots P_n a_n{}^2 I_n}{a_1{}^2 I_1 + a_2{}^2 I_2 \cdots a_n{}^2 I_n} \qquad (12.38)$$

where P_1 is the first critical load, i.e. corresponding to $y = a_1 \sin \pi x/l$, and

$$I_1 = \int_0^l \frac{dy}{dx}\, dx$$

The Is in this expression are all positive and Southwell points out that the expression is analogous with that giving the mean density of a solid from several constituents; $a_n{}^2 I_n$ is the analogue of the volume and P_n that of the density; hence the numerator is equivalent to the mass and the denominator the volume. Since the mean density is never less than that of the lightest constituent, so the value for the load will never be less than P_1.

If in the series for y, a_2, a_3, etc. are small relative to a_1, then $a_2{}^2$, $a_3{}^2$, etc. are also small compared with $a_1{}^2$; also P will differ little from P_1. Hence if the form assumed for y is a close approximation to the first characteristic mode, P as calculated from eqn. (12.38) will be close to the first critical load, with the error one of excess.

If y is taken as $a_1 \sin \pi x/l$, then it can be shown quite easily using eqn. (12.38) that $P_1 = \pi^2 EI/l^2$.

If, however, for the sake of example, y was assumed to be given by the equation

$$y = ax(l - x)$$

Then

$$\frac{dy}{dx} = a(l - 2x)$$

$$\frac{d^2y}{dx^2} = -2a$$

$$P_1 = \frac{\int_0^l EI \cdot 4a^2 \, dx}{\int_0^l a^2(l^2 - 4lx + 4x^2) \, dx} = \frac{12EIl}{l^3} = \frac{1 \cdot 22\pi^2 EI}{l^2}$$

This is in excess of the true value $\pi^2 EI/l^2$ because an incorrect form has been chosen for the strut displacement.

12.10 Buckling of Thin Plates under Loads Applied in the Plane of the Plate

The energy approach is particularly useful in the solution of this type of problem.

The governing relationship is that the loss of potential energy, or the work done by the forces in compression equals the increase of strain energy due to the bending of the plate. If the forces in compression are

N_x per unit length parallel to the x-axis,

N_y per unit length parallel to the y-axis, and a shear force N_{xy} per unit length,

then the work done is given by

$$\tfrac{1}{2} \int \int \left[N_x \left(\frac{\partial w}{\partial x}\right)^2 + N_y \left(\frac{\partial w}{\partial y}\right)^2 + 2N_{xy} \frac{\partial w}{\partial x} \cdot \frac{\partial w}{\partial y} \right] dx \, dy$$

where w is the deflexion of the plate normal to its plane.

The increase in strain energy due to bending is given by

$$\frac{D}{2} \int \int \left\{ \left(\frac{\partial^2 w}{\partial x^2} + \frac{\partial^2 w}{\partial y^2}\right)^2 - 2(1 - v)\left[\frac{\partial^2 w}{\partial x^2} \cdot \frac{\partial^2 w}{\partial y^2} - \left(\frac{\partial^2 w}{\partial x \, \partial y}\right)^2\right] \right\} dx \, dy$$

where D is the flexural rigidity of the plate per unit width = $2Eh^3/3(1 - v)$, and $2h$ is the thickness of the plate; v is Poisson's Ratio. As in our consideration of struts, an assumed deflected form is taken to satisfy the end conditions and the first critical load will be given from the solution of the equation expressing the energy relationship.

As an example, take the uniform rectangular plate simply-supported along all four edges and subjected to a thrust in the x-direction only as in Fig. 12.11.

The deflected form of the plate is given by

$$w = \sum_{m=1}^{\infty} \sum_{n=1}^{\infty} a_{mn} \frac{\sin m\pi x}{a} \cdot \frac{\sin n\pi y}{b}$$

This vanishes when $x = 0$ and $x = a$, and also when $y = 0$ and $y = b$. Since the edges are simply-supported, $\partial^2 w / \partial x^2$ and $\partial^2 w / \partial y^2$ must also be zero under the above conditions and this requirement is also satisfied.

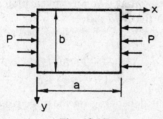

Fig. 12.11

The strain energy of bending is given by ΔU_B where

$$\Delta U_B = \frac{\pi^4 ab}{8} D \sum_{m=1}^{\infty} \sum_{n=1}^{\infty} a_{mn}^2 \left(\frac{m^2}{a^2} + \frac{n^2}{b^2}\right)^2 \qquad (12.39)$$

The work done by the forces in compression ΔU_D is given by

$$\Delta U_D = \tfrac{1}{2} P_x \int_0^a \int_0^b \left(\frac{\partial w}{\partial x}\right)^2 dx\, dy = \frac{\pi^2 b}{8a} P_x \sum_{m=1}^{\infty} \sum_{n=1}^{\infty} m^2 a_{mn}^2 \qquad (12.40)$$

Equating ΔU_B to ΔU_D gives

$$P_x = \frac{\pi^2 a^2 D \displaystyle\sum_{m=1}^{\infty} \sum_{n=1}^{\infty} a_{mn}^2 \left(\dfrac{m^2}{a^2} + \dfrac{n^2}{b^2}\right)^2}{\displaystyle\sum_{m=1}^{\infty} \sum_{n=1}^{\infty} m^2 a_{mn}^2} \qquad (12.41)$$

P_x obviously increases with n so that the smallest value occurs when $n = 1$, and also when all the coefficients a_{mn} except one are taken equal to zero, giving

$$P_{x\,cr} = \frac{\pi^2 D}{b^2} \left(\frac{mb}{a} + \frac{a}{mb}\right)^2 \qquad (12.42)$$

For a minimum value of this function $mb = a$, thus the value of m varies according to the a/b ratio.

For a square $a = b$, hence $m = 1$ for critical P_x; for $a = 2b$, $m = 2$ for critical P_x; and as a gets appreciably greater than b so m gets larger for critical P_x. For all cases however the fact that $mb = a$ means that the critical load is the same as that for the case when $a = b$, giving

$$P_{xcr} = 4\pi^2 D/b^2$$

$$= \tfrac{2}{3}\pi^2 \frac{Eh}{(1 - \nu)}\left(\frac{2h}{b}\right)^2$$

where $2h$ is the thickness of plate.

It is clear from this that, for any given plate, the load-carrying capacity depends on the $2h/b$ ratio.

This analysis was derived with the unloaded edges free. Generally it can be stated that

$$P_{cr} = \frac{k\pi^2 E(2h)}{12(1 - \nu)}\left(\frac{2h}{b}\right)^2$$

where k is a constant depending on the edge conditions.

12.11   Stiffened Plate Structures

It is seen from eqn 12.43 that the compressive load carried by a plate reduces as the width b increases and for this reason plates are often stiffened with longitudinal stiffeners as shown in Fig. 12.12 (*a*) and (*b*), the former giving the single plate construction which is largely used in civil engineering, whilst the latter gives the double plate construction used in ship and aircraft construction.

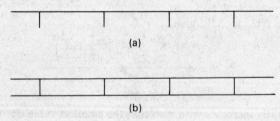

(a)

(b)

Fig. 12.12

Failure in compression of such a section may take place generally in one of three ways:

(1) By failure of the combined section as a strut as described earlier in this chapter.

(2) By failure of the panels as described in Section 12.10. These plate panels will normally be considered fixed at the edges.

(3) By failure of the longitudinal stiffening plates in a similar way. These plates, however, have one edge free.

The load-carrying capacity in a longitudinal direction will depend on the effective length of the section in that direction. The insertion of transverse stiffeners as shown in Fig. 12.13 reduces the effective

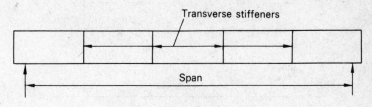

Transverse stiffeners

Span

Fig. 12.13

length resulting in the grid stiffened plate shown in Fig. 12.14.

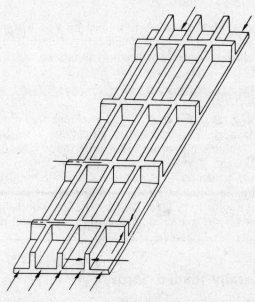

Fig. 12.14

The distribution of stress across the plate after initial buckling is as shown in Fig. 12.15 having its maximum value adjacent to the stiffeners and in predicting the ultimate strength of the plate it is this maximum stress which is the critical factor. The determination of the exact stress distribution is a difficult theoretical study but a simple approach for design purposes can be obtained by assuming that the edge stress acts over an effective width b_e and that the central portion of the plate is unstressed giving the stress distribution shown dotted in Fig. 12.15.

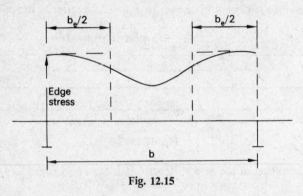

Fig. 12.15

The effective width approach was studied by Von Karman who related it to the yield stress σ_y and the modulus of elasticity E by the following expression for plates with simply supported edges

$$b_e = \pi 2h \, \frac{E}{3(1 - v^2)\sigma_y}$$

This expression gives in the case of mild steel an effective width of $55 \times$ plate thickness. Experimental investigations have, however, shown that this value is optimistic and a much lower value has been suggested by Dwight.

Stiffened plate structures of this type are generally welded. This operation results in initial stresses and deformations in the plates consequently the value to be taken for the yield stress in the material must be reduced by σ_r, the residual stress in the material resulting from the welding.

12.12 Laterally-loaded Struts

A member carrying a lateral load deflects. If to such a member an axial load is applied, the problem is analogous to that of a strut with initial curvature. If P is the magnitude of this axial load and δ the central deflection, then the member has to be designed for an axial load P together with the moment due to the lateral loads plus a moment $P\delta$. For struts where EI is large it is adequate to use the ordinary beam deflexion for δ, but this can lead to errors in more flexible struts.

12.13 Pin-jointed Strut with Uniform Lateral Load

Consider the strut shown in Fig. 12.16 which carries a uniformly-distributed lateral load of w per unit length in addition to an axial load P. With the origin at O the differential equation is

$$M = EI\,\frac{d^2y}{dx^2} = -Py - \tfrac{1}{2}w(\tfrac{1}{4}l^2 - x^2) \tag{12.43}$$

or

$$M + Py = \tfrac{1}{2}w(x^2 - \tfrac{1}{4}l^2)$$

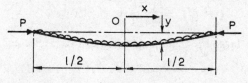

Fig. 12.16

Differentiating twice gives

$$\frac{d^2M}{dx^2} + P\,\frac{d^2y}{dx^2} = w \tag{12.44}$$

or

$$\frac{d^2M}{dx^2} + \mu^2 M = w$$

where $\mu^2 = P/EI$. A solution of this equation is

$$M = A \sin \mu x + B \cos \mu x + w/\mu^2 \tag{12.45}$$

when $x = \pm\tfrac{1}{2}l$, $M = 0$, hence $A = 0$ and $B = -w/\mu^2 \sec \tfrac{1}{2}\mu l$. Thus

$$M = (w/\mu^2)(1 - \cos \mu x \sec \tfrac{1}{2}\mu l) \tag{12.46}$$

The maximum bending moment occurs when $x = 0$ and is given by M_{max}, where

$$M_{max} = (w/\mu^2)(1 - \sec \tfrac{1}{2}\mu l)$$

or

$$M_{max} = \frac{8M_0}{\pi^2} \cdot \frac{P_E}{P}\left(1 - \sec \tfrac{1}{2}\pi\sqrt{\frac{P}{P_E}}\right)$$

where $M_0 = \frac{1}{8}wl^2$ = bending moment due to lateral loads alone, and

$P_E = \pi^2 EI/l^2$ = Euler critical load.

The maximum fibre stress σ_f is given by

$$\sigma_f = \frac{8\sigma_b}{\pi^2} \cdot \frac{\sigma_E}{\sigma}\left(1 - \sec \tfrac{1}{2}\pi \sqrt{\frac{\sigma}{\sigma_E}}\right) + \sigma \qquad (12.47)$$

where $\sigma_E = \pi^2 Ek^2/l^2$; $\sigma = P/A$; and $\sigma_b = M_0 d_2/I$ (cf. eqn. (12.25)).

If in the expression for M_{max} we write θ for $(\pi/2)\sqrt{(P/P_E)}$ and use the trignometrical series, then the expression for M_{max} becomes

$$M_{max} = \frac{8M_0}{\pi^2} \cdot \frac{P_E}{P}\left(\frac{\theta^2}{2!} + \frac{5\theta^4}{4!} + \frac{61\theta^6}{6!} + \cdots\right)$$

$$= \frac{8M_0}{\pi^2} \cdot \frac{P_E}{P}\left(\frac{\pi^2 P}{8P_E} + \frac{5\pi^4 P^2}{384P_E^2} + \cdots\right)$$

$$= M_0 + \frac{5wL^4 P}{384EI} + \cdots \qquad (12.48)$$

Where EI is large the remaining terms can be neglected, giving

$$M_{max} = M_0 + P\delta$$

where δ is the deflexion due to lateral loads alone.

An alternative solution to eqn. (12.45) is in the form

$$M = C \cos(\mu x - \epsilon) + w/\mu^2 \qquad (12.49)$$

where C and ϵ are constants of integration which can be determined from the known end conditions. Using this form of the solution H. B. Howard developed a graphical method for obtaining the bending moment on a laterally-loaded strut for any given axial and lateral load.

The load factor necessary to convert ultimate loads into working loads has to be applied to both axial and lateral loads. If the maximum load which can be carried by the member is taken as that which causes the fibre stress to reach the yield stress for the material and the load factor is N, then eqn. (12.47) becomes

$$\sigma_y = \frac{8N\sigma_b}{\pi^2} \cdot \frac{\sigma_E}{N\sigma}\left(1 - \sec \frac{\pi}{2}\sqrt{\frac{N\sigma}{\sigma_E}}\right) + N\sigma \qquad (12.50)$$

This equation, which contains the value of the working stress in the secant term in the bracket, is not easy to use and the Perry approximation is generally taken.

12.14 Perry's Approximation for Strut with Uniformly-distributed Lateral Load

This approximation is based on the assumption that the final deflected form of the strut is given by a cosine curve.

$$y = c \cos \pi x/l \qquad (12.51)$$

Then

$$M_x = EI \frac{d^2y}{dx^2} = M_x' - Py$$

where M_x' is the bending moment due to lateral loads alone. But

$$\frac{d^2y}{dx^2} = -(\pi^2/l^2)c \cos \pi x/l$$

Hence

$$(EI\pi^2/l^2)c \cos \pi x/l = Py - M_x' \qquad (12.52)$$

or

$$P_E y = Py - M_x' \qquad (12.53)$$

giving

$$y = \frac{M_x'}{P - P_E} \qquad (12.54)$$

(Since P will always be less than P_E this expression appears to give a negative value for y, but M_x' will also be negative, hence the sign of y is positive.)

Substituting in eqn. (12.53) gives

$$M_x = M_x' + \frac{PM_x'}{P_E - P} = \left(\frac{P_E}{P_E - P}\right)M_x' \qquad (12.55)$$

The maximum bending moment occurs at the centre and is given by

$$M_{max} = M_0\left(\frac{P_E}{P_E - P}\right)$$

where M_0 is the central bending moment due to lateral loads alone and equals $\frac{1}{8}wl^2$. If d_2 is the distance of the most-stressed fibre from the neutral axis and σ_f the fibre stress, then

$$\sigma_f = \frac{wl^2 d_2}{8I}\left(\frac{P_E}{P_E - P}\right) + \frac{P}{A} \qquad (12.56)$$

or, with the symbols of eqn. (12.47),

$$\sigma_f = \sigma_b\left(\frac{\sigma_E}{\sigma_E - \sigma}\right) + \sigma \qquad (12.57)$$

Applying a load factor N to the expression when the fibre stress reaches yield point gives

$$\sigma_y = \frac{Nwl^2}{8} \cdot \frac{d_2}{I} \left(\frac{\sigma_E}{\sigma_E - N\sigma} \right) + N\sigma \qquad (12.58)$$

This is a quadratic in σ which can be more easily solved than the secant equation (12.50).

The above analysis has been given for a uniformly-distributed loading. For other loading the expression for M_{max} is given approximately by

$$M_{max} = M_0 \left(\frac{P_E}{P_E - CP} \right) \qquad (12.59)$$

where C is a constant depending on the loading and has the values given in the accompanying table (Ref. 6).

Ends	Lateral load	C	Remarks	
Pin	uniform	1·000		
Pin	central point	0·894		
Pin	constant moment	1·110		
Pin	two symmetrical loads	1·064	quarter-points	Load positions
		1·030	third-points	
		1·000	$\frac{3}{4}l$ from ends	
Built-in	uniform	0·276	centre of strut	Moment positions
		0·172	ends of strut	
Built-in	central point	0·212	centre ends of strut	

12.15 Stability Functions

Equations 8.7 and 8.8 on page 224 can be written

$$M_{AB} = 4k \cdot \theta_A + 2k \cdot \theta_B - 6k \cdot \delta/L$$
$$M_{BA} = 4k \cdot \theta_B + 2k \cdot \theta_A - 6k \cdot \delta/L,$$

where 'k' is the stiffness $= EI/L$.

In the presence of an axial load the additional moments due to axial load times its eccentricity modify the equations and new functions s and c are defined as follows.

$$\left. \begin{array}{l} M_{AB} = sk \cdot \theta_A \\ M_{BA} = c(sk\theta_A) \end{array} \right\} \text{ for } \theta_B = 0 = \delta \qquad (12.68)$$

$$\left. \begin{array}{l} M_{AB} = c(sk\theta_B) \\ M_{BA} = sk \cdot \theta_B \end{array} \right\} \text{ for } \theta_A = 0 = \delta$$

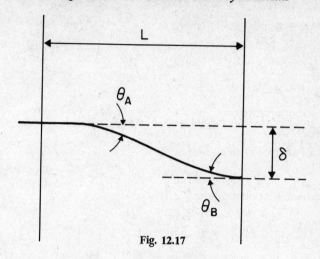

Fig. 12.17

Referring to Fig. 12.17 and re-defining θ_A and θ_B

$$\theta_A = \theta_B = -\delta/L$$

then:
$$M_{AB} = sk \cdot \theta_A + csk\theta_B$$
$$= -s(1 + c)k\delta/L$$

thus for both end rotation and element rotation (sway)

$$M_{AB} = sk\theta_A + csk\theta_B - s(1 + c)k\delta/L \qquad (12.69)$$
and
$$M_{BA} = csk\theta_A + sk\theta_B - s(1 + c)k\delta/L$$

12.16   Functions s and c

In Fig. 12.18 (*a*), taking moments about one end

$$F = -(M_{AB} + M_{BA})/L$$

Then the differential equation of equilibrium 12.1 (page 330) becomes:

$$EI\frac{d^2y}{dx^2} = -Py - M_{AB} - Fx$$

or put $P_E = \pi^2 EI/L^2$ (Eqn 12.5, page 331)
and $P = \rho P_E$

then $$\frac{d^2y}{dx^2} + \rho\frac{\pi^2}{L^2}y = \frac{1}{k_L}\left[(M_{AB} + M_{BA})\frac{x}{L} - M_{AB}\right]$$

The complementary function is

$$y = A \sin \frac{L}{\pi} \sqrt{\rho x} + B \cos \frac{\pi}{L} \sqrt{\rho x}$$

or put $\alpha^2 = \rho \pi^2/4$

$$y = A \sin \frac{2\alpha}{L} x + B \cos \frac{2\alpha}{L} x$$

using a Particular Integral

$$y = cx + D$$

and the boundary conditions

$$X = 0; \quad y = 0$$
$$X = L; \quad y = 0$$
$$X = L; \quad \dot{y} = 0$$

some troublesome manipulation (which the reader should verify) gives

$$\frac{M_{BA}}{M_{AB}} = c = \frac{2\alpha - \sin 2\alpha}{\sin 2\alpha - 2\alpha \cos 2\alpha} \qquad (12.70)$$

The condition at $\qquad x = 0; \dfrac{dy}{dx} = \theta_a$

and putting $\qquad M_{AB} = sk\theta_A$ and $M_{BA} = csk\theta_A$

after more manipulation, gives

$$s = \frac{(1 - 2\alpha \cot 2\alpha)\alpha}{\tan . \alpha - \alpha} \qquad (12.71)$$

The functions s and c are plotted in Fig. 12.18 (b).

It is seen that for $\rho = 0$, i.e. zero axial load

$s = 4; c = \frac{1}{2}$;
$s = 0$, i.e. zero stiffness;
c tends to ∞ for $\rho = 2 \cdot 046$ which is the critical load for a member fixed at one end.

For axial loads in excess of this the stiffness becomes negative, that is, the end moment M_{AB} has to reverse in order to maintain the strut in equilibrium.

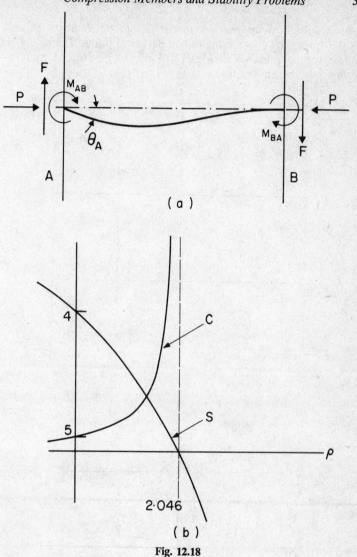

(a)

(b)

Fig. 12.18

12.17 Collapse of a Limited Frame

It will have been noticed from the moment distribution procedure of chapter 9 that the effects of applying a moment to a joint of a frame die away fairly rapidly and have little influence further than joints 'once removed' from the joint in question. It is thus possible to establish simplified approaches to the design of elements of rectangular frames by considering only a part or 'Limited Frame'. For example the column AB in Fig. 12.19 (*a*) can be studied by considering only the limited frame in Fig. 12.19 (*b*).

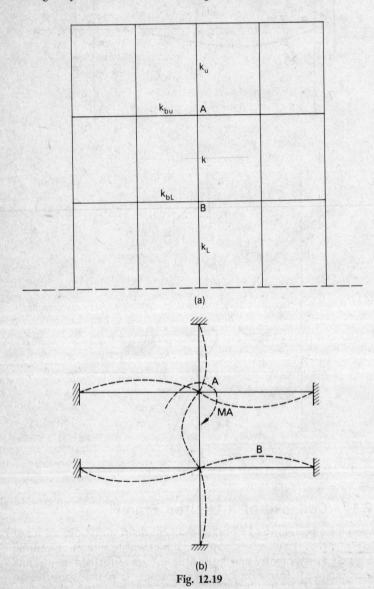

(a)

(b)

Fig. 12.19

Then at B:
$$M_B = sk\theta_B + sk_L\theta_B + \Sigma k_{bL}\theta_B + csk\theta_A$$

where Σk_{bL} = summation of beam stiffnesses at the lower end of the column B.

For $M_B = 0$
$$\theta_B = \theta_A \frac{-csk}{sk + sk_L + k_{bu}}$$

at A:
$$M_A = sk\theta_A + sk_u\theta_A + \Sigma k_{bu}\theta_A - \frac{csk(csk)}{sk + sk_L + k_{bL}}$$

and for instability the stiffness (M/θ) becomes zero.

i.e.
$$\frac{sk + sk_u + \Sigma k_{bu}}{csk} = \frac{csk}{sk + sk_L + \Sigma k_{bL}} \qquad (12.72)$$

The two ratios in (12.72) are in effect 'distribution' factors or relative stiffnesses at the lower and upper ends of AB and can be used as variables in producing charts of critical loads or 'effective lengths' to be used in the design of the column.

As a simple demonstration consider the column AB in Fig. 12.20.

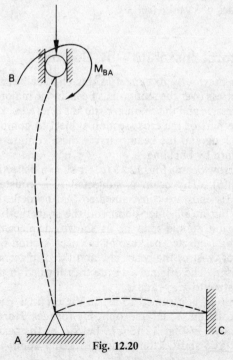

Fig. 12.20

Rotating B only $M_{BA} = sk\theta_B$; $M_{AB} = csk\theta_B$

Rotating A only $M_{BA} = csk\theta_A$; $M_{AB} = sk\theta_A$

Define the restraint offered to joint A by the built in beam AC as $qk\theta_A$, then for equilibrium of joint A

$$sk\theta_A + csk\theta_B + qk\theta_A = 0$$

i.e.
$$\theta_A = \frac{-cs}{(s + q)}\,\theta_B$$

thus
$$M_{BA} = sk\theta_B - \frac{csk(cs)\theta_B}{(s + q)}$$

or the stiffness at $B = \dfrac{M_{BA}}{\theta_B} = sk[1 - c^2/(1 + q/s)]$

and for instability of the column this becomes zero.

Direct calculation of the load at which this occurs is difficult using the eqns 12.70 and 12.71 but is readily done by trial using tabulated values of s and c (ref. 7).

12.18   Lateral Instability of Beams

The members previously discussed in this chapter have been those in which the stress over the section has been, in the majority of cases, wholly compressive and the characteristic has been a buckling failure.

Beams have part of the cross-section subject to compression and if the span is long and the beam narrow these compressive stresses can cause failure by buckling.

A deep, narrow beam (Fig. 12.13 (*a*)) is shown deflecting laterally in Fig. 12.13 (*b*). The beam is subjected to a constant bending moment *M*. Its ends are constrained so that no deflexion or twist occurs there, but at all other points on the span the beam is free. Let the deflexion at mid-span be as shown, displacements in the *x*,*y*-plane being neglected because of the deep section of the beam.

The angle of twist of the beam is θ and the displacement parallel to the *z*-axis is *u*. The original axes are transformed in the deflected (or buckled) state to x', y' and z'.

The moment *M* is assumed to act in the original *x*,*y*-plane. To prevent end rotation, equal and opposite torques *T* are applied at the ends as shown in Fig. 12.13 (*c*). Let EI_y be the flexural rigidity about the *y*,*y*-axis and *GJ* the torsional rigidity about the *x*,*x*-axis.

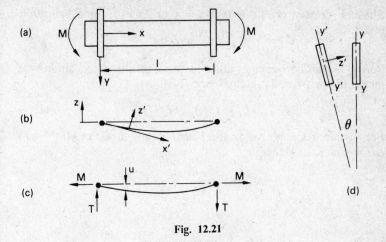

Fig. 12.21

The moment M is resolved into components along the y' and z' axes. Considering bending about the y', y'-axis gives

$$M\theta = -EI_v \frac{d^2u}{dx^2} \qquad (12.60)$$

Twisting about x' gives

$$T + M \frac{du}{dx} = GJ \frac{d\theta}{dx} \qquad (12.61)$$

Differentiating this with respect to x,

$$M \frac{d^2u}{dx^2} = GJ \frac{d^2\theta}{dx^2} \qquad (12.62)$$

Combining (12.60) and (12.62) gives

$$M\theta = -EI_v \frac{GJ}{M} \cdot \frac{d^2\theta}{dx^2}$$

or

$$\frac{d^2\theta}{dx^2} + \frac{M^2}{EI_vGJ} \cdot \theta = 0 \qquad (12.63)$$

or

$$\frac{d^2\theta}{dx^2} + k^2\theta = 0$$

where $k^2 = M^2/EI_vGJ$. A solution of this equation is

$$\theta = A \cos kx + B \sin kx$$

The end restraint gives $\theta = 0$ at $x = 0$ and $x = l$; on substitution this gives

$$A = 0 = B \sin kl$$

If $B = 0$ then no buckling takes place. The alternative $\sin kl = 0$ means that $kl = \pi, 2\pi$, etc. Hence

$$M_{cr} = (\pi/l)\sqrt{(EI_y GJ)} \tag{12.64}$$

For a thin rectangle of width b and depth d, $GJ \simeq \frac{1}{3}Gdb^3$ and $EI_y = Edb^3/12$. Hence

$$M_{cr} = \frac{\pi}{l} \cdot \frac{db^3}{6} \sqrt{(G \cdot E)}$$

If $G = E/2(1 + v)$, then

$$\sqrt{(G \cdot E)} = \frac{E}{\sqrt{\{2(1 + v)\}}}$$

leading to

$$M_{cr} = \frac{\pi}{l} \cdot \frac{db^3}{6} \cdot \frac{E}{\sqrt{\{2(1 + v)\}}}$$

The constant π applies for a uniform moment M; generally

$$M_{cr} = \frac{C}{l} \cdot \frac{db^3}{6} \cdot \frac{E}{\sqrt{\{2(1 + v)\}}} \tag{12.65}$$

where C depends on the type of loading—

> For point load at mid-span, $C = 4{\cdot}24$
> For uniformly-distributed load, $C = 3{\cdot}58$

A more exact analysis of this problem gave

$$M_{cr} = \frac{C}{l} \sqrt{\left\{\frac{EI_x \cdot EI_y \cdot GJ}{(EI_x - EI_y)}\right\}} \tag{12.66}$$

If $EI_x \gg EI_y$, as in the case of the narrow beam, this analysis leads to the earlier result which is independent of EI_x. It shows, on the other hand, that as EI_y approaches EI_x the critical moment becomes greater and that in a square-section beam, where $EI_y = EI_x$, no buckling occurs.

The analysis given applies to elastic bodies. In the inelastic range $\bar{E}$ and $\bar{G}$ must be substituted for E and G, $\bar{E}$ and $\bar{G}$ being the reduced moduli corresponding to the stress at M_{cr}.

Timoshenko and Bleich both suggest that reasonable values for the reduced moduli are the tangent moduli E_t and G_t at the fibre stress corresponding to M_{cr}. The shape of the section has an important effect on the value to be taken for the reduced moduli. The use of the tangent moduli gives reasonable results when used for I-section beams where most of the material in compression is in the flanges.

For rectangular-section beams the value to take for the reduced modulus is the secant modulus E_{sec} at the stress corresponding to the fibre stress.

The maximum fibre stress σ_{cr} corresponding to the critical moment is given by

$$\sigma_{cr} = \frac{C}{l} \cdot \frac{db^3}{6} \cdot \frac{6}{bd^2} \cdot \frac{E}{\sqrt{\{2(1+\nu)\}}} = \frac{CE}{\sqrt{\{2(1+\nu)\}}} \cdot \frac{b^2}{dl} \quad (12.67)$$

It is seen that this stress depends on two ratios, the one b/l and the other b/d. The former is sometimes referred to as the *slenderness ratio*. This is perhaps a somewhat misleading term, for whilst in struts the slenderness ratio l/k is the governing factor in the critical stress, in beams the shape factor b/d also plays a part.

12.19 Further Reading

1. TIMOSHENKO, S. P. and GERE, J. M., *Theory of Elastic Stability* (London McGraw-Hill, 1961).
2. BLEICH, F., *Buckling Strength of Metal Structures* (London, McGraw-Hill, 1952).
3. SHANLEY, F. B., "Inelastic column theory", *J. Aero Sci.* (1947).
4. MICHELL, A. G. M., "Elastic stability of long beams under transverse forces," *Phil. Mag.*, **48,** no. 292 (1899).
5. SOUTHWELL, R. V., *Theory of Elasticity* (Oxford, Clarendon Press, 1936).
6. PIPPARD, A. J. S. and BAKER, J. F., *Analysis of Engineering Structures* (London, Arnold, 1936).
7. HORNE, M. R. and MERCHANT, W., *The Stability of Frames* (Oxford, Pergamon Press).
8. GERARD, G., *Introduction to Stability Theory* (London, McGraw-Hill, 1962).
9. DWIGHT, J. B. and MAXHAM, K. E., Welded Steel Plates in Compression, *The Structural Engineer*, Vol. 47 No. 2, February 1969.
10. FAULKNER, D., "A Review of Effective Plating for Use in the Analysis of Stiffened Plating in Bending and Compression," *Journal of Ship Research*, Vol. 19, No. 1, March 1975.

12.20 Examples for Practice

(For all examples take $E = 200$ kN/mm².)
1. A column of I-section 300 mm deep is 8 m long. The cross-sectional area is 8,400 mm² and the principal second moments of area are $I_{xx} = 135 \times 10^6$ mm⁴ and $I_{yy} = 9 \times 10^6$ mm⁴.

The column may be considered pin-jointed at the ends for bending about the zz-axis and there is no other constraint against bending in this direction. For bending about the yy-axis the ends are to be considered rigidly fixed, while an additional pin-type constraint at mid-length prevents lateral displacement there.

Calculate the value of the lowest Euler critical load for the column.

If the actual load is one-fifth of the above critical value and its line of action passes through a point on the yy-axis at a distance of 6 mm from the C. of G. of the section, determine the maximum compressive stress in the section. (*Ans.* 2,250 kN; 57 N/mm².)

2. Prove that if the central axis of a column when unloaded is a sine curve with a central deflexion small compared with its length, it will remain a sine curve when carrying an end load within the elastic limit.

A tubular strut 80 mm outside diameter, 6 mm thick, 3 m long is initially out of the straight by 6 mm measured at the centre. Assuming pin-jointed ends without friction, calculate the greatest end load that may be applied if the maximum compressive stress is not to exceed 90 N/mm². (*Ans.* 80 kN.)

3. A steel tube 50 mm diameter and 2 mm thick is used as a pin-ended strut to carry an axial load of 5 kN and a uniformly distributed lateral load of 50 N per m run. The final deflected form of the strut can be assumed a cosine curve and $E = 200$ kN/mm². Calculate the maximum fibre stress. (*Ans.* 37·8 N/mm².)

4. A cantilever of length l carries a lateral load W and a central axial load P at the free end. Show that the lateral deflexion at the free end is

$$\frac{Wl}{P}\left(\frac{\tan \mu l}{\mu l} - 1\right)$$

where $\mu^2 = P/EI$. Evaluate this expression in terms of Wl^3/EI when $P = \pi^2 EI/16l^2$ and show that $(Wl^3/3EI) . P_E/(P_E - P)$ is a good approximation for this value of Δ.

5. A 100 mm diameter tube with area of 1,000 mm² and second moment of area 120×10^4 mm⁴ has a slenderness ratio $l/k = 100$. Find the permissible axial load when given the following data: hinged ends, yield stress 300 N/mm², load factor 2·0, initial curvature $\delta_{co} = 7·5$ mm.

Calculate the maximum stress caused by this load and comment on the factor of safety. (*Ans.* 70 kN; 105 N/mm².)

6. Calculate the critical load for a strip of spring steel 100 mm long and of section 12 mm by 2 mm, with the ends fixed in position and direction. Assuming the strip to be in equilibrium in a slightly bent position under the critical stress, find what lateral deflexion would accompany a maximum compressive stress of 450 N/mm². (*Ans.* 7·9 kN; 0·24 mm.)

7. The following data were taken during an experiment on a column. Determine the critical load and the initial deflexion.

Load P (kN)	15	17·5	20	23	24·5
Deflexion δ (mm)	7·6	10·0	13·5	20·0	28·0

(*Ans.* 32·4 kN; 12·3 mm.)

13

Shell Structures

13.1 Introduction

If a sheet of paper is laid to span between two supports about 500 mm apart as shown in Fig. 13.1 (*a*) and a small weight, say 0·25 kg, is placed at the centre, it will be found that the paper collapses under the load. If, however, a similar sheet of paper is taken and curved to form part of a cylinder, the ends being connected as shown in Fig. 13.1 (*b*), it will be found that it can easily carry the load.

The shell gives a form of construction by means of which very light sections can be used to cover large areas. The shells used can be subdivided into three main types—

1. Cylindrical,
2. Spherical or Domed,
3. Hyperbolic Paraboloid.

The complete theory for these shells requires a textbook in itself and all that will be attempted in this chapter is to give an elementary approach to the problem.

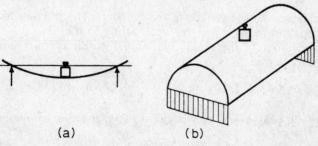

(a)   (b)

Fig. 13.1

Since the shell is thin it cannot resist large bending moments and as a first assumption it can be taken that the shell is only able to resist direct forces in its own plane. These forces are known as *membrane forces* and the theory thus derived is known as the *membrane theory*. It is proposed to give this theory for the different forms of shell. Since the theory is based on direct forces it can only be applied in cases where there is no change of shape under load. Deformation normally implies a change of curvature; such a change of curvature implies in turn that bending stresses are present and the stresses resulting from deformation must be added to the membrane stresses to give the true stress action.

13.2 The Cylindrical Shell

Consider an element of shell (Fig. 13.2) which carries a load which can be resolved into two components Z and Y per unit area normal

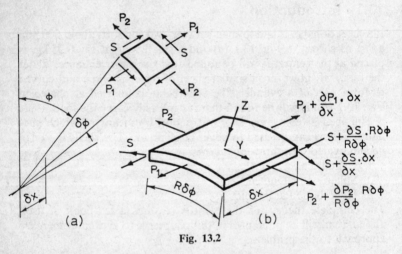

Fig. 13.2

and tangential to the surface of the cylinder. The position of the element is shown in (*a*) and the forces acting on it in (*b*).

Consideration of the equilibrium normal to the cylinder surface (*z*-direction), gives

$$Z\,\delta x R\,\delta\phi + 2p_2\,\delta x \sin\tfrac{1}{2}\delta\phi + \frac{\partial p_2}{R\partial\phi}\,R\delta\phi\,\delta x \sin\tfrac{1}{2}\delta\phi = 0$$

Taking $\sin\tfrac{1}{2}\delta\phi$ as $\tfrac{1}{2}\delta\phi$ and neglecting higher-order infinitesimals gives

$$Z\,\delta x R\,\delta\phi + p_2\,\delta x\,\delta\phi = 0$$

and hence $$p_2 = -RZ \qquad (13.1)$$

In the x-direction there are no external forces and consideration of the equilibrium gives

$$\frac{\partial p_1}{\partial x}\, \delta x R\, \delta\phi + \frac{\partial s}{R\partial\phi}\, R\, \delta\phi\, \delta x = 0$$

Thus

$$\frac{\partial p_1}{\partial x} = -\frac{\partial s}{R\,\partial\phi} \qquad (13.2)$$

Consideration of the equilibrium tangential to the shell gives

$$Y\,\delta x R\,\delta\phi + \frac{\partial p_2}{R\partial\phi}\, R\,\delta\phi\,\delta x + \frac{\partial s}{\partial x}\,\delta x R\,\delta\phi = 0$$

Hence

$$\frac{\partial s}{\partial x} = -\frac{\partial p_2}{R\,\partial\phi} - Y \qquad (13.3)$$

Integrating gives

$$s = -\int \frac{\partial p_2}{R\,\partial\phi}\, dx - \int Y\, dx + C_1 \qquad (13.4)$$

where C_1 is a constant of integration.

If the load on the shell is symmetrical about the centre where $x = 0$, then $s = 0$ at this point, which leads to $C_1 = 0$. Integrating (13.2) gives

$$p_1 = -\int \frac{\partial s}{R\,\partial\phi}\, dx + C_2 \qquad (13.5)$$

where C_2 is a constant of integration obtained by the fact that when $x = \pm l/2$, i.e. at the ends of the shell $p_1 = 0$.

The forces Y and Z depend on the loads coming on the cylinder. In the case of vertical dead load of magnitude w_0 per unit surface area,

$$Z = w_0 \cos\phi$$

and

$$Y = w_0 \sin\phi$$

giving

$$\left.\begin{aligned} p_2 &= -w_0 R \cos\phi \\ s &= -2w_0 x \sin\phi \\ p_1 &= \frac{w_0}{R}(x^2 - \tfrac{1}{4}l^2)\cos\phi \end{aligned}\right\} \qquad (13.6)$$

and

The stresses in a cylindrical shell only approximate to the membrane theory for "short" shells, i.e. those in which the ratio $L/(2R \sin \phi)$ (where L = span) is small—of the order of $\frac{1}{4}$.

Where this ratio is high—of the order of 3—the shell is known as a "long" shell and the stresses in it can be computed with reasonable accuracy by assuming that the shell acts as a beam with a span L and having a cross-section as shown in Fig. 13.3.

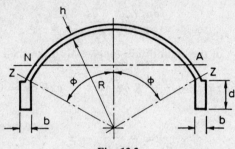

Fig. 13.3

The distance $\bar{z}$ of the neutral axis from ZZ is given by

$$\bar{z} = \frac{2R^2h(\sin \phi - \phi \cos \phi) - bd^2}{2Rh\phi + 2bd}$$

where R = radius of barrel,

h = thickness,

2ϕ = angle subtended at centre,

b = width of edge beam, and

d = depth of edge beam.

The distance $\bar{z}_S$ of the centre of area of the segmental portion alone from ZZ is given by

$$\bar{z}_S = \frac{R}{\phi} (\sin \phi - \phi \cos \phi)$$

The second moment of area I_S of the segmental portion about an axis through its centre of area is given by

$$I_S = 2R^3h\{\tfrac{1}{2}\phi(1 + 2 \cos^2 \phi) - \tfrac{3}{4} \sin 2\phi\} - \frac{2R^3h}{\phi} (\sin \phi - \phi \cos \phi)^2$$

About the neutral axis of the whole section the second moment of area of the segmental portion I_{NAS} is given by

$$I_{NAS} = I_S + 2Rh\phi(\bar{z}_S - \bar{z})^2$$

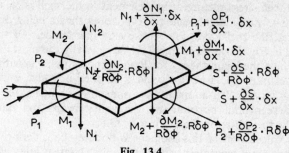

Fig. 13.4

The second moment of the valley beams I_{NAV} about the neutral axis of the whole section is given by

$$I_{NAV} = \tfrac{1}{6}bd^3 + 2bd(\bar{z} + \tfrac{1}{2}d)^2$$

The second moment of area I_{NA} of the whole section about the neutral axis is given by

$$I_{NA} = I_{NAS} + I_{NAV}$$

If the total load on the shell per ft run is w, then the bending moment is $wL^2/8$ and the compressive stress at the crown of the shell can be calculated from the standard theory of bending as given in Chap. 3.

Shells are normally made of reinforced concrete. The tensile force which the normal theory of bending gives in the lower portion will be carried by reinforcement, which can be calculated as for composite sections in Chap. 3.

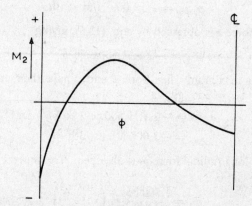

Fig. 13.5

The true stress picture at any element in the shell is as shown in Fig. 13.4, the moments and corresponding shears being due to the fact that the shell does not behave as a membrane. Of the forces shown in Fig. 13.4, the moment M_2 is generally the most important and whilst the beam theory for long cylinders predicts with reasonable accuracy the forces p_1 and s it does not give a means of calculating M_2. This moment has its maximum value at the junction of the segmental portion and the valley beam and damps out towards the centre as shown in Fig. 13.5.

The calculation of this moment is done by considering the shell and edge beam separately and setting up the compatibility equations. The resulting differential equation is one of the 8th order. The problem does, however, lend itself to a matrix solution, which can be obtained reasonably easily on the computer.

Example 13.1

Calculate the membrane forces in a shell of 4 m span built to a radius of 12 m and subtending a total central angle of 60°.

The shell is 60 mm thick and carries a load of 500 N/m² of horizontal projection in addition to its own weight, which can be taken as 1,500 N/m².

SOLUTION
Dealing first with the forces due to self-weight,

$$X = 0; \qquad Y = 1{,}500 \sin\phi; \qquad Z = 1{,}500 \cos\phi$$

Equation (13.6) gives $p_2 = 1{,}500 \times 12 \cos\phi$ and this has its maximum value when $\phi = 0$, giving

$$p_{2max} = -18{,}000 \text{ N/m width}$$

The shear force s is obtained by eqn. (13.6), giving

$$s = -2 \times 1{,}500x \sin\phi$$

This has its maximum when both ϕ and x have their maxima, i.e. $\phi = 30°$ and $x = 2$ m, giving

$$s = -2 \times 1{,}500 \times 2 \times \tfrac{1}{2}$$
$$= -3{,}000 \text{ N/m width}$$

Finally the longitudinal force p_1 is obtained. The general expression is

$$p_1 = \frac{1{,}500 \cos\phi}{12}\left(x^2 - \frac{4^2}{4}\right)$$

Since x cannot exceed 2·00 the expression within the brackets is always negative and p_1 has its maximum when $x = 0$ and $\phi = 0$, giving

$$p_1 = -500 \text{ N/m width}$$

The 500 N/m² of horizontal projection is dealt with as follows—

$$Z = 500 \cos^2 \phi; \qquad Y = 500 \cos \phi \sin \phi; \qquad X = 0$$

Hence $p_2 = -(500 \times 12) \cos^2 \phi$, which has its maximum when $\phi = 0$, giving

$$p_2 = -6,000 \text{ N/m width}$$

From eqn. (13.4)

$$s = -\int \frac{\partial p_2}{R\, \partial \phi}\, dx - \int Y\, dx$$

Substituting for p_2 and Y gives

$$s = -\int 500 \sin 2\phi\, dx - \tfrac{1}{2} \int 500 \sin 2\phi\, dx$$

$$= -\tfrac{3}{2} \int 500 \sin 2\phi\, dx$$

$$= -\tfrac{3}{2} \cdot 500x \sin 2\phi$$

This has its maximum value when x and 2ϕ have their maxima, i.e. $x = 2\cdot00$ and $2\phi = 60°$, giving

$$s = -1,300 \text{ N/m width}$$

Substitution for s in (13.5) gives

$$p_1 = \tfrac{3}{2} \cdot \tfrac{500}{12}(x^2 - 2^2) \cos^2 \phi$$

This has its maximum when $x = 0$ and $\phi = 0$, giving

$$p_1 = -250 \text{ N/m width}$$

The maximum value of the stress in the shell is that due to p_2. If this stress is σ_ϕ, then its value is given by

$$\sigma_\phi = \frac{24,000}{60 \times 1,000} = 0\cdot4 \text{ N/mm}^2$$

This is a very low value, showing the economical use of the material.

The membrane force distributions are shown in Fig. 13.6 (*a*), (*b*) and (*c*). Figure 13.6 (*a*) cannot be the final state of force since there can be no internal force perpendicular to a free edge, i.e. $p_2 = 0$ when $\phi = 30°$, giving the true value of the p_2 force as indicated by dotted lines in Fig. 13.6 (*a*).

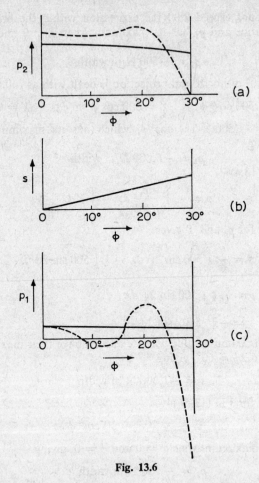

Fig. 13.6

The membrane p_1 shown in Fig. 13.6 (c) does not satisfy the conditions that at a section where there is no axial load $(X = 0)$ the total compression must equal the total tension and the true value of the p_1 force is as indicated by the dotted lines in the figure.

13.3 Membrane Stresses in Domes

Shells will be considered which are in the form of a surface of revolution loaded symmetrically with respect to an axis.

An element of the shell, as shown in Fig. 13.7 is cut out by two meridians (position defined by θ, equivalent to longitude) and two parallel or hoop circles (defined by ϕ, equivalent to latitude). The external loads are X, Y and Z acting in the longitude, latitude and normal directions respectively.

The meridian plane and the plane perpendicular to it are planes of principal curvature at any point in a surface of revolution. Let

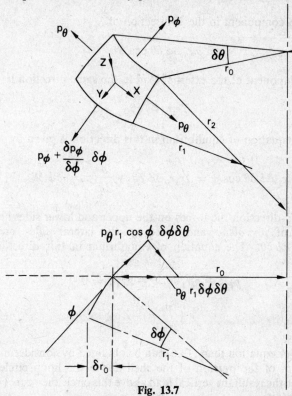

Fig. 13.7

r_1 and r_2 denote the corresponding principal radii of curvature. The radius of the hoop circle r_0 is given by $r_0 = r_2 \sin \phi$. The surface area of the element is $r_1 r_2 \sin \phi\, d\theta\, d\phi$. The symmetry of loading eliminates shearing forces.

Considering, firstly, forces in the Y-direction and neglecting second-order infinitesimals gives the resultant of the p_ϕ forces as

$$\left(p_\phi + \frac{\partial p_\phi}{\partial \phi} \cdot \partial \phi\right)(r_0 + \delta r_0)\, \delta\theta - p_\phi r_0\, \delta\theta$$

$$= p_\phi\, \delta r_0 \cdot \delta\theta + r_0 \frac{\partial p_\phi}{\partial \phi} \cdot \delta\phi\, \delta\theta$$

or

$$\frac{d}{d\phi}(p_\phi r_0)\, d\phi\, d\theta$$

The forces acting in the hoop direction have a resultant in the direction of the radius r_0 of the hoop circle of

$$2p_\theta r_1 \, \delta\phi \, \sin \tfrac{1}{2}\delta\theta = p_\theta r_1 \, \delta\phi \, \delta\theta$$

This has a component in the Y-direction of

$$p_\theta r_1 \, \delta\phi \, \delta\theta \cos \phi$$

The component of the external load in the same direction is

$$Y r_1 r_0 \, \delta\phi \, \delta\theta$$

Thus the equation of equilibrium in this direction is given by

$$p_\theta r_1 \, \delta\phi \, \delta\theta \cos \phi = Y r_1 r_0 \, \delta\phi \, \delta\theta + \frac{d}{d\phi}(p_\phi r_0) \, \delta\phi \, \delta\theta \qquad (13.6)$$

In the Z-direction the forces on the upper and lower sides have a resultant of $p_\phi r_0 \, \delta\theta \, \delta\phi$ and those on the lateral sides one of $p_\theta r_1 \sin \phi \, \delta\phi \, \delta\theta$. The equation of equilibrium in this direction is therefore

$$p_\phi r_0 + p_\theta r_1 \sin \phi + Z r_1 r_0 = 0$$

or

$$\frac{p_\phi}{r_1} + \frac{p_\theta}{r_2} = -Z \qquad (13.7)$$

A simpler equation than (13.6) can be obtained by considering the equilibrium of the portion of the shell above the hoop circle. If V_ϕ denotes the resultant vertical load above this circle then eqn. (13.6) can be rewritten

$$2\pi r_0 p_\phi \sin \phi + V_\phi = 0 \qquad (13.8)$$

Equations (13.6) or (13.8) and (13.7) enable the membrane stresses to be determined in shells of revolution. In the case of a spherical dome of radius r, carrying a self-weight of w_0 per unit area of dome (*see* Fig. 13.8)

$$r_0 = r \sin \phi$$

The resultant of the vertical forces V_ϕ is given by

$$V_\phi = 2\pi \int_0^\phi r^2 w_0 \sin \phi \, d\phi = 2\pi r^2 w_0 (1 - \cos \phi)$$

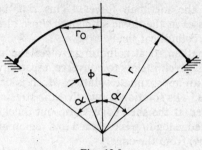

Fig. 13.8

Substituting first in (13.8) and then in (13.7) gives

$$p_\phi = -\frac{w_0 r}{1 + \cos \phi}$$

$$p_\theta = rw_0 \left(\frac{1}{1 + \cos \phi} - \cos \phi \right)$$

p_ϕ is always negative, i.e. the stresses are compressive and their magnitude increases with ϕ, being $\frac{1}{2}w_0 r$ at the apex of the dome and $w_0 r/(1 + \cos \alpha)$ at the supports.

The sign of the p_θ forces varies with ϕ, being negative, i.e. compressive, when ϕ is small. The sign changes from negative to positive when

$$\frac{1}{1 + \cos \phi} - \cos \phi = 0$$

i.e. when
$$\phi = 51°50'$$

Thus in the hoop direction the maximum compression occurs at the crown. If the supports can supply reactions tangential to the meridianal forces as in Fig. 13.8, then the stresses in the dome do not differ very much from those calculated by the membrane theory. If however the dome is only supported in a vertical direction as in Fig. 13.9, then a ring beam must be used to take the horizontal

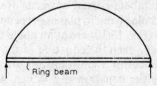

Ring beam

Fig. 13.9

component of the meridian forces. This ring beam introduces additional stresses in the dome similar to those already mentioned in dealing with cylindrical shells.

The exact mathematical solution involves the use of a hypergeometric series which is slow to converge, especially for thin shells. Geckeler gives an approximate solution which is near the truth for most thin shells. His solution shows that the stresses resulting from moments applied at the shell edge damp out quickly, and that the effect of rim or edge loading is confined to a region of width approximately $2 \cdot 5 \sqrt{(2rh)}$ from the edge.

Example 13.2

Calculate the forces in a semicircular dome 80 mm thick and 10 m radius due to a self-weight of 2,000 N/m².

SOLUTION
Due to self-weight,

$$Z = 2,000 \cos \phi; \qquad Y = 2,000 \sin \phi; \qquad X = 0$$

If p_ϕ is the force at the rim of the shell, then use of eqn. (13.8) gives

$$2\pi \times 10 \times p_\phi = -2\pi \times 10^2 \times 2,000$$

Hence $p_\phi = -20,000$ N/m width

At the crown, $p_\phi = 0$; hence by eqn. (13.7), noting that $r_1 = r_2 = 10$,

$$p_\theta = -rZ = -20,000 \text{ N/m width}$$

At the supports $p_\phi = -20,000$ and $Z = 0$; hence

$$p_\theta = +20,000 \text{ N/m width}$$

The maximum stress is that due to a force of 20,000 N/m width, i.e.

$$20,000/1,000 \times 80 = 0 \cdot 25 \text{ N/mm}^2$$

13.4 The Hyperbolic Paraboloid

A hyperbolic paraboloid is a doubly-curved surface formed by moving a vertical parabola curved upwards over one curved downwards, the two parabolae lying in planes perpendicular to each other. If the axes OX, OY (Fig. 13.10) are at an angle θ, and OZ is the axis perpendicular to them, then the equation of the surface is $z = kxy$, where k is a constant. For x constant this gives $z = k_1 y$ and for a constant value of y the equation is $z = k_2 x$, where k_1 and k_2 are constants. Thus there are two sets of straight lines forming a

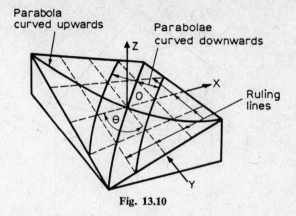

Fig. 13.10

doubly-curved surface. These straight lines are known as *ruling lines*.

If z is constant, then xy is constant, in other words all horizontal sections are hyperbolae. When $\theta = 90°$ then the surface becomes a rectangular hypar. If in Fig. 13.11 OBAH is the hypar surface and OBA′H the horizontal plane, the dimensions being as shown, then $k = d/ab$ and the equation of the hypar surface is

$$z = \frac{dxy}{ab}$$

A portion of a hypar subject to membrane stresses only is shown in Fig. 13.12 together with its projection on a horizontal plane.

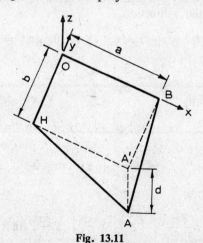

Fig. 13.11

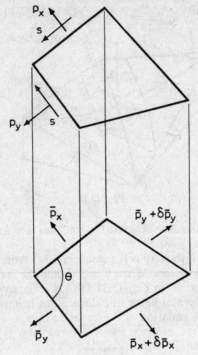

Fig. 13.12

Consider the equilibrium of the forces acting in the horizontal plane (Fig. 13.12) in the x-direction,

$$\delta\bar{p}_x \cdot \delta y + \delta s \cdot \delta x + X\,\delta y \cdot \delta x \sin\theta = 0$$

or
$$\frac{\partial\bar{p}_x}{\partial x} + \frac{\partial s}{\partial y} + X\sin\theta = 0 \qquad (13.9)$$

In the y-direction,

$$\delta\bar{p}_y \cdot \delta x + \delta s \cdot \delta y + Y\,\delta x \cdot \delta y \sin\theta = 0$$

or
$$\frac{\partial\bar{p}_y}{\partial y} + \frac{\partial s}{\partial x} + Y\sin\theta = 0 \qquad (13.10)$$

In the z-direction,

$$2s\frac{\partial^2 z}{\partial x\,\partial y} + \left(Z - X\frac{\partial z}{\partial x} - Y\frac{\partial z}{\partial y}\right)\sin\theta = 0 \qquad (13.11)$$

The expression in brackets is the z-component of the net load at the point under consideration when resolved into two components, one Z in the z-direction and the other in the tangent plane.

Thus eqn. (13.11) can be written

$$2s \frac{\partial^2 z}{\partial x \, \partial y} + Z \sin \theta = 0$$

If the applied load Z is constant at all points and $X = Y = 0$ (this state of affairs will exist in a shell which is fairly flat and carries only its own weight), then

$$2s \frac{\partial^2 z}{\partial x \, \partial y} + \text{const.} = 0$$

But the fundamental shell equation gives $\partial^2 z / \partial x \, \partial y = \text{constant}$. Hence $s = \text{constant}$ and $\partial \bar{p}_x / \partial x$ and $\partial \bar{p}_y / \partial y$ are also constant.

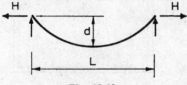

Fig. 13.13

If the shell is bounded by the ruling lines, then at the boundaries $\bar{p}_x = \bar{p}_y = 0$ and, since the rate of change is constant, then at all points on the shell $\bar{p}_x = \bar{p}_y = 0$. Hence, when there is only a vertical load the stress system consists of a shear stress along the directrices, this stress being constant throughout. Thus in this structure, the stress being constant, the material is being used in the most economical way. A uniform shear stress along the directrices implies equal and opposite principal stresses on planes bisecting the directrices. These planes are the planes of the parabolae and it means that the parabola having its apex at the top is in compression whilst that with its apex at the bottom is in tension. The shell roof thus acts as a combination of parabolic arches and suspension cables. *See* Chap. 15.

In such structures the magnitude of the horizontal force H is given by

$$H = \frac{w \times (\text{span})^2}{8 \times (\text{rise at centre})}$$

where $w = $ intensity of loading (*see* Fig. 13.13 and also Section 2.16) Applying this formula to the shell shown in Fig. 13.14 and allowing for the fact that the total load is equally shared between the tensile and compressive parabolae, we have

$$H = \frac{w_z ab}{2d}$$

where w_z is the intensity of load per unit area in the z-direction. This gives the intensity of load per unit width of shell. If the thickness is $2h$, then the stress σ is $w_z ab/4dh$.

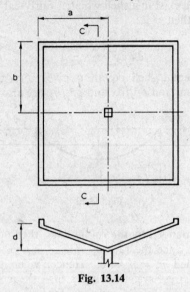

Fig. 13.14

The umbrella-type hypar shell can be analysed by considering the free-body equilibrium of half the shell. Consider the umbrella shown in Fig. 13.14, which has dimensions $2a \times 2b$ and a central dip d and carries a uniform vertical load of intensity w per unit area.

The moment across the section C–C $\quad = 2wab \times \tfrac{1}{2}a$

$\qquad\qquad\qquad\qquad\qquad\qquad\qquad = wa^2b$

The horizontal force $\qquad\qquad\qquad = wa^2b/d$

This force is a tension in each of the side beams OB and a thrust in AH. (*See* Fig. 13.11.)

The tension in each side beam $\qquad = wa^2b/2d$

The horizontal component of thrust in $AH = wa^2b/d$

The vertical component $= \dfrac{d}{a}\left(\dfrac{wa^2b}{d}\right)$

$$= wab$$

An equal force will result from each of four quadrants, giving a total vertical component of $4wab$, which is equal to the applied vertical load.

If the structure is being built of reinforced concrete then these expressions can be used to calculate the amount of reinforcement and also to check the compressive stresses in the concrete.

The membrane stresses do not give a true picture of the actual stresses in the shell since they do not allow for deformations. Experiments have, however, shown that over the major portion of the shell the stresses differ little from those calculated by the membrane theory.

Bending exists at the edges of the shell but this damps out quite quickly, leaving the membrane stresses over the central portion. As with the other types of shell, the solution involves an eighth-order differential equation, although simpler methods can be applied which give results agreeing reasonably well with experiments.

Example 13.3

The umbrella shell shown in Fig. 13.14 has dimensions $a = b = 8$ m and $d = 2$ m. It is 60 mm thick, has edge beams 150 mm by 200 mm, and carries a load of 1,250 N/m² in addition to its self-weight. Calculate the stresses in the shell and the amount of reinforcement required if the stress in this is 135 N/mm². Take the unit weight of reinforced concrete as 2.4×10^3 N/m³.

SOLUTION

The self-weight of the shell

$$= 1,440 \text{ N/m}^2$$

The edge beam has a downward load of 720 N/m run.

Assuming that this is uniformly distributed over the whole shell gives a load of

$$180 \text{ N/m}^2$$

The total load on the shell

$$= 1,250 + 1,440 + 180$$

$$= 2,870 \text{ N/m}^2$$

The force per metre width in the membrane

$$= (2,870 \times 8 \times 8)/(2 \times 2)$$
$$= 46,000 \text{ N}$$

The direct stress in the concrete

$$= 46,000/(10^3 \times 60)$$
$$= 0 \cdot 77 \text{ N/mm}^2$$

The reinforcement required

$$= 46,000/135$$
$$= 340 \text{ mm}^2 \text{ per m width}$$

The tension in the edge beam

$$= 46,000 \times 8 = 368,000 \text{ N}$$

The area of steel required

$$= 368,000/135 = 2,720 \text{ mm}^2$$

The compression in the central sloping members

$$368,000 \times 2 \times (4 \cdot 48/4 \cdot 00) = 825,000 \text{ N}$$

This member acts as a column but also as the compression flange of a beam of which the shell is the web and must be so designed.

13.5 Further Reading

1. TIMOSHENKO, S., *Theory of Plates and Shells* (New York, McGraw-Hill, 1940).
2. FLÜGGE, W., *Stresses in Shells* (Berlin, Springer, 1960).
3. JENKINS, R. S., *Theory and Design of Cylindrical Shell Structures* (London, Ove Arup, 1947).
4. GIBSON, J. E., *The Design of Cylindrical Shell Roofs* (London, Spon, 1961).
5. GECKELER, J., "Über die Festigkeit achsensymmetrischer Schalen," *Forsch-Arb. IngWes.*, **276**, 1926.
6. PARME, A. L., "Hyperbolic paraboloids and other shells of double curvature," *Proc. Amer. Soc. Civ. Engrs.*, **82**, 1,057, 1956.

14

Limit Analysis of Plane Frames

14.1 Introduction

An extensive literature has been built up dealing with the analysis of idealized materials and structures which behave in a perfectly elastic-plastic manner, and the results obtained for structures in which bending actions predominate (cf. Chap. 8) closely predict the real behaviour of such structures.

14.2 Elastic-Plastic Flexural Relations

The bending of simple beams of idealized material strained beyond the yield strain has been discussed in Chap. 3 and the moment-curvature relationship for a rectangular cross-section was shown to be similar to the curve OAB in Fig. 14.1. For $M < M_v$ the section behaves elastically, giving the straight line OA. At A, M equals the yield moment M_v, and the $M - \phi$ relation is no longer linear. Finally, as the moment tends towards the fully plastic moment, the curvature ϕ tends to infinity.

The ratio of the fully plastic moment M_p to the yield moment is called the shape factor, i.e.

$$M_p/M_v = v = 1.5 \qquad \text{for a rectangle}$$

and for an I-section the shape factor reduces to something like 1.15, giving a moment-curvature relation, shown dotted in Fig. 14.1, with a sharp knee. Finally, if the shape factor becomes unity the curve degenerates to two linear portions, a region of perfect elasticity OA'

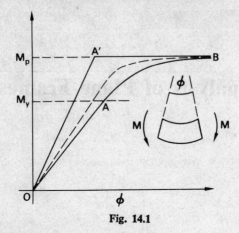

Fig. 14.1

and the perfectly plastic range A'B where the curvature can increase indefinitely at constant moment. In this case, of course,

$$M_y = M_p$$

The idealized elastic-plastic relation OA'B is assumed in the following development.

14.3 Behaviour of a Loaded Beam

In general, the bending moment on a beam is not uniform and in the centrally-loaded beam of Fig. 14.2 the maximum moment occurs at a single cross-section. If the load-point deflexion d is recorded while the load W increases from zero, the initial behaviour will be elastic to give the straight line OA. When the maximum moment reaches the fully plastic moment, a very large curvature is now possible and the deflexion increases with no change in load, i.e. A to B.

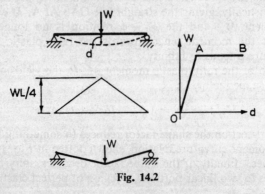

Fig. 14.2

The bending moments along the beam are everywhere less than the maximum, i.e. less than M_p, except at the load point, so that the large curvature is concentrated at the one cross-section. There is a hingeing action and the beam develops a clearly defined kink—a plastic hinge. Deformations away from the hinge are everywhere elastic and, since the plastic deformations in the hinge occur under constant moment, the elastic deformations remain unchanged and all displacements after point A is reached are due to plastic deformations in the hinge zone.

14.4 Simple Structural Behaviour

Consider the propped cantilever beam of Fig. 14.3 (*a*). As long as the beam remains elastic normal methods of analysis (e.g. as in Chap. 4) give the bending moment diagram Fig. 14.3 (*b*). The load W can

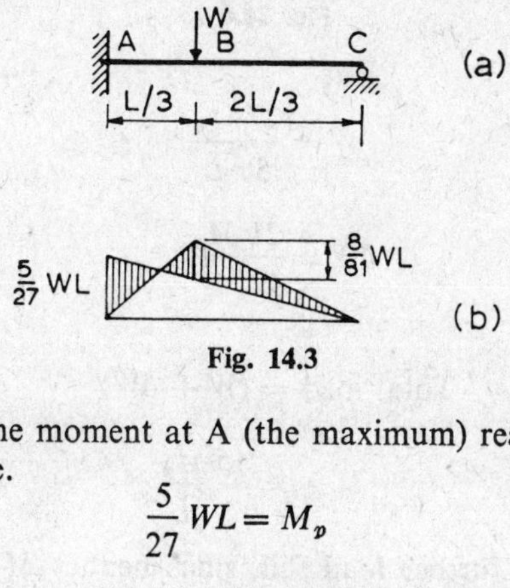

(a)

(b)

Fig. 14.3

increase until the moment at A (the maximum) reaches the plastic moment M_p, i.e.

$$\frac{5}{27} WL = M_p$$

or

$$W = \frac{27}{5} \frac{M_p}{L}$$

The moment at A cannot increase further and the curvature can increase indefinitely, so that additional load δW can cause no change in M_A, and no additional resistance to rotation is offered at A. This additional load must be carried as if the beam were simply-supported (Fig. 14.4 (*a*)); the additional moments for equilibrium are as in Fig. 14.4 (*b*). Load δW can now be increased until the gross moment at B equals the plastic moment M_p, and during this stage plastic rotation occurs at A. At $M_B = M_p$,

$$\frac{8}{81} WL + \frac{2}{9} \delta WL = M_p$$

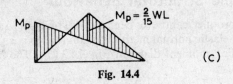

Fig. 14.4

and as

$$W = \frac{27}{5}\frac{M_p}{L}$$

$$\delta W = \frac{21}{10}\frac{M_p}{L}$$

therefore

$$\text{Total load} = (W + \delta W)$$

$$= \frac{15}{2}\frac{M_p}{L}$$

Attempts to add further load fail, since neither M_A nor M_B can increase, so that equilibrium is not possible and indefinite rotation will occur at A and B. The beam is then a two-bar mechanism and will collapse by excessive rotation at these two locations, i.e. A and B are in effect hinges of a mechanism. The physical behaviour can be demonstrated by recording the load point deflexion as in Fig. 14.5.

From O to E the beam behaves in the usual elastic manner; at E a plastic hinge occurs at end A of the beam and under additional load the structure acts as simply-supported elastic beam. Therefore

$$(d_2 - d_1) = \frac{4(\delta W)L^3}{243EI} \qquad \text{(p. 114)}$$

At F the plastic moment is reached at the load point and the structure collapses at constant load.

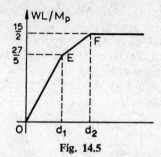

Fig. 14.5

The final bending moment diagram at collapse is recorded in Fig. 14.4 (*c*). If the location of plastic hinges had been known, the collapse load could have been calculated directly—

At F
$$M_p + \tfrac{2}{3}M_p = \tfrac{2}{9}PL$$

therefore
$$P = \frac{15}{2}\frac{M_p}{L}$$

14.5   General Structural Action

(i) A perfectly elastic-plastic structural member behaves elastically until a plastic hinge is formed at one section.

(ii) Additional load may be carried if rotation at this hinge allows diffusion of the load to other stable parts of the structure.

(iii) As each plastic hinge is formed, the moment remains constant at the fully-plastic value irrespective of deformation or additional load.

(iv) Collapse occurs when there is no remaining stable element able to carry additional load.

(v) At collapse the structure as a whole, or in part, forms a simple mechanism.

(vi) Looking ahead, it is apparent that, if the locations of plastic hinges can be predicted, the collapse load is readily calculated by simple statics.

14.6   Definition of Collapse Mechanism

The insertion of a real hinge, or pin joint, into a statically-indeterminate stiff frame reduces the number of indeterminate moments by one, so that, if the number of indeterminates is *n*, the addition of *n* hinges produces a simple statically-determinate structure. The addition of one more hinge will allow the structure to move with one degree of freedom, i.e. a mechanism is formed then the number of hinges to form a mechanism is $(n + 1)$.

This criterion, must, of course, be applied to each element of a structure as well as to the structure as a whole, because collapse of one part certainly represents practical failure. Typical collapse mechanisms are shown in Fig. 14.6 and many others can be sketched using the above criterion.

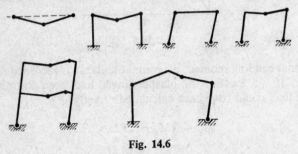

Fig. 14.6

14.7 Prediction of Collapse for Beams

The general form of the bending moment diagram for a fixed beam is readily drawn and, since it is now known that for collapse sufficient peaks of moment each equal to M_p must be obtained to form a mechanism with plastic hinges at these points, a simple trial-and-error graphical analysis is possible.

(a) Propped Cantilever (Fig. 14.7)

The final bending moment diagram is composed of two parts: (b) shows the moments on a simple beam due to the load W, and (c) the moments due to the reactant moment M_A. Combining the two diagrams in (d), it is seen that the only possible peaks of bending moment are at A and B. Then, for collapse,

$$\text{Number of redundants} = n = 1$$

therefore   Number of hinges $= (n + 1) = 2$

Thus both M_A and M_B must be equal to M_p at these sections and, for a uniform beam, $M_A = M_B = M_p$. Therefore at B

$$M_p + \tfrac{1}{2}M_p = \tfrac{1}{4}WL$$

therefore, $W = 6M_p/L$. Alternatively, for design purposes,

$$M_p = WL/6$$

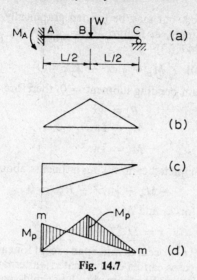

Fig. 14.7

This can, of course, be done graphically by drawing the simple beam moment diagram to scale and arranging, by trial, the reactant moment line, mm of Fig. 14.7 (*d*), to give equal peaks at A and B.

(b) Uniform Load Cases

Consider the same structure as in the previous example, but with a distributed load as in Fig. 14.8 (*a*). The full bending moment diagram is shown at (*b*) and, as before, two peaks of moment are possible, one at A and one at B somewhere in the span. The point B

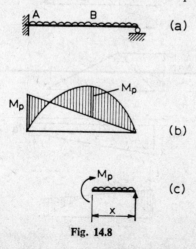

Fig. 14.8

is not well defined, but may be located graphically, by trial, or by statics. For example, at failure $M_A = M_B = M_p$, so that for the free body (Fig. 14.8 (c))—

Moments about B $M_p + \frac{1}{2}wx^2 - R_c x = 0$

Shear at maximum bending moment $= 0$, therefore

$$R_c = wx$$

therefore $M_p - \frac{1}{2}wx^2 = 0$

or $M_p = \frac{1}{2}wx^2$

Then, for the complete beam, taking moments about A,

$$-M_p + \frac{1}{2}wL^2 - R_c L = 0$$

or, substituting for R_c and M_p,

$$x = L(\sqrt{2} - 1) = 0{\cdot}41L$$

Thus point B will not occur at mid-span except for cases of symmetry, but in practical cases can be easily located; alternatively, with little resultant error, it may be assumed to be at mid-span.

(c) Fixed Beam
The two parts of the bending moment diagram for the fixed-ended beam of Fig. 14.9 (a) are shown at (b) and (c) and the combined diagram at (d).

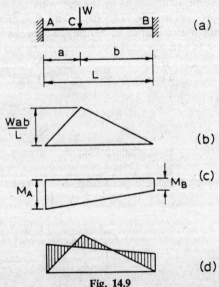

Fig. 14.9

Number of indeterminate moments $= n = 2$, therefore

Number of hinges required $= (n + 1) = 3$

From (d) it is seen that the only possible peaks of moment are at A, B and C. Thus, for collapse, hinges occur at A, B and C; and for a uniform beam

$$M_A = M_B = M_C = M_p$$

Thus at C,

$$2M_p = \frac{Wab}{L}$$

or

$$W = \frac{2M_pL}{ab}$$

or

$$M_p = \frac{Wab}{2L} = \tfrac{1}{2} \text{ simple beam moment}$$

Alternatively, it may be desirable to vary the section so that at the supports the strength is, say, $1\cdot5M_p$. Then, at C,

$$M_p + 1\cdot5M_p = \frac{Wab}{L}$$

or

$$W = 2\cdot5 \frac{M_pL}{ab}$$

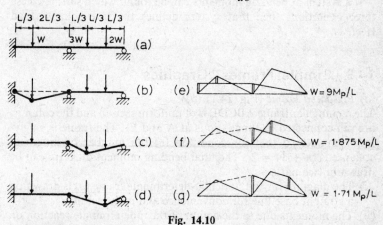

Fig. 14.10

(d) Continuous Beams

Continuous beams may be dealt with in the above manner. For example, Fig. 14.10 (a) shows a two-span beam built-in at one end and (b), (c) and (d) show possible failures in each span. The problem is then to determine the least load-value to cause collapse or,

conversely, the maximum plastic moment of resistance required for
the section. Bending moment diagrams at collapse are given in (*e*),
(*f*) and (*g*) with the associated failure loads. By varying the section
several other failures are possible.

14.8 General Collapse Conditions

The previous discussion may be summarized to give three necessary
and sufficient conditions attending collapse of a structure.

When a structure is just on the point of collapse—

(*a*) *Equilibrium condition*
The system of bending moments must be in equilibrium with the
external loads.

(*b*) *Yield condition*
The bending moments may nowhere exceed the plastic moment
values of the members.

(*c*) *Mechanism condition*
There must be sufficient plastic hinges to form a mechanism.

If a system of bending moments can be found which satisfies these
three conditions then that system defines the true collapse load
(Ref. 1).

14.9 Simple Frames: Graphics

(*a*) *Pin-based Frame* (Fig. 14.11 (*a*))
The rectangular frame ABCDE is of uniform section and the columns
are pin-connected to foundations at A and E. The frame is singly-
indeterminate so that for condition (*c*) the number of hinges at
collapse is (*n* + 1) = 2. The final bending moment diagram can be
drawn in two parts—
 (i) The diagram for the statically-determinate frame (*b*) is drawn on
     a straight base line for convenience and is shown at (*c*).
(ii) The moments due to the action of the indeterminate reaction on
     the frame are calculated. Drawn again on a straight base line,
     this gives the reactant moment diagram *abcde* of (*e*). These
     diagrams must be added to give two equal and opposite peaks of
     bending moment to satisfy the mechanism condition and so that
     nowhere is this value exceeded to satisfy the yield condition.
     This is done in (*f*); one hinge forms at D and the length BC is

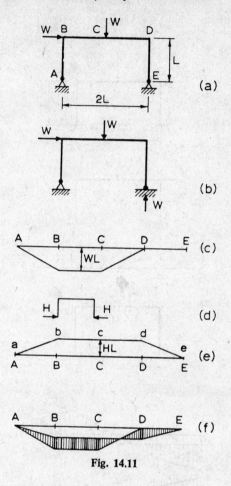

Fig. 14.11

all fully plastic. Then, considering B from (f), we have

$$2M_p = WL$$

or $$W = 2M_p/L$$

(b) Fixed-base Frame (Fig. 14.12 (a))
Considering the same frame and the same loading, but with the column bases A and E fixed, the simplest statically-determinate frame is obtained by cutting the beam as in (b) to obtain twin cantilevers ABC and CDE. The bending moment diagram is then as at (c). The reactant moment diagram is now that due to the indeterminate moment, thrust and shear at C as shown at (d). This

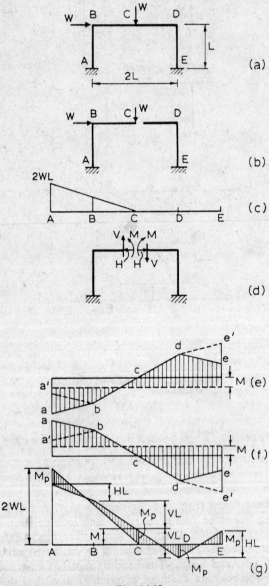

Fig. 14.12

diagram is drawn in (*e*). For *M* equal to zero the diagram *abcde* is obtained, *ab* and *de* having equal and opposite slopes, but, since the direction of *H* is not known, these slopes are not known, reversal of *H* reversing the slopes to give *a'* and *e'*; the direction of the shear *V* is also unknown. Reversal of the direction shown in (*d*) merely reverses the slope of *bcd*, as shown in (*f*). It is now necessary to fit any of the four possible diagrams in (*e*) and (*f*) to (*c*) so that the yield and mechanism conditions are satisfied.

$$\text{Number of indeterminates} = 3$$

therefore $\quad\quad$ Number of hinges required = 4

A solution can be found by trial on the drawing board as in (*g*) or algebraically. For example, try hinges at A, C, D and E; then, from the diagram,

$$\text{Change in moment along DE} = 2M_p$$

Also, from (*d*) this change $= HL$, i.e.

$$HL = 2M_p$$

In a similar manner,

Over CD, $\quad\quad\quad\quad\quad\quad VL = 2M_p$

Over ABCD, $\quad +M_p + 2VL + HL - M_p = 2WL$

and eliminating *H* and *V* gives

$$M_p = WL/3$$

or $\quad\quad\quad\quad\quad\quad\quad W = 3M_p/L$

Sketching of the diagram or simple calculation will now check that M_p is not exceeded at any other section; this therefore is the correct collapse mechanism.

(c) Pitched-roof Frame

For the pitched-roof frame of Fig. 14.13 (*a*) the same procedure is followed, splitting the frame at the ridge to give twin determinate cantilevers. The only difference is in the shape of the reactant moment diagram. This diagram is sketched in Fig. 14.13 (*b*). Line *abcde* gives moments due to *V* alone; adding the thrust *H* alters this to *a'b'cd'e'*; and the moment *M* merely moves the base line as before.

(d) Effects of Uniform Load

The presence of a distributed load has the same effect as with the beam in Fig. 14.9, that is, the hinge position is not sharply defined.

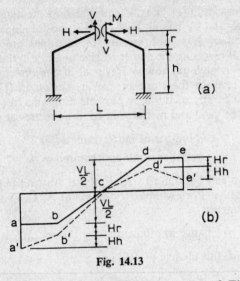

Fig. 14.13

This is demonstrated by the fixed-base frame of Fig. 14.14 (*a*) carrying a uniform load on the beam and a horizontal load at one corner. The bending moment diagram in Fig. 14.14 (*b*) is obtained by adding to the twin cantilever diagram *fghkm* the reactant moment diagram *abcde*: the span hinge is seen to have moved from the centre line. This can readily be dealt with by drawing to scale, or again, by

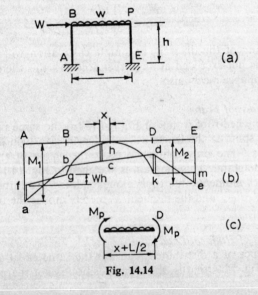

Fig. 14.14

statics. From the diagram Fig. 14.14 (b),

At A, $\qquad M_p = M_1 - \tfrac{1}{8}wL^2 - Wh$

At X, $\qquad M_p = M_1 - Hh - \tfrac{1}{2}VL\left(\dfrac{\tfrac{1}{2}L - x}{\tfrac{1}{2}L}\right) - \tfrac{1}{2}wx^2$

At D, $\qquad M_p = \tfrac{1}{8}wL^2 - (M_2 - Hh)$

At E, $\qquad M_p = M_2 - \tfrac{1}{8}wL^2$ $(M_2 = Ee,$ Fig. 14.14 (b).)

The fifth equation necessary is obtained by considering the shear at X. Then, isolating the part of the beam BD from the hinge to D, and noting that shear is zero at maximum bending moment, moments about D give

$$M_p + \tfrac{1}{2}w(x + \tfrac{1}{2})^2 + M_p = 0$$

These equations are readily solved for W in terms of M_p.

14.10 Order of Hinge Formation

The order in which plastic hinges form has a bearing on the condition of a structure under working loads and on the amount of rotation required at the hinges before the final collapse condition is reached. Once yield has occurred in a structure the relations between force and deformation are no longer simple and general analysis is difficult.

Satisfactory approximations may be obtained in the case of perfectly elastic-plastic structural members. Thus an initial elastic analysis shows, by the maximum moment value, the first hinge to form. The structure has now one less indeterminacy with respect to additional load and a second elastic analysis can be performed, and so on.

Alternatively, elastic analysis of the frame of Fig. 14.15 (a) shows the first hinge occurs at E when

$$M_p = \frac{33}{80} WL \qquad \text{or} \qquad W = \frac{80}{33}\frac{M_p}{L}$$

The B.M.D. is now as in Fig. 14.15 (b).

Assume the next hinge forms at D when the loads are increased by a factor λ_1. This will cause the moment at E to exceed M_p and it is necessary to add a reactant moment M' to reduce this excess. The

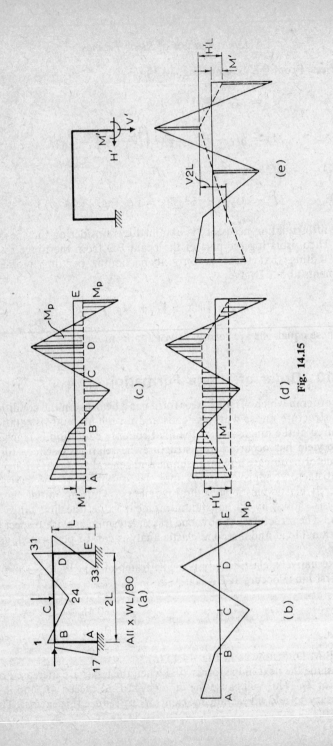

Fig. 14.15

B.M.D. is then as in (*c*)

At D, $\qquad\qquad 31\lambda_1 + M' = M_p = 33$

At E, $\qquad\qquad 33\lambda_1 - M' = M_p = 33$

$$\lambda_1 = \frac{33}{32} \quad \text{and} \quad M' = \frac{33}{32}$$

Nowhere is M_p exceeded.

If the next hinge forms at C at a load factor λ_2, moments at D and E will now exceed M_p. Reactions H' and M' must be added at E to give the B.M.D. of (*d*).

At C, $\qquad\qquad 24\lambda_2 - M' - H'L = M_p = 33$

At D, $\qquad\qquad 31\lambda_2 + M' + H'L = M_p = 33$

At E, $\qquad\qquad 33\lambda_2 - M' \qquad\;\; = M_p = 33$

This gives

$$\lambda_2 = \frac{66}{55}$$

$$M' = \frac{363}{55}$$

$$H'L = \frac{594}{55}$$

Again, nowhere is M_p exceeded.

The next hinge will form at A when the loads have been increased by a new factor λ_3 and this will increase the moments at C, D and E beyond M. M', H' and V' are added at E to restore these values. The B.M.D. is shown in Fig. 14.15 (*e*) and it follows that

at A, $\qquad 17\lambda_3 + M' \qquad\qquad + V'2L = M_p = 33$

at C, $\qquad 24\lambda_3 - M' - H'L - V'L = M_p$

at D, $\qquad 31\lambda_3 + M' + H'L \qquad\quad = M_p$

at E, $\qquad 33\lambda_3 - M' \qquad\qquad\qquad\;\; = M_p = 33$

i.e. $\qquad\qquad\qquad\qquad \lambda_3 = \frac{99}{80}$

$$M' = \frac{627}{80}$$

$$H'L = \frac{1{,}056}{80}$$

$$V'L = \frac{990}{80}$$

This gives the four hinges for collapse and nowhere is M_p exceeded.

Summarizing

First yield and first hinge at E when $W_1 = \dfrac{80}{33}\dfrac{M_p}{L}$

Second hinge at D at $W_2 = \dfrac{33}{32} W_1$

i.e. $W_2 = \dfrac{80}{32}\dfrac{M_p}{L}$

Third hinge at C at $W_3 = \dfrac{66}{65} W_1$

i.e. $W_3 = \dfrac{32}{11}\dfrac{M_p}{L}$

Fourth hinge at A at $W_4 = \dfrac{99}{80} W_1$

i.e. $W_4 = 3\dfrac{M_p}{L}$

That is, $W_1:W_2:W_3:W_4::1:1{\cdot}03:1{\cdot}20:1{\cdot}24$

14.11 Complex Frames

The graphical method described is of universal application; however, it can be found by experience that for other than the single-bay, single-storey frame, it becomes unwieldy.

Collapse loads for more complicated frames may be approximated to by the use of the following two principles. Proofs of these theorems (Refs. 1 and 2) are available, but will not be reproduced, sufficient discussion being provided to give an understanding of their application. By these means upper and lower limits can be found for the collapse load and by successive adjustment these limits can be brought as close together as desired.

14.12 Collapse Load: Lower Limit Theorem

If any bending moment distribution can be found which satisfies

(i) the equilibrium condition, and
(ii) the yield condition (i.e. bending moments nowhere exceed M_p)

that system is safe and statically sufficient and the corresponding load system is less than or equal to the true collapse load of the structure.

This theorem can be demonstrated by reference to the structure in Fig. 14.16 (*a*). An infinite number of arbitrary bending moment systems are possible which satisfy static equilibrium. For example, the previous elastic analysis provides one possible case in which full

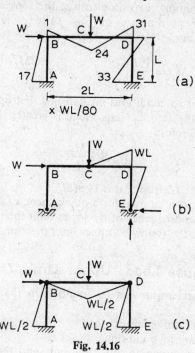

Fig. 14.16

plasticity will occur at E when

$$M_p = \frac{33}{80} WL \quad \text{or} \quad W = \frac{80}{33} \frac{M_p}{L}$$

At this load both the equilibrium and yield conditions are satisfied but, since only one plastic hinge exists, the structure will not collapse, and the true collapse load cannot be less than this value, i.e. is equal to or greater than $(80/33)(M_p/L)$.

Calculation of a lower limit to the collapse load can be obtained by selecting a very simple bending moment distribution such as that shown, for the same frame, in Fig. 14.16 (*b*). Sufficient points of zero moment are assumed to make the structure statically determinate, and the remaining moments easily calculated.

Full plasticity will be reached at D when

$$M_p = WL \quad \text{or} \quad W = \frac{M_p}{L}$$

Similarly, by assuming zero moments at the tops of the columns, the simple moment diagram of Fig. 14.16 (c) is obtained and plasticity occurs at C when

$$M_p = \frac{WL}{2} \quad \text{or} \quad W = \frac{2M_p}{L}$$

In each case the two conditions are satisfied, but only one hinge is formed, so that the actual collapse load is greater than or equal to these values. Therefore

$$W_c \geqslant 2\frac{M_p}{L}$$

(In fact $W_c = 3M_p/L$ from para. 14.9(b).)

This approach can be used for a quick check of a design but the most important consequence of the lower limit theorem is its use in conjunction with the following upper limit theorem.

14.13 Collapse Load: Upper Limit Theorem

If a collapse mechanism can be found such that the associated moments satisfy
  (i) the equilibrium condition, and
  (ii) the mechanism condition,
the mechanism is kinematically sufficient, and the corresponding load system is greater than or equal to the true collapse load.

Again consider the rectangular frame used already and reproduced in Fig. 14.17, failure could occur by either of the mechanisms (a) or (b) or by a combination such as (c). For mechanism (a) the bending moment diagram at collapse is as in Fig. 14.17 (d) and, from the beam collapse,

$$2M_p = \frac{W(2L)}{4} \quad \text{or} \quad W = 4\frac{M_p}{L}$$

This load satisfies the two conditions above, but at A the moment exceeds M_p. Therefore before W equal to $4M_p/L$ could be reached, plasticity must have occurred at A and the frame may have already failed by some other mechanism.

$$W_{collapse} \leqslant \frac{4M_p}{L}$$

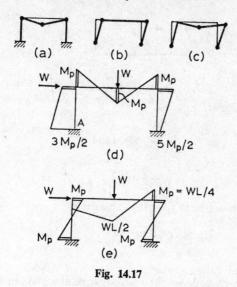

Fig. 14.17

Alternatively, consider the sidesway mechanism in Fig. 14.17 (*b*) For equilibrium in sway and taking moments about the base of the frame

$$WL = 4M_p \quad \text{or} \quad W = 4\frac{M_p}{L}$$

The bending moment diagram is drawn in Fig. 14.17 (*e*) and M_p is seen to be exceeded at the centre of the beam, so that again the collapse load has been overestimated. It is evidently possible to continue in this way trying every possible mechanism until a value for the collapse load is obtained from the lowest upper limit, but a systematic method of carrying this out is desirable.

14.14 Simple Applications of Upper Limit Theorem

The propped cantilever beam in Fig. 14.18 (*a*), when loaded with a single concentrated load W, is assumed to fail by the formation of two plastic hinges as in Fig. 14.18 (*b*). Then applying the principle of virtual work (p. 146),

$$\text{Work done by the load} = W\delta$$

(Note that W is constant at the collapse load during this deformation.) From the geometry of the deformation

$$\delta = \theta(\tfrac{1}{3}L)$$

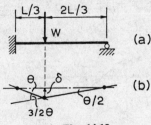

Fig. 14.18

therefore   Work done $= \frac{1}{3}W\theta L$

The work absorbed in the plastic hinges is

$$M_p \times \text{rotation} = (M_p\theta) + (M_p 3\theta/2)$$
$$= \tfrac{5}{2}M_p\theta$$

Equating work done by the load to work absorbed in the hinges,

$$W\theta L/3 = \tfrac{5}{2}M_p\theta$$

or $$W = 7{\cdot}5 M_p/L$$

This is the correct solution in this case, since (p. 378) it is already known that the assumed failure mode is the true one.

Since the work absorbed by a plastic hinge is always positive, there is no difficulty with a sign convention for moments, but it is necessary to watch that, if the geometry of the deformations "raises" a load, the work done by this load would be negative. In the case of distributed loads a little more care is necessary in calculating the work done by the load. For example, the uniformly-loaded fixed beam of Fig. 14.19 (a) fails by three hinges as shown at (b). The deflexion of the mid-span is now $\frac{1}{2}\theta L$ but the centroid of the load on each link of the mechanism is at the centre of the length, i.e. $\frac{1}{4}L$ from

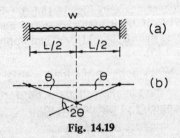

Fig. 14.19

the supports, and will translate only $\frac{1}{4}\theta L$. Then

$$\text{External work} = 2(\tfrac{1}{2}wL)(\tfrac{1}{4}\theta L)$$
$$= \tfrac{1}{4}wL^2\theta$$
$$\text{Internal work} = M_p\theta + M_p2\theta + M_p\theta$$
$$= 4M_p\theta$$

Therefore from virtual work,

$$\tfrac{1}{4}wL^2\theta = 4M_p\theta$$

or
$$w = 16M_p/L^2$$

Again this failure mode is known, from symmetry, to be correct.

14.15 Portal Frame Application

The simple rectangular portal frame has already been discussed, but in this case consider separately the possible local collapse mechanisms of Fig. 14.20. Figure 14.20 (*b*) shows the way in which the beam could collapse on its own. Applying the principle of virtual work,

$$\text{External work} = 3W(\theta L)$$
$$\text{Internal work} = 4M_p\theta$$

Therefore equating,
$$W = 4M_p/3L$$

In the same way the sway mechanism is shown on its own in Fig. 14.20 (*c*). Applying the principle of virtual work gives

$$\text{External work} = 2W(\theta L)$$
$$\text{Internal work} = 4M_p\theta$$

so
$$W = 2M_p/L$$

From these two steps it is seen that the collapse load cannot be greater than $W = 4M_p/3L$, but it would be necessary to check to ensure that all three collapse conditions (p. 383) are satisfied before deciding that this is the true collapse load. It is simpler to continue by adding these two mechanisms as in Fig. 14.20 (*d*), noting that at the left-hand eaves joint the two mechanisms cancel. The principle of virtual work can be applied directly to Fig. 14.20 (*d*), but it is simpler, and saves time, to combine the previous results noting that one hinge has been eliminated.

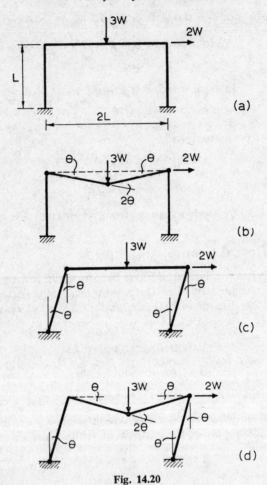

Fig. 14.20

Thus, adding the previous results and subtracting the work which was done in each case at the lost hinge,

External work $= 3W\theta L + 2W\theta L = 5W\theta L$
Internal work $= 4M_p\theta + 4M_p\theta - 2M_p\theta = 6M_p\theta$

Therefore equating,
$$W = 6M_p/5L$$

The resulting load is smaller than that to cause the separate beam collapse and hence more nearly correct. Since no other combinations are possible in this case, this is the correct result. Check of the three collapse conditions (p. 383) is simple since the forces are statically determinate.

14.16 Elementary Mechanisms

Considering a structure loaded by concentrated loads so that the bending moment diagram consists of straight lines, if there are m peak values these m values completely define the moment profile and each peak is a possible plastic hinge location. If the structure has r redundancies these would, for an elastic structure, be determined from r compatibility equations, that is, the remaining $(m - r)$ values must be obtained from $(m - r)$ independent equations of equilibrium.

Each independent elementary mechanism with one degree of freedom defines an equilibrium equation, therefore

Number of possible hinge locations $= m$

Number of redundants $= r$

Therefore

Number of elementary mechanisms $= m - r$

Example 14.1 Combined Mechanisms
The bending moment diagram for the uniform rectangular plane frame ABCDE shown in Fig. 14.21 (a) is defined by the five values at A, B, C, D and E (Fig. 14.21 (b)) and the frame has 3 redundants,

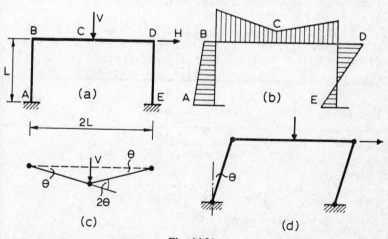

Fig. 14.21

therefore

$$m = \text{Number of possible hinge locations} = 5$$
$$r = \text{Number of redundants} = 3$$

therefore   Number of elementary mechanisms $= 2$

Collapse of the beam BCD could occur as in (c) with 3 hinges to form a typical beam mechanism, and by virtual work,

$$V\theta L = M_p(\theta + 2\theta + \theta)$$
$$= 4M_p\theta$$

i.e. $$VL = 4M_p \qquad (14.1)$$

Similarly, the typical sway mechanism shown in Fig. 14.21 (d) gives

$$HL\theta = M_p(\theta + \theta + \theta + \theta)$$

i.e. $$HL = 4M_p \qquad (14.2)$$

Including load factors λ_1 and λ_2,

Beam mechanism $\lambda_1 VL = 4M_p$ or $\lambda_1 = 4M_p/VL$
Sway mechanism $\lambda_2 HL = 4M_p$ or $\lambda_2 = 4M_p/HL$

If $V = 1.5H$, then

$$\lambda_1 = (4/1.5)M_p/HL = (8/3)M_p/HL$$
$$\lambda_2 = 4M_p/HL = 3\lambda_1/2$$

Thus failure will occur by a beam mechanism when the load factor equals λ_1, and λ_2 cannot be reached.

The elementary mechanisms may combine; if both occur together at the same load factor (λ_3) the virtual work equation is obtained by adding eqns. (14.1) and (14.2), but noticing that the beam mechanism closes joint B while the sway opens the joint, the movement at hinge B is cancelled. Thus

Eqn. (14.1) $\lambda_3 VL = 4M_p$

Eqn. (14.2) $\lambda_3 HL = 4M_p$
                    _____
                    $\lambda_3 VL + \lambda_3 HL = 8M_p$

Cancel hinge B $- 2M_p$
                    _____
                    $\lambda_3 VL + \lambda_3 HL = 6M_p$

For $V = 1.5H$, the new load factor is

$$\lambda_3 = 2.4M_p/HL < \lambda_1$$

Thus failure will occur by this combined mechanism before the beam failure and, if the true load factor at collapse is λ, the upper limit theorem states

$$\lambda \leqslant \lambda_3$$

It remains to check the yield condition, i.e. $M = M_p$, at all points. Using $\lambda_3 = 2\!\cdot\!4M_p/HL$, the B.M.D. is readily obtained by statics or, more conveniently, use of the elementary mechanism equilibrium equations. For the beam mechanism, using virtual work, we have

$$\lambda_3 VL\theta = M_B\theta + M_C2\theta + M_D\theta$$
$$\lambda_3 VL = M_B + 3M_p$$

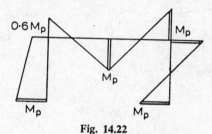

Fig. 14.22

Therefore, since $\quad V = 1\!\cdot\!5H = 1\!\cdot\!5 \times 2\!\cdot\!4M_p/\lambda_3L,$

$$M_B = 0\!\cdot\!6M_p$$

i.e. safe, all three collapse conditions are satisfied and the collapse bending moment diagram is shown in Fig. 14.22.

14.17 General Procedure

It is advantageous to establish a systematic selection of elementary mechanisms: three groups are clearly defined—
   (a) Laterally-loaded members may collapse by the typical *beam mechanism* (Fig. 14.23 (b)).
   (b) Parts of a structure may give a *sway mechanism* (Fig. 14.23 (c)).
   (c) There may be joint rotations to give a *joint mechanism* (Fig. 14.23 (d)).
Each independent mechanism is first investigated separately and then the mechanisms may be combined in any way, but it is usually useful to start combinations with that elementary mechanism giving the lowest load factor. It is apparent from the above example that the load factor for a combined mechanism cannot be smaller than the separate factors unless the combination eliminates a hinge, i.e reduces the internal virtual work.

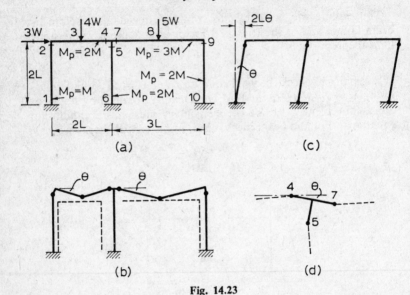

Fig. 14.23

Example 14.2 Two-bay Frame

A more complicated frame can now be considered and again the collapse will be broken down into fundamental possible local collapse modes or elementary mechanisms, each one examined separately and then combined. The frame discussed is shown in Fig. 14.23. (*a*) gives the dimensions, loading, and possible plastic hinge locations; (*b*) shows the beam collapse modes; and (*c*) the sideways or sway collapse. There is in addition, one other elementary mechanism. It is the joint 4, 5, 7 and is shown in Fig. (*d*). This is not a possible local collapse and it is evident that, to satisfy the principle of virtual work, the effect of (*d*) on the whole frame would have to be considered. However, this mechanism is required in combination with others.

Considering the separate mechanisms—

Beam 2–3–4 Internal work $= M_p\theta(1 + 4 + 2) = 7M_p\theta$
External work $= 4W\theta L$

Therefore $\qquad\qquad W = 1\cdot75M_p/L$ $\qquad\qquad$ (14.3)

Beam 7–8–9 Internal work $= M_p\theta(3 + 6 + 2) = 11M_p\theta$
External work $= 5W(3L\theta/2) = 7\cdot5W\theta L$

Therefore $\qquad\qquad W = 1\cdot47M_p/L$ $\qquad\qquad$ (14.4)

Table 14.1

Hinge Point	1	2	3	4	5	6	7	8	9	10		
M_p Value	1	1	2	2	2	2	3	3	2	2	$WL\theta$	$M_p\theta$
Mechanism 1 (Beam)		$-\theta$	$+2\theta$	$-\theta$							4·0	7·0
Mechanism 2 (Beam)							$-\theta$	$+2\theta$	$-\theta$		7·5	11·0
Mechanism 3 (Sway)	$-\theta$	$+\theta$			$-\theta$	$-\theta$			$-\theta$	$+\theta$	6·0	10·0
Mechanism 4 (Joint)				$-\theta$	$+\theta$		$+\theta$				0	7·0

(Note that the corner hinges occur in the weaker member.)

Sway   Internal work $= M_p\theta(2 + 8) = 10M_p\theta$

                 External work $= 6W\theta L$

Therefore $W = 1\cdot67M_p/L$ (14.5)

These results can be summarized in Table 14.1 where, in order to note possible cancelling of hinges, a sign convention has been adopted. Hinges at which tensile stresses occur on the same side of a member as the dotted reference lines in Fig. 14.23 (*b*) are positive. Also included in the table are the external and internal virtual work for each mechanism—including the joint mechanism of Fig. 14.23 (*d*).

To find combinations giving lower collapse loads start with the lowest elementary collapse load and try and reduce it further. Adding eqns. (14.4) and (14.5) (cf. Fig. 14.24 (*a*)) will not do, as no hinge is eliminated and the collapse of beam 7–8–9 is not affected.

If the joint mechanism Fig. 14.23 (*d*) is now included, the rotation in the column hinge at 5 and in the beam hinge at 7 can both be eliminated at the expense of introducing a beam hinge at 4, as in Fig. 14.24 (*b*).

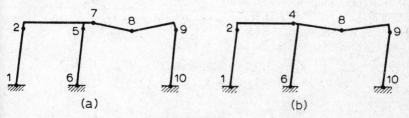

Fig. 14.24

Mechanism 2	$7.5WL = 11M$
Mechanism 3	$6.0WL = 10M$
	$13.5WL = 21M$
Mechanism 4	$7M$
	$13.5WL = 28M$
Hinges 5 and 7	$-(2(2M) + 2(3M))$
	$13.5WL = 18M$

Therefore
$$W = \frac{18}{13.5}\frac{M}{L} = 1.33\frac{M}{L}$$

This is lower than any previous result, but, proceeding, mechanism 1 can be included to eliminate hinge 2 (Fig. 14.25 (*a*)).

Thus		
	Earlier result	$13.5WL = 18M$
	Mechanism 1	$4.0WL = 7M$
		$17.5WL = 25M$
	Hinge 2	$-2M$
		$17.5WL = 23M$

Therefore $W = 1.315M/L$

This is still lower and, as no further possibilities are obvious, this mechanism is checked to see if it satisfies the yield condition ($M \leqslant M_p$).

Mechanism 1 gives
$$4\lambda WL\theta = M_2\theta + 2M_3\theta + M_4\theta$$
$$4\lambda WL = M_2 + 4M + 2M$$

and since $\lambda = 1.315M/WL$,
$$M_2 = 5.26M - 6M$$
$$= -0.74M < M$$

The negative sign indicates M_2 to be of opposite sign to that assumed for the mechanism (Fig. 14.23 (*b*)), i.e. it was assumed to produce tension on the outer fibres. Therefore the final value of $-0.74M$ gives tension on the inside.

Mechanism 2 gives
$$7.5\lambda WL\theta = M_7\theta + M_8 2\theta + M_9\theta$$
$$7.5\lambda WL = M_7 + 6M + 2M$$

and, substituting for λ,
$$M_7 = 1.86M < 2M$$

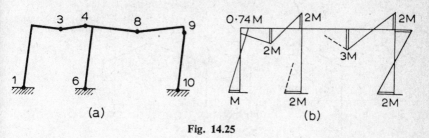

(a)   (b)

Fig. 14.25

Equilibrium of joint 4, 5, 7, i.e. mechanism 4, then gives

$$M_5 = 0.14M < 2M$$

and the final bending moment diagram (Fig. 14.25 (b)) satisfies the yield conditions so that Fig. 14.25 (a) is the true collapse mode.

14.18 Pitched-roof Frames

Analysis of pitched-roof frames introduces two useful points—
   (a) The *gable* mechanism,
   (b) The instantaneous centre.
   The frame in Fig. 14.26 (a) has seven possible hinge locations, lettered A to G and is initially three times redundant, so that

$$p = 7$$
$$r = 3$$

and m = number of independent mechanisms
$$= 7 - 3 = 4$$

There are two beam mechanisms BCD and DEF, (b) and (c), and one sway mechanism (d). The fourth mechanism (e) peculiar to such frames is the *gable* mechanism. Virtual work equations for the beam and sway mechanisms are as before, but the geometry of the gable mechanism is simplified by using the instantaneous centre of rotation. The leg GF rotates about point G so that F translates only in a direction normal to GF. Similarly BD rotates about point B so that D translates normal to BD. It follows that DF must rotate about a point in the line of both GF and BD, i.e. where they intersect. Let the deformation be such that GF rotates through an angle θ about G then F moves $2L\theta$ to the right, and the rotation about the instantaneous centre O is $2L\theta/OF = \theta$. The vertical motion of D can be found as the rotation about O times the horizontal distance from O to D, i.e. $\theta \times 2L = 2L\theta$. This vertical motion is also given by BD

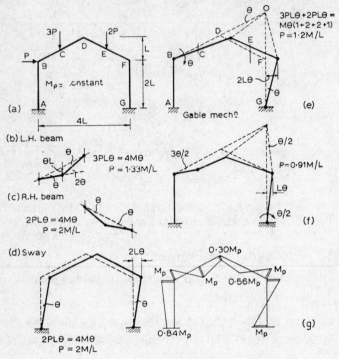

Fig. 14.26

rotating about B, i.e.

$$\theta_B \times \text{horizontal distance to } D = \theta \times 2L$$

Therefore $\qquad\qquad \theta_B = \theta$

Movements of the load points are also easily established; in this case the load $3P$ drops due to BD rotating about B and the load $2P$ by E rotating about O. In each case the vertical movement is given by the product of the angle of rotation and the horizontal distance to the centre of rotation.

The virtual work equations are set out in Fig. 14.26 and it is seen that using the left-hand beam mechanism (b) and combining this with one-half of the deformation due to the gable mechanism (e) will eliminate the hinge at D, as in (f). Thus

$$\text{Internal work} = 4M\theta + 6M\theta/2 - M\theta\,(1 + 2/2)$$
$$= 5M\theta$$
$$\text{External work} = 3PL\theta + 5PL\theta/2$$
$$= 5{\cdot}5PL\theta$$

therefore $$P = 0.91M/L$$

Checking the bending moment diagram (g), no moment exceeds M_p, so that this is the correct collapse mode.

14.19 Deflexions

From a practical point of view it is always necessary to ensure that the deformations of a structure at working loads are not excessive.

For elastic structures methods are available, but in the case of limit analysis where, after some yielding, there is no longer a linear relationship between loads and deflexions the problem is more difficult. Also final deformations may depend on the earlier load history of the structure and may not be unique functions of the existing load.

Simple structures of ideal elastic-plastic material such as mild-steel beams can be treated with a fair degree of rigour, e.g. the numerical methods of Chap. 9.

However, the most important practical problems are still more difficult and a sufficiently accurate approach (Ref. 3) will be discussed.

14.20 Slope-Deflexion Equations

The slope-deflexion eqns. (8.7) and (8.8) express the joint rotations and member rotations in terms of the bending moments. Referring to Fig. 14.27, they are written as

$$M_{AB} = \frac{2EI}{L}\left(2\theta_{AB} + \theta_{BA} - \frac{3\delta}{L}\right) + M_{AB}^{F}$$

$$M_{BA} = \frac{2EI}{L}\left(2\theta_{BA} + \theta_{AB} - \frac{3\delta}{L}\right) + M_{BA}^{F}$$

All terms are positive when clockwise on the member, as in Fig. 14.27, and M^F is the end fixing moment due to lateral loads. In the

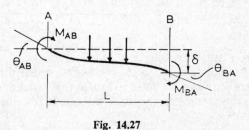

Fig. 14.27

present context, it is more convenient to rearrange the terms and write

$$\theta_{AB} = \frac{\delta}{L} + \frac{L}{6EI}\,[2(M_{AB} - M_{AB}^F) - (M_{BA} - M_{BA}^F)] \quad (14.6)$$

$$\theta_{BA} = \frac{\delta}{L} + \frac{L}{6EI}\,[2(M_{BA} - M_{BA}^F) - (M_{AB} - M_{AB}^F)] \quad (14.7)$$

The application of these equations to the plastic collapse of a structure requires several assumptions—

(*a*) The plastic zone does not extend along the member, i.e. the shape factor is unity.

(*b*) Strain hardening is ignored.

(*c*) Rotations are assumed not to reverse during collapse.

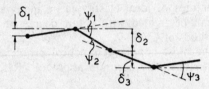

Fig. 14.28

Then, at the instant before formation of a plastic hinge, elastic conditions apply and the conventional slope-deflexion equations are valid and will define the end slopes of the members on either side of the hinge position. During collapse the bending moments remain unaltered and rotation at the hinge is due entirely to plastic deflexion—the shape of the elastic part between hinges remains unchanged. Thus the hinge rotation will consist of two parts—

(1) An elastic part due to end rotations, and

(2) A change of angle due to elastic and plastic deflexion, i.e. δ/L. The elastic end rotations θ_e are all known since the moments are all known and (Fig. 14.28)

$$\psi_1 = \text{hinge rotation} = \text{function of } (\theta_e, \delta_1/L_1, \delta_2/L_2)$$
$$\psi_2 = \text{hinge rotation} = \text{function of } (\theta_e, \delta_2/L_2, \delta_3/L_3)$$
$$\text{etc.}$$

thus giving n equations with $(n + 1)$ unknowns.

The additional equation necessary is obtained if the position of the last hinge to form is known; then, at that point and just before the plastic rotation starts, the hinge rotation is zero, i.e.

$$\psi_n = 0$$

and we have $(n + 1)$ equations for $(n + 1)$ unknowns. There are two ways of getting round the difficulty of not knowing which hinge is in fact the last to form, but the easier to grasp immediately is to try each possibility; then that giving the largest deflexion is the correct one.

14.21 Propped Cantilever

The collapse condition for the propped cantilever in Fig. 14.29 (a) gives

$$M_p = 2WL/15$$

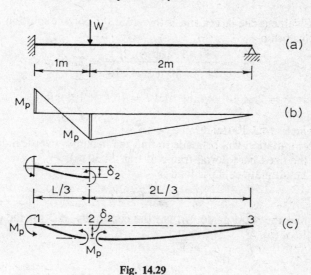

Fig. 14.29

Using a beam section with

$$M_p = 45 \text{ kN-m}$$

and assuming that the last hinge forms at 1, then just prior to hinge formation,

$$\theta_{12} = 0$$

From eqn. (14.6),

$$0 = \frac{\delta_2}{\frac{1}{3}L} + \frac{\frac{1}{3}L}{3EI}(-M_p + \tfrac{1}{2}M_p)$$

Therefore

$$\delta_2 = M_p L^2/54EI$$

(Since there are no lateral loads on this segment $M^F = 0$.)

Assuming that the last hinge forms at 2 and because of continuity at the last hinge,

$$\theta_{21} = \theta_{23}$$

$$\theta_{21} = \frac{\delta_2}{\frac{1}{3}L} + \frac{\frac{1}{3}L}{3EI}\left(-M_p + \tfrac{1}{2}M_p\right)$$

$$\theta_{23} = -\frac{\delta_2}{\frac{2}{3}L} + \frac{\frac{2}{3}L}{3EI}\left(+M_p\right)$$

i.e. $$\delta_2 = \frac{5M_pL^2}{27EI}$$

This value is the larger and is therefore the correct solution, so that in this beam

$$\delta_2 = \frac{5 \cdot 45 \cdot 10^3 \cdot 9 \cdot 10^6}{27 \cdot 2 \cdot 10^2 \cdot 16\cdot4 \cdot 10^6} = 22\cdot9 \text{ mm}$$

where $E = 2 \times 10^5$ N/mm^2 and $I = 16\cdot4 \times 10^6$ mm^4.

Example 14.3 Portal Frame
A computation will be made to find the maximum vertical deflexion for the fixed-base portal frame of Fig. 14.30 (*a*).
   The ultimate collapse load is

$$W_u = 6M_p/L$$

To find the ratio δ_V to δ_H, use the continuity at 2 for the collapse mode of (*b*),

$$\theta_{23} = \theta_{21}$$

$$\theta_{23} = \frac{\delta_V}{\frac{1}{2}L} + \frac{\frac{1}{2}L}{3EI}\left(0 + \tfrac{1}{2}M_p\right)$$

$$\theta_{21} = \frac{\delta_H}{\frac{1}{2}L} + \frac{\frac{1}{2}L}{3EI}\left(0 + \tfrac{1}{2}M_p\right)$$

Equating

$$\frac{2\delta_V}{L} + \frac{M_pL}{12EI} = \frac{2\delta_H}{L} + \frac{M_pL}{12EI}$$

or $$\delta_V = \delta_H$$

Assume last hinge to form at 1 (Fig. 14.39 (*c*)). Here

$$\theta_1 = 0$$

i.e. $$0 = \frac{\delta_H}{\frac{1}{2}L} + \frac{\frac{1}{2}L}{3EI}\left(-M_p + 0\right)$$

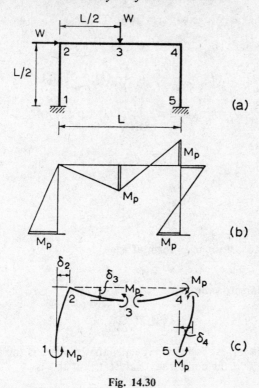

Fig. 14.30

Therefore

$$\delta_H = \frac{M_p L^2}{12EI} = \delta_V$$

Assume the last hinge to form at 3. Then

$$\theta_{32} = \theta_{34}$$

$$\theta_{32} = \frac{\delta_V}{\frac{1}{2}L} + \frac{\frac{1}{2}L}{3EI}(-M_p + 0)$$

$$\theta_{34} = -\frac{\delta_V}{\frac{1}{2}L} + \frac{\frac{1}{2}L}{3EI}(M_p - \tfrac{1}{2}M_p)$$

Therefore

$$\delta_V = \frac{M_p L^2}{16EI}$$

Assume the last hinge to form at 4. Then

$$\theta_{43} = \theta_{45}$$

$$\theta_{43} = -\frac{\delta_V}{\frac{1}{2}L} + \frac{\frac{1}{2}L}{3EI}(M_p - \tfrac{1}{2}M_p)$$

$$\theta_{45} = \frac{\delta_H}{\frac{1}{2}L} + \frac{\frac{1}{2}L}{3EI}(-M_p + \tfrac{1}{2}M_p)$$

i.e.

$$\delta_V + \delta_H = \frac{M_p L^2}{12EI}$$

Since

$$\delta_V = \delta_H$$

Therefore

$$\delta_V = \frac{M_p L^2}{24EI}$$

Assume the last hinge to form at 5

$$\theta_{54} = 0$$

By a similar procedure,

$$\delta_V = \frac{M_p L^2}{24EI}$$

Thus the largest solution indicates that hinge 1 is the last one to form, and that the correct vertical deflexion at 3 is

$$\frac{M_p L^2}{12EI}$$

If the last hinge is wrongly assumed, then the continuity that is assumed is false and rotation does occur. The wrong assumption is equivalent to assuming that the section bends backward and that the frame mechanism is backward to the collapse mode. This discrepancy does not occur when the correct assumption is made: therefore the largest deflexion is found when the last hinge is properly assumed.

14.22 Deflexions at Working Loads

The method explained above for the calculation of deflexion is suitable for finding the maximum deflexion of a structure at the point of collapse. The designer is more interested in the state of the structure at working loads and a safe estimate of the working load deflexion will be obtained as follows.

Figure 14.31 represents the observed and idealized load-deflexion curves for some structure. Where the two curves cross is the point which has been calculated in the past two examples—the deflexion of the structure at imminent collapse. If the deflexion at this point is divided by the load factor, the result will be an approximation of the working load deflexion and will be greater than the deflexion exhibited by the actual structure at that load. This working load

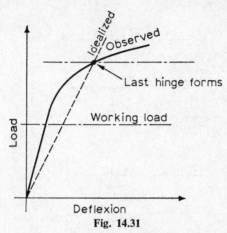

Fig. 14.31

deflexion may be in error as much as 100 per cent, but will give the designer a figure on which to decide whether or not the deflexions may be critical and whether or not to attempt a more rigorous calculation.

14.23 Practical Loadings

In general there is not one single relationship between the several loads acting on a real structure. For example, wind loading may reverse, whereas other loads may remain constant; again, live loads on floors may vary independently.

Previous chapters have considered only one loading system and it has been assumed that all loads retain constant proportions; while this is not a serious defect for normal design where each possible load system can be investigated separately and the strongest structure selected, in exceptional conditions it may lead to other forms of failure than the simple *static* mechanism.

14.24 Alternating Plasticity

Alternating plasticity occurs when a member is loaded up to the plastic moment at a section and, after some rotation has occurred in the plastic hinge, the deformation is reversed to cause plastic rotation in the opposite sense.

The fixed beam of Fig. 14.32 (a) has a static load at the third point. The collapse bending moment diagram is shown at (b). The collapse load seen to be

$$W_c = \frac{M_p}{0 \cdot 111L}$$

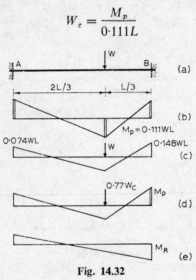

Fig. 14.32

Consider a load just sufficient to cause some plastic rotation at B. Thus from the elastic moment diagram shown in (c) the plastic moment of $0 \cdot 111 W_c L$ will be reached at B when W equals $(0 \cdot 111/0 \cdot 148) W_c$. Now allow W to increase a little more, say to $0 \cdot 77 W_c$, when the bending moment diagram is as in (d). Removal of the load is equivalent to applying an equal and opposite load and the relief of moments will occur elastically. Thus the relief of moment at B is

$$M_B = 0 \cdot 148 \times 0 \cdot 77 W_c \times L = 0 \cdot 114 W_c L$$

and the final residual moments due to the permanent deformation at B are shown in (e) where the residual moment at B is

$$M = (0 \cdot 111 - 0 \cdot 114) W_c L$$
$$= 0 \cdot 003 W_c L$$

If the load is now reversed and reapplied, it is seen that the moment induced at B is of the same sign as the residual moment left by the preceding load cycle, so that the plastic moment, in the reversed direction will be reached at a load W' such that

$$0{\cdot}148W'L + 0{\cdot}003W_cL = M_p$$
$$= 0{\cdot}111W_cL$$

Therefore
$$W' = \frac{0{\cdot}108}{0{\cdot}148}\,W_c = 0{\cdot}729W_c$$

This is less than the preceding loading and it is clear that reversed rotation will occur at B and again leave a residual moment, this time unfavourable to the original downward loading. In this trivial case it is clear that in order to prevent reversal of yielding the load must be restricted to that which leaves no residual moments, that is, to the elastic loading; for this condition

$$0{\cdot}148WL = M_p = 0{\cdot}111W_cL$$
or
$$W = 0{\cdot}75W_c$$

It is now possible to generalize and state that, in order to prevent reversal of yielding, a system of residual moments must be reached such that all subsequent loadings occur elastically, and, of course, nowhere is M_p exceeded. Then when m is the residual moment at any section, three necessary conditions are—

$$m + M_{max} \leqslant M_p$$
$$m + M_{min} \geqslant -M_p$$
$$M_{max} - M_{min} \leqslant 2M_p$$

The third inequality is, in fact, contained in the other two. If these conditions are established the structure is said to have "shaken down."

14.25 Incremental Collapse

Incremental collapse of a structure will occur due to gradual build-up of deformations if successive load cycles produce small amounts of yielding at one or more sections, all of the same sense. This may be illustrated by considering the fixed beam in Fig. 14.33 to carry, alternatively, equal loads at the right and left third points as in (*a*) and (*b*). As before, the static collapse load is

$$W_c = M_p/(0{\cdot}111L)$$

and the collapse moment diagram is as in Fig. 14.32 (*b*). Removal of the load causes the elastic moments of Fig. 14.32 (*c*) but with reversed signs, and leaves the residual moments in Fig. 14.33 (*c*).

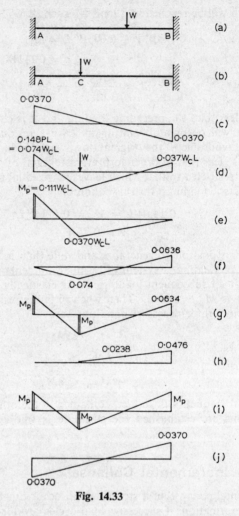

Fig. 14.33

This stage is reached after some plastic deformation has occurred at B. Application and removal of the left-hand load can now be followed in the same way. Initially, normal elastic fixed beam moments (Fig. 14.33 (*d*)) are added until M_p is reached at A to give the moments of Fig. 14.33 (*e*). Under additional load we have an effective pin at A and moments of Fig. 14.33 (*f*) occur, until M_p is reached at C and the bending moment diagram is as in (*g*). Final increments of load are now carried as by a simple cantilever CB to give moments of (*h*) and a final hinge at B, when the collapse moments are shown at (*i*). Unloading, that is, adding the elastic moments due to an equal and an opposite load at C, gives the residual moments in (*j*). It is now apparent that re-application of the right-hand load will produce M_p at B and before plastic hinges can form elsewhere some plastic rotation must occur at B in the same sense as under the first loading; as this cycle of loading is continued deformations will build up and lead to failure of the structure's usefulness, due to excessive deflexion. It is also evident that this situation will continue as long as all plastic hinges do not form simultaneously, that is, to prevent incremental build up of deformations all moments must occur elastically and the same generalization for Section 14.24 applies and, if satisfied, the structure is said to shake down to a stable state.

14.26 Shakedown Theorem

It is now possible to state, without proof, the shakedown theorem (Refs. 1 and 3.): "If any particular system of residual moments exists which would enable all variation of the applied loads between their prescribed limits to be supported in a purely elastic manner, then the structure will shake down, although the actual system of residual moments existing in the structure when it has shaken down will not necessarily be the particular system which has been found."

14.27 Graphical Analysis

As a simple example consider the pin-base portal frame shown in Fig. 14.34, which is of uniform section and is required to support a vertical and a horizontal load to be of equal maximum value, any lower positive value, and applied in any order and in any combination. Since we are finally concerned that bending moment

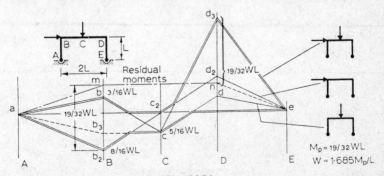

Fig. 14.34

variations shall occur elastically, it is necessary to obtain all possible maximum values of the elastic moments, that is, a conventional maximum moments diagram. Three possible load systems are shown in Fig. 14.34 and they give, respectively, the moment diagrams *abcde*, $ab_2c_2d_2e$, and ab_3cd_3e. Irrespective of individual values of the loads, bending moments cannot exceed the envelope of the above diagrams.

By trial on a scale drawing, a system of residual moments represented by *amne* in Fig. 14.34 is possible, and gives two equal peaks of moment occurring at B and D, but with different load combinations. For this system the conditions of Section 14.23 are satisfied and

$$W = 1 \cdot 685 M_p / L$$

The static collapse load with both loads at full value is $3M_p/L$ as in Section 14.9.

It is also seen, in this case, that no other system of residual moments will allow more than two equal peaks of moment. It may then be inferred that, in general, the true system will be that which gives the largest number of equal moment peaks and that a full collapse mechanism will not be developed under any of the alternative combinations of load involved.

14.28 Further Reading

1. BAKER, HORNE and HEYMAN, *The Steel Skeleton*, Vol. II, Chap. 7 (Cambridge, University Press, 1956).
2. PRAGER, W., *An Introduction to Plasticity* (London, Addison-Wesley, 1959).
3. NEAL, B. G., *The Plastic Methods of Structural Analysis* (London, Chapman and Hall, 1956).

15

The Arch and the
Suspension Bridge

15.1 Introduction

An arch can be defined as a member curved upwards in elevation
whose ends are restrained in position but not necessarily in direction.
The restraint at the ends introduces a horizontal thrust similar to
that which occurs in a portal (*see* Section 2.15), and the upward
curvature means that the bending moment at any section resulting

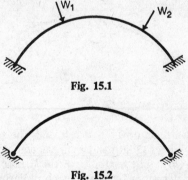

Fig. 15.1

Fig. 15.2

from this thrust is of opposite sign from that due to the lateral
loading; consequently the net bending moment at the section is
much less than that which would occur if a simply-supported beam
were carrying the same loading system. Arches can be subdivided
into the statically-determinate three-pinned, the fixed-ended (Fig.
15.1) and the two-pinned (Fig. 15.2).

The fixed-ended arch shown in Fig. 15.1 has three redundancies, e.g. the reactions at one of the supports, whilst the two-pinned arch of Fig. 15.2 has only one redundancy.

15.2   The Two-pinned Arch

An arch is a redundant structure for which the flexibility method of Chap. 7 gives the better form of solution. Consider the two-pinned arch shown in Fig. 15.3 (*a*), where the horizontal and vertical reactions are H_A, V_A, and H_B, V_B, at A and B respectively.

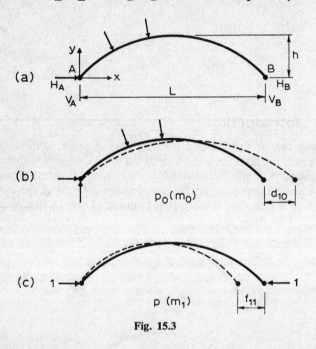

Fig. 15.3

If the horizontal reaction at B is released, the reduced structure becomes statically-determinate and a displacement d_{10} occurs at the release (Fig. 15.3 (*b*)). The flexibility of the structure corresponding to the releases is determined by applying a unit horizontal load corresponding to the release (Fig. 15.3 (*c*)).

Then for compatibility $d_{10} + f_{11}r_1 = 0$; and from Chap. 5, if only bending deformations are considered.

$$d_{10} = \int \frac{m_1 m_0}{EI}\, ds \qquad f_{11} = \int \frac{m_1^{\,2}}{EI}\, ds$$

giving
$$\int \frac{m_1 m_0}{EI} ds + H \int \frac{m_1{}^2}{EI} ds = 0 \qquad (15.1)$$

Taking a set of coordinates with an origin at A, and noting that y is positive upwards (cf. Fig. 15.3 (a)),

$$m_1 = +y$$

and eqn. (15.1) becomes

$$+\int \frac{m_0 y}{EI} ds + H \int \frac{y^2}{EI} ds = 0 \qquad (15.2)$$

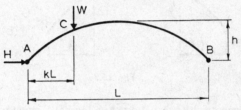

Fig. 15.4

There are two common forms of arch rib where this equation can be solved analytically. They are—
(1) The segmental arch of uniform section throughout, i.e. $EI =$ constant. In this case $ds = R\,d\alpha$, where R is the radius of the arch rib and y can be expressed in terms of R, the half-angle of the arch and the variable α.
(2) The parabolic arch rib, where $EI_x = EI_0 \sec \alpha$, EI_0 is the flexural rigidity at the crown and α is the inclination of the tangent to the rib to the x-axis at the point under consideration. In this case,

$$\int \frac{ds}{EI_x} = \int \frac{dx}{EI_0} \qquad (15.3)$$

Such an arch is shown in Fig. 15.4. The equation of the parabolic curve is

$$y = \frac{4h}{L}\left(x - \frac{x^2}{L}\right)$$

The determination of the horizontal thrust H due to a single asymmetrical load W at a distance kL from the origin is best carried out by using the principle of superposition and replacing the asymmetrical system by a symmetrical plus a skew-symmetrical one, as shown in Fig. 15.5.

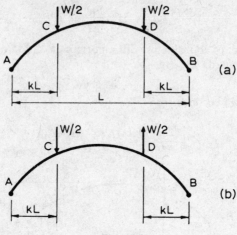

Fig. 15.5

From symmetry, the thrust due to a load at C must be the same as that due to an equal load acting in the same direction at D. Thus for the loading shown in Fig. 15.5 (*a*) the horizontal thrust will be the same as that for the original loading shown in Fig. 15.4, whilst for the skew-symmetrical loading in Fig. 15.5 (*b*) the thrust is zero. (This result is to be expected since the loading of Fig. 15.5 (*b*) has no resultant vertical component.)

Following the flexibility approach, the m_0 diagram for (*a*) is as shown in Fig. 15.6 and

$$\int_0^L \frac{m_0 y}{EI_x} \, ds = \frac{2}{EI_0} \int_0^{L/2} m_0 y \, dx$$

since $EI_x = EI_0 \sec \alpha$.

$$\frac{2}{EI_0} \int_0^{L/2} m_0 y \, dx = \frac{2}{EI_0} \left[\int_0^{kL} m_0 y \, dx + \int_{kL}^{L/2} m_0 y \, dx \right]$$

$$= -\frac{2}{EI_0} \left[\int_0^{kL} \frac{W}{2} xy \, dx + \int_{kL}^{L/2} \frac{W}{2} kLy \, dx \right] \quad (15.4)$$

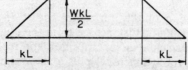

Fig. 15.6

Substituting for y gives

$$\frac{1}{EI_0} \int m_0 y \, dx = -\frac{1}{EI_0} \cdot \frac{WL^2h}{3} [k - 2k^3 + k^4] \qquad (15.5)$$

and

$$\int \frac{y^2}{EI_x} \, ds = \frac{1}{EI_0} \int_0^L y^2 \, dx = \frac{8h^2L}{15EI_0} \qquad (15.6)$$

Substituting in eqn. (15.2) gives

$$H \cdot = \frac{5WL}{8h} k(1 - 2k^2 + k^3) \qquad (15.7)$$

The final moments are

$$m = m_0 + m_1 r_1$$
$$= m_0 + Hy \qquad (15.8)$$

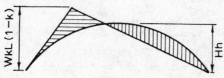

Fig. 15.7

The sign of m_0 is negative and the resultant diagram is as shown in Fig. 15.7.

For a uniformly-distributed load extending from $k_1 L$ to $k_2 L$, an element of load is $w \, d(kL)$; for the total load

$$H = \frac{5L}{8h} \int_{k_1}^{k_2} wk(1 - 2k^2 + k^3) \, d(kL)$$

$$= \frac{5wL^2}{8h} \int_{k_1}^{k_2} k(1 - 2k^2 + k^3) \, dk \qquad (15.9)$$

If the load covers the whole span, $k_1 = 0$ and $k_2 = 1$, giving

$$H = \frac{wL^2}{8h} \qquad (15.10)$$

The moment m_0 has a parabolic distribution with a maximum value of $-\frac{1}{8}wL^2$, and

$$m_1 r_1 = Hy = \frac{wL^2}{8h} \cdot \frac{4h}{L} \left(x - \frac{x^2}{L} \right)$$

This also has a maximum value of $\frac{1}{8}wL^2$ and a parabolic distribution; hence the bending moment at all points on the arch rib is zero.

This interesting result has been obtained by considering bending energy only, but since there is no moment there is no bending energy and the results to be developed in Section 15.4 are important in such a case.

15.3 Graphical Solution of Two-pinned Arch

Where the eqn. (15.2) is not integrable, a graphical method has to be used. Consider the arch shown in Fig. 15.8, which is pinned to the supports A and B and carries the vertical loads shown.

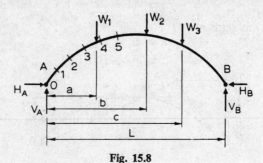

Fig. 15.8

The vertical reactions V_A and V_B can be obtained from the equations of static equilibrium, leaving the redundant force H to be determined from the equation

$$\int \frac{m_0 y}{EI_x} ds + H \int \frac{y^2}{EI_x} ds = 0 \qquad (15.11)$$

The bending moment m_0 is given from

$$m_0 = -V_A x + [W_1(x - a)] + [W_2(x - b)] + [W_3(x - c)] \qquad (15.12)$$

the [ ] signifying that the terms are only taken when they are positive in sign. Thus

$$\int \frac{m_0 y}{EI_x} ds = -V_A \int \frac{xy}{EI_x} ds + W_1 \int \frac{(x - a)y}{EI_x} ds - \cdots \text{ etc.}$$

and the equation for determining H becomes

$$H \int \frac{y^2}{EI_x} ds = V_A \int \frac{xy}{EI_x} ds - W_1 \int \frac{(x - a)y}{EI_x} ds - \cdots \text{ etc.}$$

In order to integrate the terms graphically, the rib is divided into a number of equal parts, marked 1, 2, 3, etc. on Fig. 15.8. At each point EI_x, x and y are measured and the values of y^2/EI_x, xy/EI_x, $y(x-a)/EI_x$, $y(x-b)/EI_x$ and $y(x-c)/EI_x$ calculated. Curves of these functions against the length s along the arch rib are plotted in Fig. 15.9. The areas under the curves are obtained by planimeter

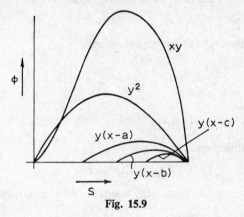

Fig. 15.9

or Simpson's Rule. Let these be as follows—

$$\int \frac{y^2}{EI_x}\,ds = A_1 \qquad \int \frac{y(x-a)}{EI_x}\,ds = A_3$$

$$\int \frac{xy}{EI_x}\,ds = A_2 \qquad \int \frac{y(x-b)}{EI_x}\,ds = A_4$$

$$\int \frac{y(x-c)}{EI_x}\,ds = A_5$$

Then

$$HA_1 - V_A A_2 + W_1 A_3 + W_2 A_4 + W_3 A_5 = 0 \qquad (15.13)$$

leading to the value for H.

As an alternative to determining the areas the tabulation method given by Chettoe and Adams (Ref. 2) may be employed.

The method of solution given in Chap. 11, using indirect model analysis, is also very suitable for arches.

15.4 Lack of Fit

Both the rib and the abutments of an arch will strain and change in length. This change may be important in the analysis and it is convenient to regard it as a lack of fit so that the compatibility statement becomes

$$d_{10} + f_{11}r_1 + \sum p_1 u_e = 0$$

where for the abutments

$$u_e = \pm\lambda \text{ (movement, adjustment or shortening)}$$

or $\qquad = +He \text{ ($e$ = yield for unit load)}$

and $\qquad p_1 = +1$

This gives the compatibility equation as

$$\int \frac{m_0 y}{EI_x}\, ds + H \int \frac{y^2}{EI_x}\, ds \pm \lambda + He = 0 \qquad (15.14)$$

15.5 Rib Shortening

The compatibility equation

$$d_{10} + f_{11}r_1 = 0$$

has in the previous sections only taken bending effects into account.

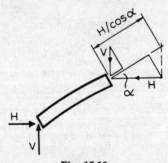

Fig. 15.10

If axial effects are included the additional terms in this equation are

$$\int \frac{p_1 p_0}{AE}\, ds + H \int \frac{p_1^{\,2}}{AE}\, ds = 0$$

Reference to Fig. 15.10 shows that

$$p_1 = \cos \alpha$$
$$p_0 = V \sin \alpha$$

and $\qquad ds = dx/\cos \alpha$

Putting $AE_\alpha = AE_0 \sec \alpha$, where AE_0 is the extensibility at the crown, the compatibility equation becomes

$$\int \frac{V \sin \alpha \cos \alpha}{AE_0} \, dx + H \int \frac{\cos^2 \alpha}{AE_0} \, dx = 0 \qquad (15.15)$$

i.e.
$$\int \frac{(V \sin \alpha + H \cos \alpha) \cos \alpha}{AE_0} \, dx = 0 \qquad (15.16)$$

Fig. 15.10 shows that, approximately, $H/\cos \alpha = V \sin \alpha + H \cos \alpha$. Therefore

$$\int \frac{(V \sin \alpha + H \cos \alpha) \cos \alpha}{AE_0} \, dx = H \int \frac{dx}{AE_0} = \frac{HL}{AE_0} \qquad (15.17)$$

The complete arch equation then becomes

$$\int \frac{m_0 y}{EI_0} \, dx + H \int \frac{y^2}{EI_0} \, dx + \frac{HL}{AE_0} = 0 \qquad (15.18)$$

The rib-shortening effect has been shown only to be important for arches which are flat, those, for example, with a span/rise ratio greater than 10.

15.6 Temperature Changes

The effect of a change of temperature on the length of an arch is similar to that of a lack of fit.

$$d_{10} = \int p_1 u$$

where
$$\begin{aligned} u &= \text{change in length} \\ &= \pm ct \, ds \\ &= \frac{\pm ct}{\cos \alpha} \, dx \end{aligned}$$

$$\begin{cases} \text{Elongation} \equiv \text{rise in temperature} \ (+) \\ \text{Shortening} \equiv \text{fall in temperature} \ (-) \end{cases}$$

$$p_1 = -\cos \alpha$$

Therefore
$$\int p_1 u = \mp \int ct \, dx = \pm cLt$$

This gives the compatibility equation as

$$\int \frac{m_0 y}{EI_0} \, dx + H \int \frac{y^2}{EI_0} \, dx \pm cLt = 0 \qquad (15.19)$$

15.7 The Fixed-ended Arch

The fixed-ended arch shown in Fig. 15.11 (a) has three redundancies. The most convenient way of solving is to take these as M_A, V_A, and

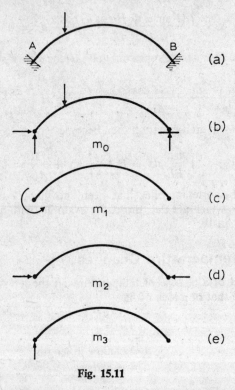

Fig. 15.11

H_A; the corresponding releases are r_1, r_2, and r_3 as shown in Fig. 15.11 (c), (d) and (e). The m_0 diagram is, however, best obtained by taking the redundancies as M_A, H_B and M_B, giving the released structure shown in Fig. 15.11 (b). This problem then becomes one of the "mixed release" type given in Section 7.8.

Using the fundamental equation

$$d_0 + fr = 0$$

gives in this case

$$d_{10} + f_{11}r_1 + f_{12}r_2 + f_{13}r_3 = 0 \tag{15.20}$$

$$d_{20} + f_{21}r_1 + f_{22}r_2 + f_{23}r_3 = 0 \tag{15.21}$$

$$d_{30} + f_{31}r_1 + f_{32}r_2 + f_{33}r_3 = 0 \tag{15.22}$$

where, taking bending terms only,

$$d_{10} = \int \frac{m_1 m_0}{EI} \, ds; \quad f_{11} = \int \frac{m_1^2}{EI} \, ds; \quad f_{12} = \int \frac{m_1 m_2}{EI} \, ds, \text{ etc.}$$

and $\quad m_1 = +1; \qquad m_2 = +y; \qquad m_3 = -x$

leading to

$$+\int \frac{m_0}{EI} \, ds + M_A \int \frac{1}{EI} \, ds - V_A \int \frac{x}{EI} \, ds + H_A \int \frac{y}{EI} \, ds = 0$$

$$-\int \frac{m_0 x}{EI} \, ds - M_A \int \frac{x}{EI} \, ds + V_A \int \frac{x^2}{EI} \, ds - H_A \int \frac{xy}{EI} \, ds = 0$$

$$+\int \frac{m_0 y}{EI} \, ds + M_A \int \frac{y}{EI} \, ds - V_A \int \frac{xy}{EI} \, ds + H_A \int \frac{y^2}{EI} \, ds = 0$$

$$(15.23)$$

These are the equations to be solved for the fixed-ended arch.

Before the digital computer was applied in structural analysis much thought was given to finding a convenient method of carrying out the integrations for hingeless arches in order to solve these equations.

Culmann was the first to show that these integrations could be done more easily if the redundant reactions are considered to act at the centroid of the elastic areas. (The element of elastic area $\delta A = \delta s/EI$.)

This method is referred to as the *elastic centre* or *neutral point* method.

The essence of it is to select as the origin of coordinates that point which gives

$$\int x \, dA = 0 = \int y \, dA$$

Then if x and y are principal axes, the product of inertia $\int xy \, dA$ will also be zero. This origin is known as the *elastic centre*.

With this set of axes the problem reduces to three equations each of which contains only one unknown instead of the three simultaneous eqns. (15.23).

The application of superposition also simplifies the fixed-arch problem. A symmetrical load as in Fig. 15.5 (*a*) has two redundancies—the horizontal thrust and the fixing moment—whilst the skew-symmetrical load has only one—the vertical reaction—there being no horizontal thrust and no central bending moment.

15.8   The Suspension Bridge

A parabolic arch rib carrying a uniformly-distributed load will have
a very small bending moment in it and the force will be largely a
thrust. Thus an economical bridge could be built (*see* Fig. 15.12)
from a girder which rested on a large number of columns which in
turn were supported on a parabolic arch rib. In such a structure the
arch rib and the columns it carries would be in compression.

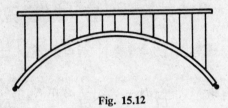

Fig. 15.12

An even more economical structure could be obtained by inverting
the arch rib, giving the suspension bridge shown in Fig. 15.13, where
a girder carrying the roadway is suspended by hangers from a cable.
In this example the cable and hangers are in tension. Since normally
a tension member requires less area than a compression member
(Chap. 12) carrying the same load, the suspension bridge is a more
economical structure than the arch.

The structure consists in essence of three parts: the cable, the
hangers and the stiffening girder. The analysis must allow for the
interconnexion of these parts and satisfy the equilibrium and
compatibility conditions.

There are a number of theories for the analysis; the outlines of
four of them will be given.

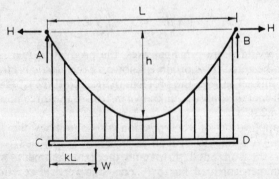

Fig. 15.13

15.9   The Rankine Theory

The assumptions made in this theory are as follows—
 (i) That under the total *dead* loading the cable has a parabolic shape and the stiffening girder is unstressed.
 (ii) That any live loading applied to the girder produces a uniform load in all of the hangers, i.e. w_h is uniform across the whole span.
 (iii) That the intensity of $w_h = W/L$, where W = applied load and L = span.

The third assumption is only necessary when the stiffening girder is two-pinned, i.e. supported at C and D. If a third pin is provided at some intermediate point, then the relationship between w_h and W can be determined in terms of kL (the load position) and L without assumption (iii) being made.

This theory satisfies the conditions of equilibrium of forces but makes no attempt to satisfy the compatibility conditions. It cannot therefore be used in analysis and is only of historic interest.

15.10   The Linear Elastic Theory

The cable itself is a statically-determinate structure; so is a simply-supported beam: interconnecting them by means of hangers gives a redundant structure.

The linear elastic theory reduces the problem to a single redundancy by using the first two assumptions of the Rankine theory and also assuming that the loaded structure has a linear-load displacement diagram.

The single redundancy is best taken as H, the horizontal tension in the cable. The released structure is shown in Fig. 15.14, with the live load consisting solely of a single load W at a distance kL from one support. The beam only is stressed and, since a linear load-displacement relationship is assumed, strain-energy can be used in the compatibility equation

$$d_{10} + f_{11}r_1 = 0 \qquad (15.24)$$

Applying a unit load corresponding to the release gives the internal forces shown in Fig. 15.15 (*a*), with a free-body diagram shown in Fig. 15.15 (*b*). The moment m_1 at any section of the beam is given by

$$m_1 = y$$

and the shear and direct forces in the beam can be neglected.

The parts of the compatibility eqn. (15.24) can be treated in just the same way as the arch.

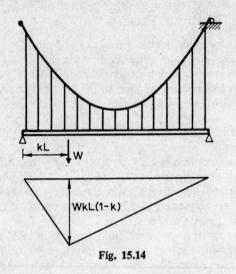

WkL(1-k)

Fig. 15.14

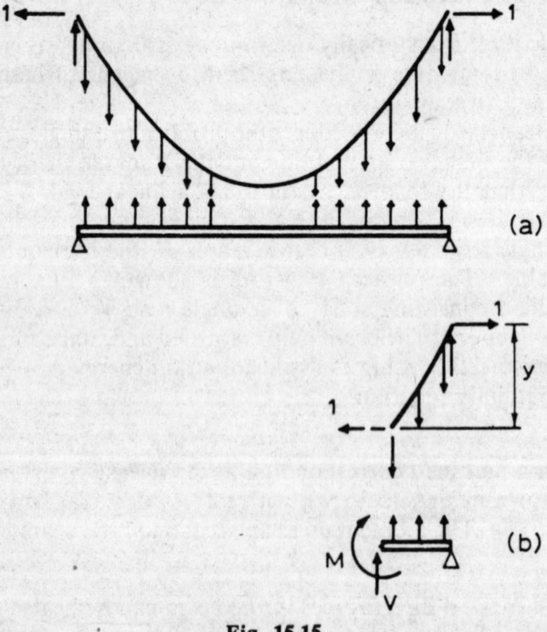

(a)

y

M

V

(b)

Fig. 15.15

(a) Bending of the stiffening girder (assumed of uniform section throughout)

$$d_{10} = \int_0^L \frac{m_1 m_0}{EI}\, dx$$

$$= \int_0^L \frac{m_0 y}{EI}\, dx$$

$$= \frac{WL^2 h}{3EI} (k - 2k^3 + k^4) \qquad (15.25)$$

$$f_{11} = \int \frac{m_1^2}{EI}\, ds = \int_0^L \frac{y^2}{EI}\, dx \qquad (15.26)$$

(b) Tension in the cable

$$d_{10} = 0$$

$$f_{11} = \int \frac{p^2}{AE}\, ds = \int_0^L \frac{\cos\theta}{AE}\, dx$$

$$\simeq \frac{L}{AE} \qquad (15.27)$$

(c) The effect of the hangers and towers may be included in the same way, but in normal bridges these elements contribute about 1 per cent to the value of H and, in view of the doubtfulness of the initial assumptions, they can well be neglected, giving as the compatibility equation

$$\int_0^L \frac{m_0 y}{EI}\, dx + H \int_0^L \frac{y^2}{EI}\, dx + \frac{HL}{AE} = 0 \qquad (15.28)$$

where EI = flexural rigidity of stiffening girder, and
AE = extensibility of the cable.

15.11 The Deflexion Theory

This theory deals with the compatibility relationship by considering the differential equations of the cable and stiffening girder. It makes the same basic assumptions as the previous approach but differs in that it takes into consideration the deflexion of the beam, symbolized by v. It assumes that the hangers are inextensible; thus the deflexion of the cable will be the same as that of the beam.

The bending moment on the girder $M_x = m_0 + Hy$ when v can be neglected, but if this is not the case then

$$M_x = m_0 + Hy + (H_D + H)v \qquad (15.29)$$

where H_D is the horizontal force in the cable due to the dead weight of the structure. In many long-span bridges the multiplier $(H_D + H)v$ may not be negligible compared with Hy and must be taken into account.

If the girder carries a load of intensity w per unit length, the differential equation of flexure of the beam is

$$EI \frac{d^4v}{dx^4} = w - w_h \qquad (15.30)$$

where $w_h =$ load in hangers due to live load.

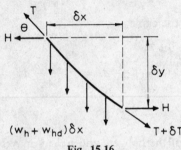

Fig. 15.16

Consideration of the vertical equilibrium of the section of the cable shown in Fig. 15.16 gives

$$(\delta T \sin \theta) + (w_h + w_{hd}) \, dx = 0 \qquad (15.31)$$

where w_{hd} is the load in the hangers due to the dead load of the bridge. But $T \sin \theta = H \tan \theta$. The horizontal pull H is constant, hence

$$\frac{d(T \sin \theta)}{dx} = H \frac{d}{dx} (\tan \theta)$$

$$= H \frac{d^2y}{dx^2}$$

Thus
$$w_h + w_{hd} + H \frac{d^2y}{dx^2} = 0 \qquad (15.32)$$

Applying this general equation to the particular case under consideration and remembering that y has increased to $(y + v)$ we have

$$(H_D + H_L) \frac{d^2}{dx^2} (y + v) = -w_h - w_{hd} \qquad (15.33)$$

For dead load only

$$H_D \frac{d^2y}{dx^2} = -w_{hd} \qquad (15.34)$$

Eliminating w_{hd} from eqns. (15.33) and (15.34) gives

$$H_L \frac{d^2y}{dx^2} + (H_D + H_L)\frac{d^2v}{dx^2} = -w_h \qquad (15.35)$$

Eliminating w_h from eqns. (15.30) and (15.35) gives

$$EI \frac{d^4v}{dx^4} - (H_D + H_L)\frac{d^2v}{dx^2} = w + H_L \frac{d^2y}{dx^2} \qquad (15.36)$$

Alternatively, eqn. (15.36) can be written

$$\frac{d^2v}{dx^2} = \frac{m_0 + H_L y}{EI} + \frac{(H_L + H_D)v}{EI}$$

$$= c^2 v + \frac{c^2}{(H_L + H_D)}(M_x' + H_L y) \qquad (15.37)$$

where $c^2 = (H_L + H_D)/EI$.

Whichever form is adopted, either equation contains two unknowns and a further relationship is required. This is obtained by considering the extension of the cable.

To a first approximation the change in length Δl is given by

$$\Delta l = \frac{H_L L}{AE} \qquad (15.38)$$

This extension must be consistent with the shortening which results from the vertical movements v at the points along the cable. This can be shown to be given by

$$\Delta l = \frac{w_d}{H_D} \int_0^L x \, dv \qquad (15.39)$$

Integrating by parts and noting that $v = 0$ at the upper and lower limits of x gives

$$\Delta l = -\frac{w_d}{H_D} \int_0^L v \, dx \qquad (15.40)$$

Combining eqns. (15.38) and (15.40) gives, since the cable ends are fixed,

$$\Delta l = \frac{H_L L}{AE} - \frac{w_d}{H_D} \int_0^L v \, dx = 0 \qquad (15.41)$$

In many practical cases the elastic extension of the cable is negligible and the first term in eqn. (15.41) can be omitted, giving

$$\int_0^L v \, dx = 0 \tag{15.42}$$

The problem, therefore, in the analysis of a suspension bridge is the solution of the simultaneous eqns. (15.36) or (15.37), and (15.41) or (15.42). The solution of these equations can best be carried out by the use of the Fourier series expressing the deflexion v in terms of a trignometrical series.

15.12 The Flexibility Coefficient Method

The fundamental difficulties in the analysis of suspension bridge structures are the change in shape of the cable curve and the non-linear deformation relations of the cable. The linear elastic theory

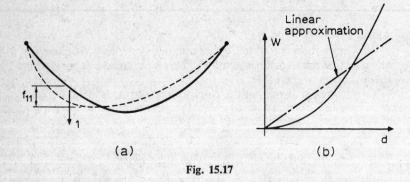

(a)   (b)

Fig. 15.17

given in Section 15.10 avoids the difficulties by the basic assumption of no change in the form of the cable, while the deflexion theory given in Section 15.11 attempts to include that effect and results in difficult differential equations.

Pugsley has given a useful intermediate form of solution recognizing the change of shape of the cable under load (*see* Fig. 15.17 (*a*), but assuming that the load-deflexion relationship is linear (*see* (*b*)) over the normal working ranges. This is often justified because the live load which is the cause of the change in shape is small compared with the dead load.

The recognition of the change of shape of the cable means that the simple relationship between cable tension and hanger forces of the linear-elastic theory is no longer valid.

Taking the hanger forces as the unknowns, the released structure of Fig. 15.18 (*a*) results. These are the same number of releases as unknown hanger forces, say *n*, so that the general compatibility condition

$$d_0 + fr = 0$$

contains *n* equations. The flexibilities at the releases (Fig. 15.18 (*b*)) are then conveniently evaluated by considering separately the three

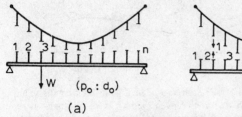

Fig. 15.18

elements of cable, hangers and stiffening girder, i.e.

$$(f_{ij}) = (f_{ij})_{cable} + (f_{ij})_{hanger} + (f_{ij})_{girder}$$

The flexibilities of the cable can be determined analytically, or experimentally by using a model. The cable in practice will have some bending stiffness, which has so far been neglected, but this can be included if an experimental approach is used.

The flexibilities of the hangers are simply

$$f_{ii} = L/AE$$
$$f_{ij} = 0$$

The stiffening girder is analysed by the orthodox beam approach.

15.13 Further Reading

1. PUGSLEY, A. G., *The Theory of Suspension Bridges* (London, Edward Arnold 1957).
2. CHETTOE, C. S., and ADAMS, H. C., *Reinforced Concrete Bridge Design* (London, Chapman and Hall, 1933).

15.14 Examples for Practice

1. Derive a formula for the horizontal thrust of a parabolic, two-pinned arch (Fig. 15.19) when a single, concentrated, vertical load *W* acts at a point *kL* from the left-hand abutment.

   [Take an origin of coordinates as shown, $y = (4h/L)(x - x^2/L)$, and assume that at any section $I = I_c \sec \alpha$, where I_c is the second moment of

area of the arch rib at the crown and α is the angle the arch axis makes with the horizontal. In this way, putting $ds = dx/\cos \alpha$, integrations are made with respect to x. It is also convenient to notice that for two loads symmetrically placed the horizontal reaction is simply doubled.]
(*Ans.* $H = 5WL/8h(k - 2k^3 + k^4)$.)

2. Show that, for the arch in Fig. 15.19 the horizontal thrust due to a uniformly-distributed load of w/ft run over the whole span is $H = wL^2/8h$.

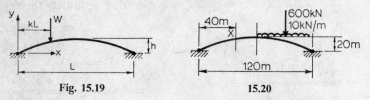

Fig. 15.19   15.20

3. In the two-hinged parabolic arch shown in Fig. 15.20, find the value of the B.M. at the section X due to the given loading. One abutment is rigid, but the other yields 0.1 mm/kN of horizontal thrust. Include a fall in temperature of 8 degC. ($EI_c = 100 \times 10^6$ kN-m²; coefficient of lin. exp. $= 12 \times 10^{-6}$/degC; and $I = I_c \sec \alpha$.)
(*Ans.* $H = 870$ kN, $M_x = 1,800$ kN-m.)

Appendix
Elements of Matrix Algebra

A.1 Definition

The way in which a matrix is built up and defined is shown in Chap. 6, so that a suitable formal definition is: "A matrix is simply an ordered array of numbers arranged in m rows and n columns."

This would be described as an $m \times n$ matrix and the numbers are called the elements, e.g. a_{ij} is the element in the ith row and jth column.

A.2 Vector

A vector is a matrix with one column, or one row, i.e. a $m \times 1$ or a $1 \times m$ matrix or simply a *column matrix* or a *row matrix*.

A.3 Unit and Null Matrices

The unit matrix is one which has unit elements on the diagonal only, all other elements being zero, that is—

$$I = \begin{bmatrix} 1 & 0 & 0 \\ 0 & 1 & 0 \\ 0 & 0 & 1 \end{bmatrix}$$

The null matrix has zero elements only, that is—

$$\begin{bmatrix} 0 & 0 & 0 \\ 0 & 0 & 0 \\ 0 & 0 & 0 \end{bmatrix}$$

A.4 Addition (and Subtraction)

Matrices are added by adding corresponding elements, e.g.

$$\begin{bmatrix} a_{11} & a_{12} & a_{13} \\ a_{21} & a_{22} & a_{23} \end{bmatrix} + \begin{bmatrix} b_{11} & b_{12} & b_{13} \\ b_{21} & b_{22} & b_{23} \end{bmatrix}$$

$$= \begin{bmatrix} (a_{11} + b_{11}) & (a_{12} + b_{12}) & (a_{13} + b_{13}) \\ & \text{etc.} & \end{bmatrix}$$

and written $\qquad A + B = C$

It is seen that the two matrices A and B must be of the same order and also that

$$A + B = B + A = C$$

A.5 Scalar Multiplication

Scalar multiplication is the multiplication of a matrix by a single constant, each element being multiplied by the same scalar, e.g.

$$\begin{bmatrix} a_{11} & a_{12} \\ a_{21} & a_{22} \end{bmatrix} \times s = \begin{bmatrix} sa_{11} & sa_{12} \\ sa_{21} & sa_{22} \end{bmatrix}$$

A.6 Matrix Multiplication

This operation may be described as *transforming* one matrix D into another C, by the operation (or transformation)

$$A \times B = C$$

and the matrix A is called the transformation matrix.

Consider the set of equations defining y in terms of x—

$$y_1 = a_{11}x_1 + a_{12}x_2 + a_{13}x_3$$
$$y_2 = a_{21}x_1 + a_{22}x_2 + a_{23}x_3$$

and defining z in terms of y

$$z_1 = b_{11}y_1 + b_{12}y_2$$
$$z_2 = b_{21}y_1 + b_{22}y_2$$
$$z_3 = b_{31}y_1 + b_{32}y_2$$

These can be written in matrix notation as, for example,

$$\begin{bmatrix} y_1 \\ y_2 \end{bmatrix} = \begin{bmatrix} a_{11} & a_{12} & a_{13} \\ a_{21} & a_{22} & a_{23} \end{bmatrix} \begin{bmatrix} x_1 \\ x_2 \\ x_3 \end{bmatrix}$$

or

$$y = A \times x$$

and the operation of multiplication must be defined carefully as
for the element in the first row and first column of the product:
multiply each element of the first row of A by the corresponding
element in the first column of X and add the products:—
It is clearly seen that this is only possible if there are the same
number of columns in A as there are rows in X; in other words, two
matrices must be *conformable* for multiplication.

In the case of the above equations for y and z, substituting for y
in the second set gives

$$z_1 = (b_{11}a_{11} + b_{12}a_{21})x_1 + (b_{11}a_{12} + b_{12}a_{22})x_2 + (b_{11}a_{13} + b_{12}a_{23})x_3$$
$$z_2 = (b_{21}a_{11} + b_{22}a_{21})x_1 + (b_{21}a_{12} + b_{22}a_{22})x_2 + (b_{21}a_{13} + b_{22}a_{23})x_3$$
$$z_3 = (b_{31}a_{11} + b_{32}a_{21})x_1 + (b_{31}a_{12} + b_{32}a_{22})x_2 + (b_{31}a_{13} + b_{32}a_{23})x_3$$

The matrix of the coefficients of these equations can naturally be
regarded as the product of the matrices B and A, i.e.

$$y = Ax$$

and

$$z = By$$

therefore

$$z = B \times A \times x$$

By writing out the matrices B and A it is seen that the multiplication rule already stated is used, that is,

$$\begin{bmatrix} b_{11} & b_{12} \\ b_{21} & b_{22} \\ b_{31} & b_{32} \end{bmatrix} \times \begin{bmatrix} a_{11} & a_{12} & a_{13} \\ a_{21} & a_{22} & a_{23} \end{bmatrix}$$

$$= \begin{bmatrix} (b_{11}a_{11} + b_{12}a_{21}) & (b_{11}a_{12} + b_{12}a_{22}) & \text{etc.} \\ & \text{etc.} & \end{bmatrix}$$

i.e.

$$c_{ij} = \sum_{k=1}^{n} b_{ik}a_{kj}$$

e.g.

$$c_{21} = b_{21}a_{11} + b_{22}a_{21}$$

The order of multiplication is critical; it is easy to see that $AB \neq BA$, e.g.

$$\begin{bmatrix} 1 & -2 & 3 \\ -4 & 2 & 5 \end{bmatrix} \times \begin{bmatrix} 1 & 3 \\ -1 & 0 \\ 2 & 4 \end{bmatrix} = \begin{bmatrix} 9 & 15 \\ 4 & 8 \end{bmatrix}$$

$$\begin{bmatrix} 1 & 3 \\ -1 & 0 \\ 2 & 4 \end{bmatrix} \times \begin{bmatrix} 1 & -2 & 3 \\ -4 & 2 & 5 \end{bmatrix} = \begin{bmatrix} -11 & 4 & 18 \\ -1 & 2 & -3 \\ -14 & 4 & 26 \end{bmatrix}$$

Special cases of multiplication are—

(a) $AI = A = IA$; and (b) $A0 = 0A = 0$

A.7 Transpose of a Matrix

The transpose of a matrix A is a new matrix B such that the columns of B are the rows of A and is often written

$$B = A'$$

i.e. $A = \begin{bmatrix} a_{11} & a_{12} & a_{13} \\ a_{21} & a_{22} & a_{23} \end{bmatrix}$ then $B = A' = \begin{bmatrix} a_{11} & a_{21} \\ a_{12} & a_{22} \\ a_{13} & a_{23} \end{bmatrix}$

A.8 Transpose of a Product

It is easy to demonstrate for oneself that

$$(A \times B)' = B'A'$$

i.e. the transpose of a product is equal to the reverse order product of the transposes: this is a *reversal rule*. Thus

$$\begin{bmatrix} 1 & 3 \\ 6 & 2 \end{bmatrix} \begin{bmatrix} 2 & 1 & 0 \\ 7 & 0 & 4 \end{bmatrix} = \begin{bmatrix} 23 & 1 & 12 \\ 26 & 6 & 8 \end{bmatrix}$$

and the transpose of the product is

$$\begin{bmatrix} 23 & 26 \\ 1 & 6 \\ 12 & 8 \end{bmatrix}$$

Then, multiplying the transposes in the reverse order,

$$
\begin{bmatrix} 2 & 7 \\ 1 & 0 \\ 0 & 4 \end{bmatrix}
\begin{bmatrix} 1 & 6 \\ 3 & 2 \end{bmatrix} =
\begin{bmatrix} 23 & 26 \\ 1 & 6 \\ 12 & 8 \end{bmatrix}
$$

A.9 Partitioned Matrices

A matrix may be divided for convenience, into blocks of elements, these blocks are called sub-matrices. Consider for example,

$$
\begin{bmatrix} a_{11} & a_{12} & \vdots & a_{13} & a_{14} \\ a_{21} & a_{22} & \vdots & a_{23} & a_{24} \end{bmatrix}
$$

This matrix is subdivided by the dotted line into two groups of elements

$$
\begin{bmatrix} a_{11} & a_{12} \\ a_{21} & a_{22} \end{bmatrix} \quad \text{and} \quad \begin{bmatrix} a_{13} & a_{14} \\ a_{23} & a_{24} \end{bmatrix}
$$

these may be designated A_1 and A_2, so that we can write

$$
A = [A_1 A_2]
$$

where the elements of the new matrix A are themselves matrices.

All the foregoing rules of matrix algebra still apply even though the elements are themselves matrices.

A.10 Reciprocal or Inverse Matrix

The reciprocal of a matrix, written A^{-1}, is defined by the relation $AA^{-1} = I$. This is equivalent to division in scalar arithmetic, but as with matrix multiplication the order is important. Thus, while

$$
AA^{-1} = I = A^{-1}A
$$
$$
BA^{-1} \neq A^{-1}B
$$

This definition allows us to write the solution of linear algebraic equations in matrix form, namely

$$
y_1 = a_{11}x_1 + a_{12}x_2 + a_{13}x_3
$$
$$
y_2 = a_{21}x_1 + a_{22}x_2 + a_{23}x_3
$$

or $\qquad y = Ax$

and the solution is obtained by premultiplying by A^{-1}, i.e.

$$
A^{-1}y = A^{-1}Ax = Ix = x
$$

The various methods employed for inverting a matrix are not developed here, it being desirable for the reader who wishes to study the subject further to refer to some standard work on matrices. However two points are of present importance: in order to have an inverse,

(a) a matrix must be square, and
(b) a matrix must be non-singular, that is the determinant of the elements must be non-zero, i.e.

$$|A| \neq 0$$

It is also of interest to note the reversal rule, that the inverse of a product is the reversed product of the inverses, i.e.

$$(AB)^{-1} = B^{-1}A^{-1}$$

Index